T0189236

Advances in Intelligent Systems and Computing

Volume 463

Series editor

Janusz Kacprzyk, Polish Academy of Sciences, Warsaw, Poland
e-mail: kacprzyk@ibspan.waw.pl

About this Series

The series "Advances in Intelligent Systems and Computing" contains publications on theory, applications, and design methods of Intelligent Systems and Intelligent Computing. Virtually all disciplines such as engineering, natural sciences, computer and information science, ICT, economics, business, e-commerce, environment, healthcare, life science are covered. The list of topics spans all the areas of modern intelligent systems and computing.

The publications within "Advances in Intelligent Systems and Computing" are primarily textbooks and proceedings of important conferences, symposia and congresses. They cover significant recent developments in the field, both of a foundational and applicable character. An important characteristic feature of the series is the short publication time and world-wide distribution. This permits a rapid and broad dissemination of research results.

More information about this series at http://www.springer.com/series/11156

Phayung Meesad · Sirapat Boonkrong
Herwig Unger
Editors

Recent Advances in Information and Communication Technology 2016

Proceedings of the 12th International Conference on Computing and Information Technology (IC^{2}IT)

Editors
Phayung Meesad
Faculty of Information Technology
King Mongkut's University of Technology North Bangkok
Bangkok
Thailand

Herwig Unger
Lehrgebiet Kommunikationsnetze
FernUniversität in Hagen
Hagen
Germany

Sirapat Boonkrong
Faculty of Information Technology
King Mongkut's University of Technology North Bangkok
Bangkok
Thailand

ISSN 2194-5357 ISSN 2194-5365 (electronic)
Advances in Intelligent Systems and Computing
ISBN 978-3-319-40414-1 ISBN 978-3-319-40415-8 (eBook)
DOI 10.1007/978-3-319-40415-8

Library of Congress Control Number: 2016942013

Printed on acid-free paper

This Springer imprint is published by Springer Nature
The registered company is Springer International Publishing AG Switzerland

Preface

Data mining, machine learning as well as data networks and communication are the key areas of information technology—a field in which massive changes and challenges have occurred within the last few years while significantly influencing other areas of daily life. Progress is made not only in designated, separate research areas, but also by interdisciplinary influences among them. Of course, these rapid developments in research and daily life came about by the massive utilisation of the Internet, which enables to create and store an already huge and permanently growing pile of (raw) data generated by millions of users from different locations. Obtaining knowledge or wisdom from them becomes an increasingly tedious task and depends on the opportunities to combine user activities and requests with the data representing the requested information from the right locations at desired times. If the necessary network and data mining activities can successfully be carried out, great achievements for businesses, governments and academia can be realised on the one hand, but on the other hand dangers for people and society at large may arise.

The chapters of this book contain the main, well-selected and reviewed contributions of scientists who met at the 12th International Conference on Computing and Information Technology in recognition of the importance to expand and improve the current technologies. There are three main sections in this book representing the three main directions in the recent developments of data mining and computer networking.

Since most activities originate are made on behalf of or directly controlled by users, and since 80 % of all contents in the World Wide Web are text-based, aspects of User Centric Data Mining and Text Processing are the topic of the first chapter, besides classical text-mining, picture and video processing, handwriting recognition and audio processing methods, which become more and more important in our multimedia world.

Data Mining Algorithms and Their Application constitute the key technology to elicit desired information in a reasonable amount of time and stand, and therefore become the focus of this book's considerations. Applications from industry, business, medicine and engineering give an idea of the always-present and everything-penetrating character of these technologies.

Finally, as the big breakthrough of information technology has been caused by the Internet and modern communication technologies, which allow to access data anywhere in the world in extremely short times, Optimisation of Complex Networks must ensure that the required speed of data access and transfer can be also guaranteed in the future. In this context, Cloud Computing and Security are the central issues of recent developments and need to be understood and applied in a right and efficient manner.

With going through the chapters of this book, we hope that readers, especially researchers, will be able to find novel contributions on state-of-the-art technologies for data mining, machine learning and data networking. Beginners or research students should be able to grasp basic ideas, as well as to gain fundamental knowledge.

The book was prepared with the combined effort of the staff of the Faculty of Information Technology at King Mongkut's University of Technology North Bangkok. We also would like to thank Ms. Watchareewan Jitsakul, whose work has been very valuable to the success of the publishing process. Finally, we are grateful to Springer-Verlag for the support provided and for agreeing to publish this book.

Bangkok, Thailand
March 2016

Phayung Meesad
Sirapat Boonkrong
Herwig Unger

Program Committee

A. Agarwal, UoHyd, India
T. Anwar, UTM, Malaysia
S. Auwatanamongkol, NIDA, Thailand
G. Azzopardi, University of Groningen, The Netherlands
T. Bernard, Syscom CReSTIC, France
T. Boehme, TU Ilmenau, Germany
M. Caspar, TU Chemnitz, Germany
N. Ditcharoen, UBU, Thailand
K. Djemame, University of Leeds, UK
T. Eggendorfer, HS Weingarten, Germany
L. Fung, Murdoch University, Australia
M. Ghazali, UTM, Malaysia
W. Halang, FernUni, Germany
C. Haruechaiyasak, NECTEC, Thailand
S. Hengpraprohm, NPRU, Thailand
U. Inyaem, RMUTT, Thailand
M. Kaenampornpan, MSU, Thailand
P. Kirdwichai, KMUTNB, Thailand
A. Kongthon, NECTEC, Thailand
M. Kubek, FU Hagen, Germany
P. Kucharoen, NIDA, Thailand
K. Kyamakya, University of Klagenfurt, Austria
U. Lechner, UniBwM, Germany
A. Mingkhwan, KMUTNB, Thailand
C. Namman, UBU, Thailand
H.H.C. Nguyen, QNU, Vietnam
S. Nitsuwat, KMUTNB
S. Nuanmeesri, RSU, Thailand
P. Palawisut, NPRU, Thailand
J. Phuboonob, MSU, Thailand
N. Porrawatpreyakorn, KMUTNB, Thailand

P. Prathombutr, NECTEC, Thailand
S. Puangpronpitag, MSU, Thailand
T. Quan, HCUT, Vietnam
P. Saengsiri, TISTR, Thailand
J. Shen, University of Wollongong, Australia
S. Smanchat, KMUTNB, Thailand
M. Sodanil, KMUTNB, Thailand
S. Sodsee, KMUTNB, Thailand
G. Somprasertsri, MSU, Thailand
T. Srikhacha, TOT, Thailand
W. Sriurai, UBU, Thailand
T. Sucontphunt, NIDA, Thailand
S. Suranauwarat, NIDA, Thailand
D. Thammasiri, NPRU, Thailand
S. Tongngam, NIDA, Thailand
D.H. Tran, HNUE, Vietnam
K. Treeprapin, UBU, Thailand
D. Tutsch, University of Wuppertal, Germany
S. Valuvanathorn, UBU, Thailand
M. Weiser, OSU, USA
N. Wisitpongphan, KMUTNB, Thailand
K. Woraratpanya, KMITL, Thailand
P. Wuttidittachotti, KMUTNB, Thailand

Organizing Partners

In Cooperation with

King Mongkut's University of Technology North Bangkok (KMUTNB)
FernUniversitaet in Hagen, Germany (FernUni)
Chemnitz University, Germany (CUT)
Oklahoma State University, USA (OSU)
Edith Cowan University, Western Australia (ECU)
Hanoi National University of Education, Vietnam (HNUE)
Gesellschaft für Informatik (GI)
Mahasarakham University (MSU)
Ubon Ratchathani University (UBU)
Kanchanaburi Rajabhat University (KRU)
Nakhon Pathom Rajabhat University (NPRU)
Mahasarakham Rajabhat University (RMU)
Phetchaburi Rajabhat University (PBRU)
Rajamangala University of Technology Krungthep (RMUTK)
Rajamangala University of Technology Thanyaburi (RMUTT)
Prince of Songkla University, Phuket Campus (PSU)
National Institute of Development Administration (NIDA)
Council of IT Deans of Thailand (CITT)

Contents

Part IV Optimisation of Complex Networks

Part I
Invited Paper

Cloud Computing: New Paradigms and Challenges

Djamshid Tavangarian

Abstract Cloud computing is an Internet-based distributed system architecture to provide the customers with services and resources in a scalable and self-manageable fashion. Whatever the user needs and wherever he demands it—Cloud can deliver the regarding software services, platform or infrastructural resources. Services are furnished by software whereas resources are a complex of virtualized computing facilities, storages, and networking hardware provided by the Cloud for end users and enterprises. Therefore, the Cloud computing principle leads to a new paradigm in computer architecture. We can divide the whole system into two parts:

- the front-end which consists of different kind of mobile devices like tablets, smart phones, PCs, or simply embedded systems, and regarding software as well as
- the back-end, which is the Cloud infrastructure consisting of resources and services (as public, private, hybrid, or federated Cloud) with a centralized view for users.

The front-end and back-end are connected to each other through a broadband network, usually the Internet. The front-end is what the computer user, or customer, sees. The back-end works in the background to satisfy user demand. In this presentation, in a short introduction the principle of cloud computing, its philosophy, and methods will be discussed. As main topics of the talk we will discuss some paradigms and challenging topics of cloud computing like the new architectural view, its benefits, advantages and disadvantages. Introduction of virtualization conceptions for an energy efficient cloud as well as the specification of security and security problems are further subjects of the talk. In this part we will introduce also a new security algorithm to store the big data in a cloud. It will then also illustrate how the solution can be leveraged in context of different applications to identity business, strategic, technical, and implementation challenges and find solutions for different kind of cloud back-end structures (like private, public, and hybrid cloud).

D. Tavangarian (✉)
Faculty of CS and EE Research Group of Computer Architecture,
University of Rostock, Rostock, Germany
e-mail: Djamshid.tavangarian@uni-rostock.de

P. Meesad et al. (eds.), *Recent Advances in Information and Communication Technology 2016*, Advances in Intelligent Systems and Computing 463,
DOI 10.1007/978-3-319-40415-8_1

Part II
User Centric Data Mining and Text Processing

User-Triggered Structural Changes in OSN-Alike Distributed Content Networks

Hauke Coltzau and Mario Kubek

Abstract We continue the evaluation of our model to describe regularly occurring phenomena in online social networks. The influence of link removals triggered by dislikes on the network structure is analyzed. Additionally, two scenarios for link replacement are investigated. It is shown that structured link replacement in the local neighborhood has acceptable impact on the graph structure and substantial benefits for the average like rate.

Keywords Online social networks · Modeling

1 Introduction

The remarkable sizes of today's online social networks (OSNs) allow for a unique practical view on the development of global behavior patterns in user-controlled decentralized systems. Interesting characteristics of these systems are the distributions of relationships between users, the diameter of the structure as well as the perseverative occurrence of superpopular contents.

In previous works, a simple model for the distribution of contents in OSNs was discussed. It could be shown that the occurrence of superpopular contents is very robust against parameter changes in this model [4]. Ongoing work will focus on the influence of network changes on the distribution and occurrence of superpopular contents. The work described in this article connects these two topics, the focus therefore lies on the enhancement of the basic model to integrate user triggered changes in the network structure based on the evaluation of recommendations received from other users.

H. Coltzau (✉) · M. Kubek
Fernuniversität in Hagen, Hagen, Germany
e-mail: hauke.coltzau@fernuni-hagen.de

M. Kubek
e-mail: mario.kubek@fernuni-hagen.de

P. Meesad et al. (eds.), *Recent Advances in Information and Communication Technology 2016*, Advances in Intelligent Systems and Computing 463,
DOI 10.1007/978-3-319-40415-8_2

The remaining article is organized as follows: In Sect. 2, an overviewing extract of existing works on OSN analysis and modeling is given. Our model is described in Sect. 3. Simulations in Sect. 4 discuss the results for three different link-adding scenarios.

2 Related Work

A remarkable amount of scientific works already exists in the field of online social network analysis, covering various fields like computer science and mathematics, but also physics and sociology. Almost all relevant online social networks already have been analyzed with regard to information flow, network structure and user behaviors.

Kleinberg [8], Watts and Strogatz [13], and Albert [1] and Barabási [2] independently have described or predicted reoccuring structural properties relying only on knowledge of networks with much smaller scale. They proposed different models for the evolution of large graphs, but merely focused on the distribution of the nodes degrees, their connectivity and the graph diameters. The distribution of contents was not part of their analysis, neither were the influences of dynamic user-triggered changes.

Cha et al. [3] found popularity growth patterns for pictures in the Flickr OSN. Lerman and Gosh [9] analyzed news spreading in Digg and Twitter from a user activity point-of-view. They deduce that in both platforms, news spread in an almost deterministic way similar for both platforms. Doerr et al. [5] confirm that information in social networks is propagated very quickly due to influential users. Zhang et al. [14] developed mechanisms to efficiently identify these influential users in a mobile online social network environment.

As for the analysis of social network structures, Schiöberg et al. [10] tracked the growth of Google+. Viswanath et al. [12] found that most of the links between users in Facebook are only seldomly used, while few links show a high level of activity. Additionally, it could be shown that the activity on a single link remarkably decreases over time.

Only few works exist regarding abstract modeling of online social networks. Kleinberg [8], Easly [6], as well as Watts and Strogatz [13] describe core principles on the efficiency of social distributed systems. But again, the perspective of their works is descriptive and does not discuss individual actions of users and their influence on the networks. Tovionen et al. [11] propose a network creation model that resembles some existing social networks e.g. in terms of degree correlations and clustering. Jamali et al. [7] derived a user rating-behavior model especially of the Flickr network but only with regard to the dynamics of the rating system and not as a general model.

3 Model

The model is an extension of the still unpublished works of Böhme and Airmarn. Let there be a set of users U and a set of files C that represent content distributed over the network. For now, we let C be $\in \mathbb{N}$, although future works will focus on more realistic document corpora. The dissimilarity $\delta(c_i, c_j)$ between any two contents $c_i, c_j \in C$ shall simply be their Euclidean distance. As usual, the structure of the distributed system is defined by the neighborhood relationships of their users. For each user, a neighborhood $N(u) \in U$ is defined, containing all users, a directed connection exists to.

Users can recommend content they already know to other users in their neighborhood. For this purpose, each user $u \in U$ maintains a recommendation stack $R(u)$. To recommend a content to a neighbor $n \in N(u)$, a user pushes the content to $R(n)$. In a subsequent step, the target user of a recommendation can either accept ("like"), ignore or actively reject ("dislike") the recommendation. To evaluate recommendations received from other users, each user defines a set of interest areas I_u containing tuples (c_i, d) with $i_c \in C, d \in \mathbb{R}^+$ during initialization, where d is a system-wide parameter and all c_i are chosen randomly from C. In the same way, a set of avoided topics $A_u = \{(c_a \in C, d \in R)\}$ is defined.

An element $c \in R(u)$ is evaluated using a decision function $f : C \rightarrow -1, 1$ that maps c to a value from -1.0 (dislike) over 0.0 (ignore) to 1.0 (like) according to the minimum dissimilarity to the user's interest and avoidance areas. Recommendations that are within a distance of d to any of the users interests $c_i \in I$ are mapped to 1.0 (like). All distances between d and $2 \cdot d$ are linearly mapped to a value in [1.0..0.0] and all distances above $2 \cdot d$ are mapped to 0.0. In the same way, the dissimilarity to the closest avoided topic is taken to map the recommendation to a dislike-value in [-1.0..0.0] If a content is within the radius of $2 \cdot d$ of both an interest area and an avoidance area, the results of the evaluation functions for both areas are added.

The elements, a user likes (i.e., those, who evaluate to 1.0), are put to the users *like-stack* $L(u)$, which serves as memory with a maximum size of $k_u \in \mathbb{N}^+$, where a new element will replace the oldest one in case the maximum size is reached. Each element in $L(u)$ can be recommended to the user's neighbors (Table 1).

Initially, each user's like stack $L(u)$ is randomly filled with elements from the user's interest areas. Each timestep, a user can only process the top element from the recommendation stack $R(u)$ and then decide, if the element is accepted and transferred to $L(u)$, dropped (ignored) or rejected. In the same timestep, the user may also recommend one element from $L(u)$ as described above.

Each user u keeps track of evaluation results received from and given to other users. For each user $e \in U$, a recommendation was given to or received from, a limited set $M(u)_e$ exists containing the last m recommendation results for/from user e to/from user u. Should the average value of these m entries in $M(u)_e$ fall below a system wide threshold Δ, the user u will initiate a removal of any existing connection $u \rightarrow e$ or $e \rightarrow u$ between itself and e. As long as $M(u)_e$ does not yet contain of m elements, no action is performed at all.

Table 1 Model parameters

$\|U\|$	Number of users in the network. Here, this number is fixed to 1000
$\|C\|$	Number of available distinguishable contents, also fixed to 1000 here
p_{int}	The probability for each $c \in C$ that c is of interest for $u \in U$
p_{avoid}	The probability for each $c \in C$ that c is to be avoided by $u \in U$
d	The maximum dissimilarity to any of a user's interest (avoidance) areas, recommended content may, to be liked (disliked) by the user
k	The size of the recommendation memory of a user, i.e. the maximum for $\|L(u)\|$
Δ	Threshold for average result for the evaluation of recommendations to remove a connection between two users
r	The depth of the neighborhood search for best-fitting neighbors

4 Simulation and Results

We investigate three different scenarios for how users deal with the removal of a link to another user:

1. In a *remove-only* scenario, no further action occurs, when a link is removed.
2. In the second scenario, a removed connection is replaced with a *random link* to any user in the network.
3. In the *best-fitting neighbors* scenario, the user parses through their neighborhood to find a neighbor, whose interest areas differ as little as possible from the interest area of the user itself. The deepness of the search is given by the simulation parameter r.

All simulations displayed in this article have been set up with the same number of users (1000) and contents (1000). In all cases, the graph properties of the user connection graph where tracked as well as the average value over all user's recommendation memories, which could be interpreted as average acceptance over all recommendations.

4.1 Scenario 1—Remove Only

The underlying initial network is a Watts/Strogatz with $k = 4$ and $\beta = 5\,\%$, hence the clustering coefficient is rather high and the average path length relatively low in the beginning of the simulation. The clustering coefficient decreases after the first users begin to remove links, the time of which being almost independent from the chosen threshold Δ (see Fig. 1a). At the same time, the average path length starts to increase (Fig. 1b). Nevertheless, after a rather short period of time, both graph properties stablilize at a value depending on Δ, where higher values for Δ lead to a

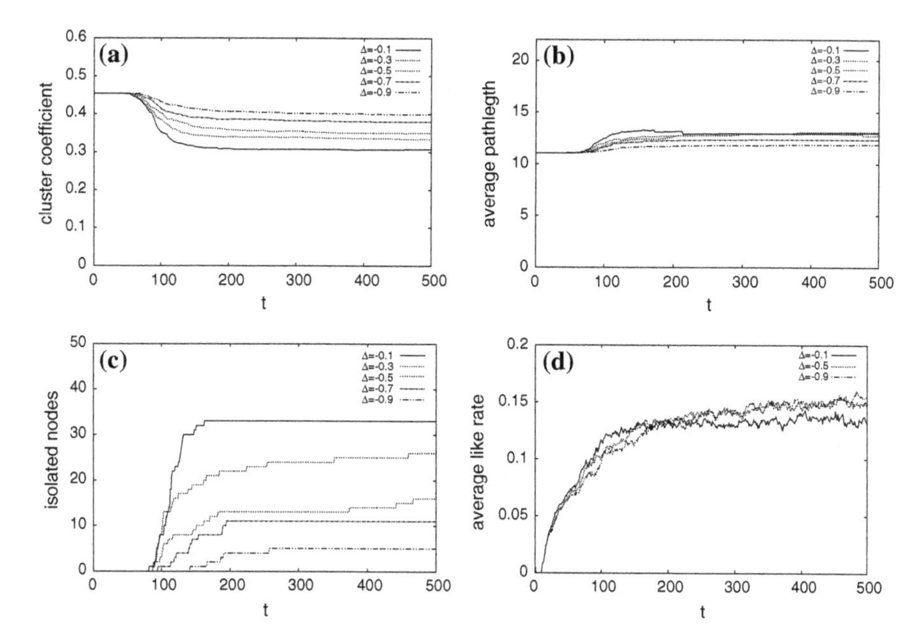

Fig. 1 Scenario 1—remove links only. **a** Clustering coefficient. **b** Average path length. **c** Isolated nodes. **d** Average like rate

higher clustering coefficient and lower average path length, because fewer links are actually removed.

By removing links, the user graph becomes disconnected. Some nodes do not have any incoming or outgoing neighbours with sufficiently overlapping interest areas any more. These nodes become isolated over time (see Fig. 1c). Not surprisingly, the number of isolated nodes also depends on the value of Δ, such that higher values for Δ result in lower numbers of isolated nodes.

As a general result for this scenario, it can be concluded that removing links depending on the average like value of recommendations does influence the connectivity of the graph as well as its clustering coefficient negatively. Nevertheless, both graph properties converge after a short period of time.

4.2 *Scenario 2—Random Links*

Adding random links after deletion of neighborhood relations transforms the user graph into a mixture between a Watts/Strogatz small world and an Erdös-Rényi random graph. Therefore, it is not surprising that the clustering coefficient decreases remarkably, as links are changed from the Watts/Strogatz local neighborhood to random targets. On the other hand, the average path length also decreases remarkably.

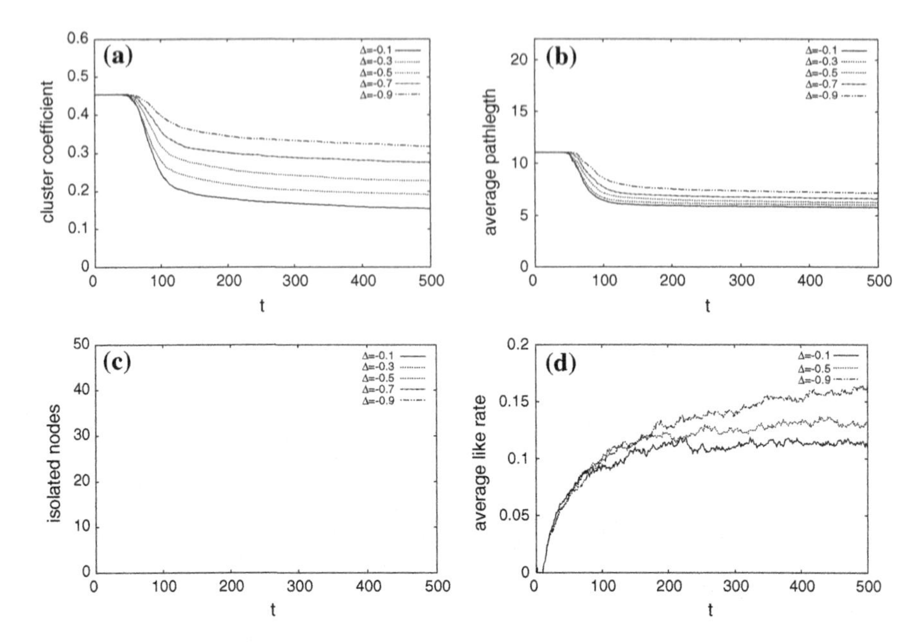

Fig. 2 Scenario 2—add random links after removal. **a** Clustering coefficient. **b** Average path length. **c** Isolated nodes. **d** Average like rate

Both changes are dependent on the threshold Δ, where higher values generally lead to lesser decreasing of the clustering coefficient but also only lesser positive impact on the path length (Fig. 2).

Although no more nodes are isolated, the user graph still regularly looses its connectivity over time without notable influence of the threshold Δ. This can also easily be explained by the underlying transformation to a random graph, which has no guarantee to be connected at all. The average like rate for Δ closer to 0 is even worse than it was in scenario 1 without adding new links at all. Again, the obvious explanation lies in the randomness of the newly added links, meaning that the newly added neighbor may as well be "worse" (in the sense of similarity to the user's interests) than the previously deleted link.

This approach is of course not realistic, since real-world users will neither simply select new contacts randomly and are not able to browse a whole OSN randomly to do so. Instead, new contacts will more likely be found in a user's local neighbourhood and only very seldomly in more distant areas of the network.

4.3 Scenario 3—Best Fitting Neighbors

In the third scenario, we therefore let a user try to replace removed link with a link to a user more similar to itself from within the local neighbourhood. When a link is removed, the affected users actively parse its neighborhood to find the neighbor with the highest similarity to the user's interest areas. The interest areas of all neighbors within r hops are compared to the interest area of the current user. The neighbor with the highest similarity is added to the local neighborhood of the user who initiated the search. Since all neighbours within 1 hop are already neighbors, r starts with a value of 2. In the simulations discussed in this article, the maximum value for r was chosen to be 4, which also seems to be a reasonable practical limit.

This approach still breaks up some local connections, so that the clustering coefficient still decreases in comparison to the initial Watts/Strogatz small world. Nevertheless, it remains much closer to the initial value than it did in the previous scenario. The influence of Δ is similar to the random scenario, where values close to 0 lead to stronger decrease of the clustering coefficient than values closer to -1.0 (see Fig. 3a). The impact increases for larger values of r, because in this case, new links can be added to more distant users, which will most likely not have any connections to the current user's neighbors and hence lead to a lower local clustering coefficient (Fig. 3b).

The influence on the average path length of both Δ and r is similar to the influence on the clustering coefficient (Fig. 4c, d). The closer Δ is to -1.0, the lesser the path length decreases. The higher the value of r is, the more the path length decreases. The average like rate increases in all cases (Fig. 4a, b), converging almost to the same value for all Δ and only to minor differences over different values of r.

Although loosing graph connectivity still remains as unsolved problem, the approach to add links by performing a local similarity search seems to be an appropriate solution for handling necessary removals of connections to other users.

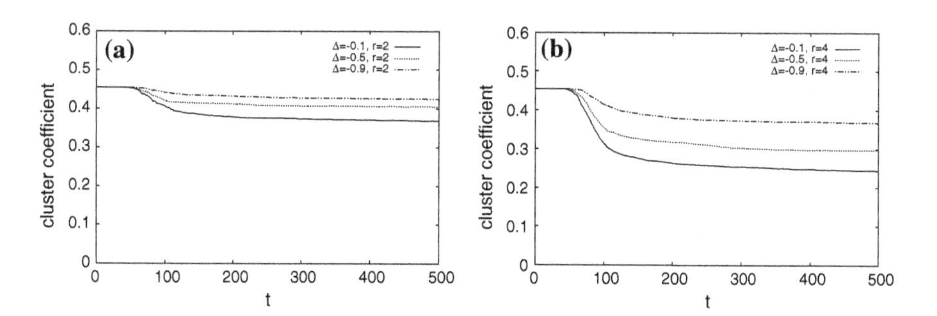

Fig. 3 Scenario 3—add best fitting neighbour, cluster coefficients. **a** Clustering coefficient $r = 2$. **b** Clustering coefficient $r = 4$

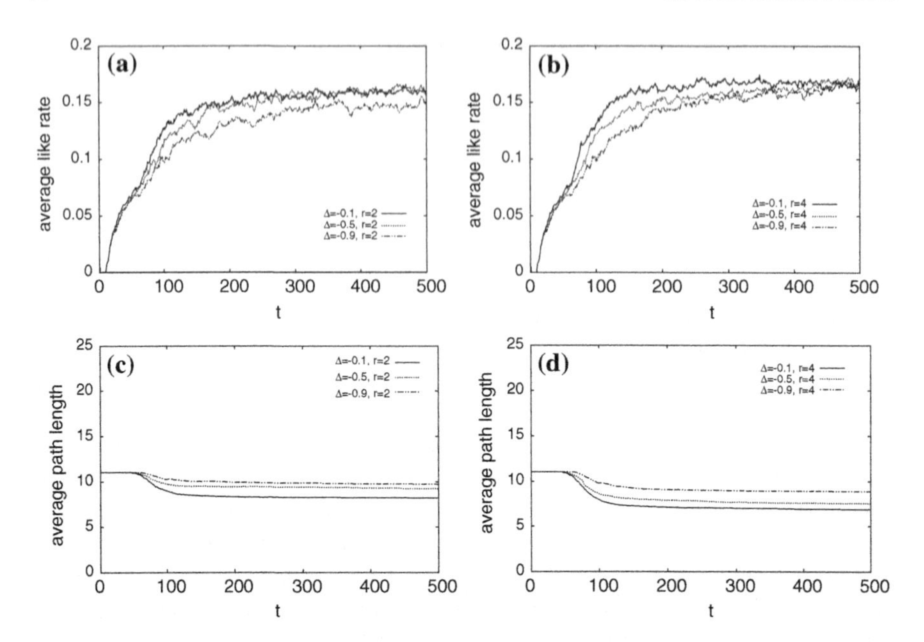

Fig. 4 Scenario 3—add best fitting neighbour, like rate and path length. **a** Average like rate $r = 2$. **b** Average like rate $r = 4$. **c** Average path length $r = 2$. **d** Average path length $r = 4$

5 Summary and Outlook

In this article, we have discussed an extension of our basic user centric model for content distribution in online social networks. It was shown that the introduction of link removal and link replacement based on the ongoing evaluation of recommendations from the user's perspective allows for an increasing average like-rate but negatively impacts the user graph's connectivity and local clustering. These effects can be moderated but not completely eliminated, when removed links are replaced using a best-fitting strategy in a user's local neighborhood. Future works will analyze the influence of these structural changes on the content distribution in the network, especially with regard to occurrence and distribution of superpopular contents.

References

1. Albert, R., Barabási, A.: Statistical mechanics of complex networks. Rev. Mod. Phys. **74**(1), 47–97 (2002)
2. Barabási, A., Albert, R.: Emergence of scaling in random networks. Science **286**(5439), 509–512 (1999)
3. Cha, M., Mislove, A., Gummadi, K.P.: A measurement-driven analysis of information propagation in the flickr social network, In: 18th International Conference on World Wide Web, pp. 721–730. ACM, New York (2009)

4. Coltzau, H., Unger, H.: A model based approach on superpopular contents in online social networks. In: 8th GI Conference on Autonomous Systems, pp. 145–156. VDI Verlag, Düsseldorf (2015)
5. Doerr, B., Fouz, M., Friedrich, T.: Why Rumors spread so quickly in social networks. CACM **55**(6), 70–75 (2012)
6. Easley, D., Kleinberg, J.: Networks, crowds, and markets: reasoning about a highly connected world. Cambridge University Press (2010)
7. Jamali, M., Haffari, G., Ester, M.: Modeling the temporal dynamics of social rating networks using bidirectional effects of social relations and rating patterns. In: 20th International Conference on World Wide Web, pp. 527–536. ACM, New York (2011)
8. Kleinberg, J.: The small-world phenomenon: an algorithmic perspective, In: Thirty-second annual ACM symposium on Theory of computing, pp. 163–170. ACM, New York (2000)
9. Lerman, K., Ghosh, R.: Information contagion: an empirical study of the spread of news on Digg and Twitter social networks. In: 4th International Conference on Weblogs and Social Media. AAAI (2010)
10. Schiöberg, D., Schmid, S., Schneider, F., Uhlig, S., Schiöberg, H., Feldmann, A.: Tracing the birth of an OSN: social graph and profile analysis in Google+. In: 4th Annual ACM Web Science Conference, pp. 265–274. ACM, New York (2012)
11. Toivonen, R., Onnela, J., Saramäki, J., Hyvönen, J., Kaski, K.: A model for social networks. In: Dawson, K.A., Indekeu, J.O., Stanley, H.E., Tsallis, C. (eds.) Physica A, vol. 371, no. 2, pp. 851–860. Elsevier (2006)
12. Viswanath, B., Mislove, A., Cha, M., Gummadi, K.P.: On the evolution of user interaction in Facebook, In: 2nd ACM Workshop on Online Social Networks, pp. 37–42. ACM, New York (2009)
13. Watts, D.J., Strogatz, S.H.: Collective dynamics of 'small-world' networks. Nature **393**, 440–442 (1998)
14. Zhang, M., Yang, P., Tian, C., Xiang, C., Xiong, Y.: Walk globally, act locally: efficient influential user identification in mobile social networks. In: First International Workshop on Mobile Sensing, Computing and Communication, pp. 29–34. ACM, New York (2014)

Modeling Disease Spread at Global Mass Gatherings: Data Requirements and Challenges

Sultanah M. Alshammari and Armin R. Mikler

Abstract Spread of infectious diseases at global mass gatherings can pose health threats to both the hosting country and the countries where participants originate. The travel patterns at the end of these international events may result in epidemics that can grow to pandemic levels within a short period of time. Computational models are essential tools to estimate, study, and control disease outbreaks at mass gatherings. These models can be integrated in the planning and preparation process of mass gatherings. In this paper, we present a review of the key data requirements and the challenges encountered when modeling infectious diseases epidemics initiated by global mass gatherings. This review can assist epidemic modeling at global mass gatherings providing researchers with insights of the main aspects and possible big data opportunities in this emerging research area.

Keywords Mass gatherings (MGs) · Disease spread · Infectious diseases · Epidemic · Outbreak · Computational modeling

1 Introduction

There are several recurrent global mass gatherings (MGs) occurring on different places in the world such as sport, religious, and political events. Participants attending these events may be exposed to serious health threats. The gathering of high population density in a close proximity during MGs has the potential to facilitate the transmission of communicable diseases [1–5]. Furthermore, the global

S.M. Alshammari (✉)
Department of Computer Science & Engineering, University of North Texas,
Denton, TX, USA
e-mail: SultanahAlshammari@my.unt.edu

A.R. Mikler
Center for Computational Epidemiology and Response Analysis (CeCERA),
University of North Texas, Denton, TX, USA
e-mail: Armin.Mikler@unt.edu

P. Meesad et al. (eds.), *Recent Advances in Information and Communication Technology 2016*, Advances in Intelligent Systems and Computing 463,
DOI 10.1007/978-3-319-40415-8_3

events (such as the Olympics, the FIFA World Cup, and the Hajj; the Muslim pilgrimage to Makkah) where people gather from different countries present serious health threats and challenges for both the hosting countries and the countries of origin of the participants. These global MGs allow the mixing of various infectious pathogens due to the diverse disease exposure history and demographics of the participants. When these global events are over, the travel patterns of the participants could cause a rapid spread of infectious diseases affecting large number of people and causing global epidemic within a short period.

There was a rising concern in 2009, as MGs occurring that year such as the Hajj and the Southeast Asian Games could further contribute to global spread of the 2009 pandemic influenza A (H1N1) [6]. In fact, after the 2009 Hajj season several studies [7–10] were conducted to trace H1N1 pandemics among the returning pilgrims and their contacts. Also, activities of the 2009 pandemic influenza A (H1N1) virus have been reported during several music festivals [11]. Several outbreaks have been reported in different MGs of varying sizes with the most common reported infections being respiratory viruses including different strains of influenza [12]. As shown in the Fig. 1 below, there was a global spread of N meningitidis W135 after recurrent outbreaks of W135 during the Hajj seasons of 2000 and 2001. Also, influenza outbreaks have been reported in 2002 and in 2008 at the Winter Olympics in Salt Lake City, UT, USA [13], and the World Youth Day in Sydney [14] respectively. In 2008, at the European Football Championship several measles outbreaks occurred in both hosting countries Austria and

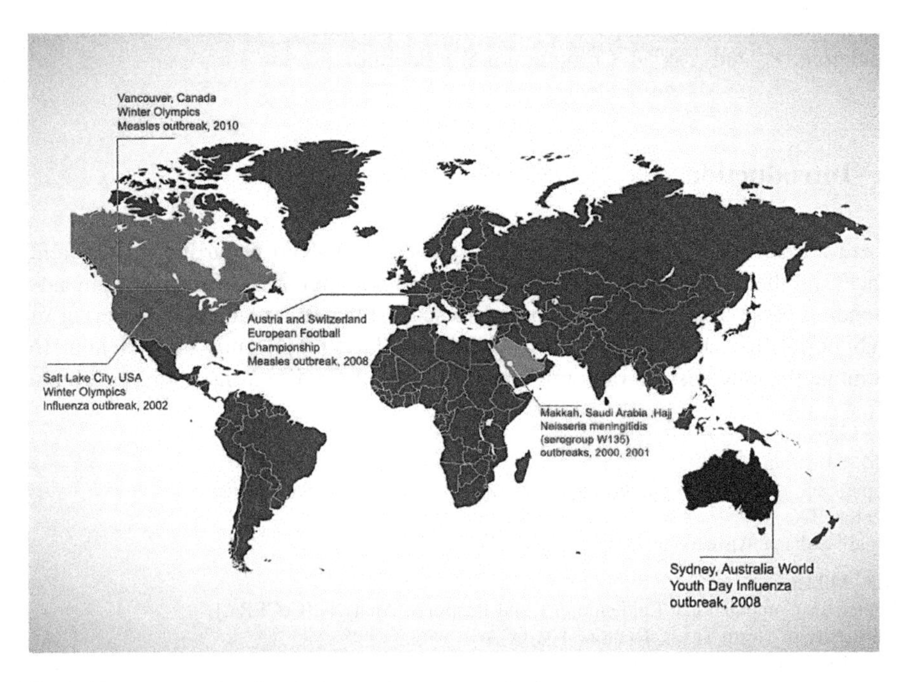

Fig. 1 Some reported outbreaks in previous mass gatherings

Switzerland, with other measles cases reported in the neighboring countries of France, Germany, and Spain [15].

There are increasing research efforts addressing the health hazards in the context of mass gatherings and their impacts on global health. Many studies [2, 5, 16–19] draw attention to major threats presented in mass gatherings, and provide guidelines and recommendations to manage possible health hazards. In their study, Nsoesie et al. [20] reviewed the use of new technologies to improve infectious diseases surveillance in different types of MGs as a part of planning and preparation for these events. These technologies include web-based systems, smartphone applications, wireless sensor networks and syndromic surveillance systems. While, applying all the different measures and efforts can be effective in preventing and controlling the spread of infectious diseases during mass gatherings, it may not be sufficient to provide insights of the potential risk of global epidemics after these events. Computational methods and models can play an important role not only in the prevention and control of disease outbreaks in mass gatherings, but also in predicting a global spread of infectious diseases.

In this paper, we aim to summaries the main aspects of modeling disease spread at mass gatherings, and provide a review over relevant literature. Based on this review, the key data requirements to model disease spread at global MGs and the encountered challenges related to big data are presented. The paper is organized as follows; the next section summaries several recent computational models of infectious disease in mass gatherings. Next, data requirements and challenges in modeling epidemics at global MGs are covered. Then, some big data aspects and opportunities when modeling disease spread at MGs are highlighted. Finally, the last section summaries the review presented in this paper.

2 Modeling Disease Outbreaks at MGs

The success of computational modeling in epidemiology and public health motivates applying its use in the context of mass gatherings to provide efficient tools to study the spread of diseases at mass gatherings and assess the risk of global epidemics. There are several studies proposing computational models for analyzing and simulating epidemics of different diseases in different settings and populations. These models vary from the simple mathematical SIR [21] (Susceptible–Infected–Recovered) model to more complicated spatial-temporal simulations that allow the introduction of prevention and mitigation measures. However, few studies are devoted to computational models of epidemics at mass gatherings. In their study, Chowell et al. [22] examined different aspects of epidemics that must be addressed in computational models including; the environmental, behavioral, health-related, and demographics factors and highlighted several data requirements and modeling challenges in these settings. Their article concluded with recommendations of integrating disease spread models as a part of the preparedness and the planning of some of the global events such as the Olympics and the Hajj.

In an event-based study, Khan et al. [23] proposed a conceptual model to integrate data of international air travels from different sources and reported cases of infectious diseases as a preparation for potential infectious diseases threats at the Vancouver 2010 winter Olympics. This work suggested that the hosting country of such global event could use this integrated knowledge model to estimate emerging health threats. First, the authors gathered and analyzed the previous global travel patterns, focusing on the Olympic Winter Games for the past decade. Then, based on results of that analysis, the 25 cities with the highest numbers of travellers were identified, and the infectious disease surveillance in those cities was conducted. Next, the information from the HealthMap (www.healthmap.org) was integrated to the global infectious disease surveillance. As described by Khan et al. [23], HealthMap is an online tool that tracks outbreaks reports from different sources and transforms these reports into interactive maps. HealthMap provided a real time analysis before and during the 2010 Olympic Winter Games. Although, applying their model to the 2010 Olympic Winter Games did not reveal any serious threats, it provided the Canadian health officials with advanced knowledge to be prepared to any threats.

While Khan et al. [23] targeted one of the global mass gatherings to identify possible incoming health hazards, Stehle et al. [24] proposed a simulation of infectious disease transmission for a relatively small gathering among the attendees of the 2009 Annual French Conference on Nosocomial Infections. They used a stochastic SEIR (Susceptible–Exposed-Infected–Recovered) model to demonstrate the impact of using different contact patterns on the dynamic of the infection. The authors were able to measure the interactions between the individuals within the two days of the conference using wearable sensors (RFID; radiofrequency identification devices). Then, the collected data was used to compare three different contact networks: dynamic, homogeneous, and heterogeneous. These contact networks differ in the type and the duration of the interactions between individuals. Their results of the contact duration and distribution showed a high probability of pathogen spread for a small number of infected people. This probability was smaller in homogeneous networks as compared to heterogeneous and dynamic networks.

Shi et al. [25] developed a SEIR agent-based simulation model using data from the state of Georgia to study the impacts of the contact patterns and social mixing on influenza pandemic in different settings including mass gatherings and holiday traveling. In their simulation, they divided the simulation period into a “regular” period and a “traveling” or “mass gathering” period. In the “regular” period, Agents are moving to and from households, schools or workplaces, and other public places. The proportion of the created population was selected to mix within a large group to model temporal mass gatherings. To validate their model, Shi et al. [25] executed the simulation using several scenarios of different populations interacting in mass gatherings or traveling periods varying in length and size. The results obtained from the simulation experiments identified several factors that affect the dynamics of the influenza epidemic, including the timing of mass gatherings or travel in relation to the epidemic peak and the pathogen’s infectious period. Also, the simulation indicated little impact of the changes in the social mixing on the course of an epidemic. The authors suggested that using such predictive model could provide public health officials with insights to take proper actions and control measures during a mass gathering.

3 Data Requirements

As stated by Chowell et al. [22], the risk of disease spread at MGs is related to the participants attending these events and the environment where the event took place. Thus, the key data requirements to model outbreaks in MGs include the characteristics of the participating populations and the event itself.

3.1 Participating Populations

The main attributes of the population attending MGs that should be used to guide modeling disease spread, include the population demographic and epidemiological disease-specific aspects such as susceptibility levels and vaccination status. The most important demographic data about the participants attending a global event are *age*, *gender*, and the *country of origin*. In fact, based on the epidemic modeling approach these demographic data if known, can be used to identify, estimate, or make reliable assumptions about other modeling attributes such as the contact patterns, the susceptibility levels, and the vaccination status. Knowing the age distribution of the population can provide insights into the contact rates and the susceptibility levels of different age groups. In fact, the correlation between the different age groups and the contact rates was investigated in several studies. In a recent study [26], Cowling et al. revealed a strong age-based contact rate based on analyzing contact patterns in different settings in several European countries. It was concluded that compared to other age groups, children and adolescents tend to have a higher rate of contacts within their age group. Thus, children and teenagers play an important role in disease spread of close contact infections especially influenza. As a result it is important to conduct studies for age-based risk groups for different diseases in the context of MGs.

Gender is another important factor when modeling disease transmission among individuals. Gender differences can contribute in understanding and controlling the transmission of infectious diseases [27]. However, most studies investigated gender differences on emerging infectious diseases focused mainly on sexual transmitted diseases. During mass gatherings, the gender distribution of the population is important to study and model disease spread especially when combined with gender-based behavior analysis within the event. As in some MGs especially the religious events, there is a gender separation either for the entire event or in some stages of the event. For example, at the Hajj, men and women are only segregated in their sleeping arrangements, as opposed to another annual religious event of Attukal Pongala festival in Kerala, India where only women are participating.

Countries of origins of the participants might provide insights of their susceptibility to some diseases, possible disease history, and vaccination status. As different countries have different vaccination policies and populations from different geographical areas vary in their disease exposure history. The incoming travel patterns can be used to determine the country of origin of the population during global MGs. In some MGs like Hajj, vaccination requirements for entry visa are issued based on the country of origin. For example, arrivals from polio-endemic countries are required to provide a proof of Poliomyelitis vaccination 6 weeks prior attending Hajj [28]. The vaccination coverage of different age groups can be acquired from different sources based on the country of origin such as official public health websites. The National Center for Health Statistics (NCHS) (www.cdc.gov/nchs/) provides detailed health measures and statistics including age-based vaccination coverage of different diseases. Also, there are several published reports and surveys provide information about the immunizations status in several countries either for specific age groups or the whole population. For example, the publicly available information provided in several reports [29, 30] of the seasonal influenza vaccination in Europe can be used to estimate the levels of immunity and susceptibility when modeling a MGs occurring in Europe.

3.2 Mass Gatherings Event

The different characteristics of the event such as timing, size, incoming travel patterns, setting, schedule of events, and the spatial considerations needed to be identified and included in the epidemic model. There is evidence of associations between climatic conditions and some infectious diseases such as seasonal influenza [7]. Thus, it is important to have a better understanding of climate contribution to the disease(s) under study and to integrate climate data into epidemic modeling. The expected weather patterns at MGs can be estimated based on the timing and the geographical location of the event. For example, in the past few years the Hajj season shifted from summer to winter [31]. As a result in the last seasons of Hajj, the summer related diseases such as heat stroke and food poisoning are not reported and the more expected diseases are related to cold weather such as influenza and asthma.

Arrival and departure are the most important phases of a global MGs, as the movement of large numbers of people from and to different countries to attend an event can pose public health threats to the hosting country. The incoming travel patterns can be used to determine the countries of origin of the participants, the expected diseases, and the possible vaccinations. Thus, identifying and analyzing the incoming travel patterns can help hosting countries to estimate the risks of importing infectious diseases. Moreover, when participants are travelling back to their countries after a global event this might contribute to a global epidemic within a short period of time. For example, in their study Khan et al. [32] provided detailed analysis of the outgoing travel patterns from Mexico after the identification of the novel 2009 A (H1N1) influenza in Mexico and California. Their study confirmed H1N1 introductions within a few weeks in 20 countries where the highest number of arriving passengers from Mexico. Also, as stated previously after the Hajj season on 2009 several studies [7–10] were conducted to trace H1N1 pandemics among the returning pilgrims arriving from Saudi Arabia. A recent study by Khan et al. [33] analyzed 2012 international air travel data and Hajj data to predict pilgrims' movements after Hajj and estimate the potential novel coronavirus (MERS-CoV) importations.

Also, it is important to know and include the setting of the event; whether the event is held in one location or several locations; in a confined space or an open area, and whether it is a seated or mobile event. For example, in the Olympics Games and the Soccer World Cups several locations are involved. However, for most religious events, the rituals are performed at specific holy sites. Also, there is a need to determine the possible movements of the participants during the event; is there any specific regulations on their movements or their interactions with the local population within the hosting country. For instance, the movement of pilgrims attending Hajj is restricted and they are not allowed to extend their stay in the country after the completion of Hajj. All these aspects can play an important role in the dynamics of the disease spread and should be used to guide the modeling process of infectious disease in MGs.

4 Big Data and Modeling Disease Spread at MGs

There are several challenges related to the required data when attempting to model disease spread in global MGs. The variety of data to be captured from several sources is extensive, and includes demographics, airline traffic data, climate data, spatial data, and social and mobile phone network. Moreover, gathering detailed data about the expected participating populations at MGs prior to the event is very challenging if not infeasible. But, it is essential to have at least an adequate approximation of the demographics of the expected populations. For example, data from previous occurrences of MGs can be collected and analyzed to estimate the features of future participants attending similar events. In addition, as suggested by Nsoesie et al. [20] and Chowell et al. [22] using advanced monitoring and sensing devices of a representative sample provides higher levels of details about participants at MGs and their interactions and movements. With such advanced methods extensive amounts of data will be generated and need to be analyzed to extract useful information about the participants, their demographics, contact patterns, and behaviors during the event. Social media such as twitter can provide an alternative data sources about these events. Analyzing tweets posted by the participants can provide insights about their behaviors, movements and the event itself, which can be used when modeling disease spread. Moreover, simulating vastly increasing and high dimensional heterogeneous data representing the populations at global MGs provide a great big data challenge. As there is a need to integrate individual differences when modeling disease transmission at each stage of the event. In fact, some recent studies are aiming to include genetic variation among individuals when modeling infectious diseases and to use pathogens' genetic profiles to construct epidemics [33], which will result in having to process vast amount of genetic data.

Moreover, representing populations attending MGs in the epidemic model is a challenging task, as we want to capture differences on individual levels, we also need to represent specific subgroups among the entire population. For example, participants need to be grouped based on their countries of origin to model disease spread in the arrival and departure stages of the event. In addition, modeling epidemics in global MGs needs to handle the multiple spatial scales as the participants in these events arrive from different geographical locations and gathered in a specific local space. Even within the setting of the event the participants can go back and forth to different locations either randomly (e.g. audience at Olympic games) or following a specific rituals (e.g. pilgrims at Hajj). Modeling epidemics at MGs requires advanced spatial-temporal simulations that allow the introduction of prevention and mitigation measures. Therefore, it is important to capture and represent the interactions and movements of participants at these different spatial-temporal levels.

5 Summary

Global mass gatherings can pose public health threats on a large scale. Thus, it is important to apply computational epidemic modeling to estimate potential global epidemics. Based on the comprehensive review of the literature presented in this paper, we were able to summarize the key data requirements to model disease spread in these events. While there are several required data related to both the participants attending the event and the event itself, the most important aspect is the incoming and outgoing travel patterns before and after the event. Travel patterns of the participants can be used to predict the path and the dynamics of a potential global epidemic. There is a need for more advanced studies in the context of modeling disease spread in MGs. These studies should provide reliable methods to determine or adequately estimate the incoming and outgoing travel patterns, the event size and setting, and the characteristics of the expected population. These studies should also identify the best methodologies to utilize and integrate these data into disease-spread models. The diversity of the data sources provides opportunities to apply big data methods to assist developing epidemic models in the context of MGs. These challenges and opportunities include large data gathering, simulating enormous heterogeneous populations, approximate the underlying contact patterns among the participants using different data sources, and capturing the multiple spatial-temporal layers.

References

1. World Health Organization: Communicable Disease Alert and Response for Mass Gatherings: Key Considerations. WHO, Geneva (2008)
2. Memish, Z.A., Stephens, G.M., Steffen, R., Ahmed, Q.A.: Emergence of medicine for mass gatherings: lessons from the Hajj. Lancet. Infect. Dis **12**(1), 56–65 (2012)
3. Ebrahim, S.H., Memish, Z.A., Uyeki, T.M., Khoja, T.A., Marano, N., McNabb, S.J.: Pandemic H1N1 and the 2009 Hajj. Science **326**(5955), 938 (2009)
4. Al-Tawfiq, J.A., Smallwood, C.A., Arbuthnott, K.G., Malik, M.S., Barbeschi, M., Memish, Z.A.: Emerging respiratory and novel coronavirus 2012 infections and mass gatherings/infections respiratoires émergentes, nouveau coronavirus 2012 et rassemblements de masse. East. Mediterr. Health J. **19**, S48 (2013)
5. Tabatabaei, S.M., Metanat, M.: Mass gatherings and infectious diseases epidemiology and surveillance. Int. J. Infection **2**(2) (2015)
6. Gautret, P., Parola, P., Brouqui, P.: Risk factors for H1N1 influenza complications in 2009 Hajj pilgrims. Lancet **375**(9710), 199–200 (2010)
7. Al-Jasser, F.S., Kabbash, I.A., AlMazroa, M.A., Memish, Z.A.: Patterns of diseases and preventive measures among domestic hajjis from Central Saudi Arabia. Saudi Med. J. **33**(8), 879–886 (2012)
8. Ziyaeyan, M., Alborzi, A., Jamalidoust, M., Moeini, M., Pouladfar, G.R, Pourabbas, B., et al.: Pandemic 2009 influenza A (H1N1) infection among 2009 Hajj pilgrims from Southern Iran: a real-time RT-PCR based study. Influenza Other Respir.Viruses **6**(6), e80–e84 (2012)
9. Kandeel, A., Deming, M., Elkreem, E.A., El-Refay, S., Afifi, S., Abukela, M., Earhart, K., El-Sayed, N., El-Gabay, H.: Pandemic (H1N1) 2009 and hajj pilgrims who received predeparture vaccination, Egypt. Emerg Infect Dis. **17**(7), 1266–1268 (2011)
10. Gautret, P., Vu Hai, V., Sani, S., Doutchi, M., Parola, P., Brouqui, P.: Protective measures against acute respiratory symptoms in French pilgrims participating in the Hajj of 2009. Journal of travel medicine **18**(1), 53–55 (2011)
11. Gutiérrez, I., Litzroth, A., Hammadi, S., Van Oyen, H., Gerard, C., Robesyn, E., Bots, J., Faidherbe, M.T., Wuillaume, F.: Community transmission of influenza A (H1N1) V virus at a rock festival in Belgium, 2–5 July 2009. Euro surveillance: bulletin Europeen sur les maladies transmissibles = Eur. Commun. Dis. Bull. **14**(31), 2202–2206 (2009)
12. Abubakar, I., Gautret, P., Brunette, G.W., Blumberg, L., Johnson, D., Poumerol, G., Memish, Z.A., Barbeschi, M., Khan, A.S.: Global perspectives for prevention of infectious diseases associated with mass gatherings. Lancet. Infect. Dis **12**(1), 66–74 (2012)
13. Gundlapalli, A.V., Rubin, M.A., Samore, M.H., Lopansri, B., Lahey, T., McGuire, H.L., Winthrop, K.L., Dunn, J.J., Willick, S.E., Vosters, R.L., Waeckerle, J.E.: Influenza, winter olympiad, 2002. Emerg. Infect. Dis. **12**(1), 144 (2006)
14. Jorm, L.R., Thackway, S.V., Churches, T.R., Hills, M.W.: Watching the games: public health surveillance for the Sydney 2000 olympic games. J. Epidemiol. Community Health **57**(2), 102–108 (2003)
15. Steffens, I., Martin, R., Lopalco, P.: Spotlight on measles 2010: measles elimination in Europe—a new commitment to meet the goal by 2015. Euro. Surveill. **15**(50), 19749 (2010)
16. Steffen, R., Bouchama, A., Johansson, A., Dvorak, J., Isla, N., Smallwood, C., Memish, Z.A.: Non-communicable health risks during mass gatherings. Lancet Infect. Dis. **12**(2), 142–149 (2012)
17. Johansson, A., Batty, M., Hayashi, K., Al Bar, O., Marcozzi, D., Memish, Z.A.: Crowd and environmental management during mass gatherings. Lancet Infect. Dis. **12**(2), 150–156 (2012)
18. Khan, K., McNabb, S.J., Memish, Z.A., Eckhardt, R., Hu, W., Kossowsky, D., Sears, J., Arino, J., Johansson, A., Barbeschi, M., McCloskey, B.: Infectious disease surveillance and modelling across geographic frontiers and scientific specialties. Lancet. Infect. Dis. **12**(3), 222–230 (2012)

19. Tam, J.S., Barbeschi, M., Shapovalova, N., Briand, S., Memish, Z.A., Kieny, M.P.: Research agenda for mass gatherings: a call to action. Lancet. Infect. Dis. **12**(3), 231–239 (2012)
20. Nsoesie, E.O., Kluberg, S.A., Mekaru, S.R., Majumder, M.S., Khan, K., Hay, S.I., Brownstein, J.S.: New digital technologies for the surveillance of infectious diseases at mass gathering events. Clin. Microbiol. Infect. **21**(2), 134–140 (2015)
21. Daley, D.J., Gani, J., Gani, J.M.: Epidemic Modelling: An Introduction. Cambridge University Press, Cambridge (2001)
22. Chowell, G., Nishiura, H., Viboud, C.: Modeling rapidly disseminating infectious disease during mass gatherings. BMC Med. **10**(1), 159 (2012)
23. Khan, K., Freifeld, C.C., Wang, J, Mekaru, S.R., Kossowsky, D., Sonricker, A.L., Hu, W., Sears, J., Chan, A., Brownstein, J.S.: Preparing for infectious disease threats at mass gatherings: the case of the Vancouver 2010 olympic winter games. Can. Med. Assoc. J. **182** (6), 579–583 (2010)
24. Stehlé, J., Voirin, N., Barrat, A., Cattuto, C., Colizza, V., Isella, L., Régis, C., Pinton, J.F., Khanafer, N., Van den Broeck, W., Vanhems, P.: Simulation of an SEIR infectious disease model on the dynamic contact network of conference attendees. BMC Med. **9**(1), 87 (2011)
25. Shi, P., Keskinocak, P., Swann, J.L., Lee, B.Y.: The impact of mass gatherings and holiday traveling on the course of an influenza pandemic: a computational model. BMC Public Health **10**(1), 1 (2010)
26. Cowling, B.J., Chan, K.H., Fang, V.J., Lau, L.L., So, H.C., Fung, R.O., Ma, E.S., Kwong, A.S., Chan, C.W., Tsui, W.W., Ngai, H.Y.: Comparative epidemiology of pandemic and seasonal influenza a in households. N. Engl. J. Med. **362**(23), 2175–2184 (2010)
27. World Health Organization. Taking sex and gender into account in emerging infectious disease programme: an analytical framework. WHO Regional Office for the Western Pacific, Manila (2011)
28. The European Centre of Disease Prevention and Control (ECDC). Seasonal influenza vaccination in EU/EEA, influenza season 2010–11. http://ecdc.europa.eu
29. The European Centre of Disease Prevention and Control (ECDC). Seasonal influenza vaccination in EU/EEA, influenza season 2011–12. http://ecdc.europa.eu
30. Al-Jasser, F.S., Kabbash, I.A., AlMazroa, M.A., Memish, Z.A.: Patterns of diseases and preventive measures among domestic hajjis from Central Saudi Arabia. Saudi Med. J. **33**(8), 879–886 (2012)
31. Khan, K., Arino, J., Hu, W., Raposo, P., Sears, J., Calderon, F., Heidebrecht, C., Macdonald, M., Liauw, J., Chan, A., Gardam, M.: Spread of a novel influenza A (H1N1) virus via global airline transportation. New England journal of medicine **361**(2), 212–214 (2009)
32. Khan, K., Sears, J., Hu, V.W., Brownstein, J.S., Hay, S., Kossowsky, D., Eckhardt, R., Chim, T., Berry, I., Bogoch, I., Cetron, M.: Potential for the international spread of middle East respiratory syndrome in association with mass gatherings in Saudi Arabia. PLoS Currents **5** (2013)
33. Kao, R.R., Haydon, D.T., Lycett, S.J., Murcia, P.R.: Supersize me: how whole-genome sequencing and big data are transforming epidemiology. Trends Microbiol. **22**(5), 282–291 (2014)

Concept-Based Sentiment Analysis for Opinion Texts with Multiple-Languages

Jantima Polpinij, Natthakit Srikanjanapert and Chetarin Wongsin

Abstract Today, millions of message posted daily contain opinions of users in a variety of languages, including emoticon. Sentiment analysis becomes a very difficult task, and the understanding and knowledge of the problem and its solution are still preliminary. Therefore, this work presents a new methodology, called Concept-based Sentiment Analysis (C-SA). The main mechanism of the C-SA is Msent-WordNet (Multilingual Sentiment WordNet), which is used to prove and increase the results accuracy of sentiment analysis. By using the Msent-WordNet, all words in opinion texts having similar sense or meaning will be denoted and considered as a same concept. Indeed, concept-level sentiment analysis aims to go beyond a mere word-level analysis of text and provide novel approaches to sentiment analysis that enables a more efficient solution from opinion text. This can help to reduce the inherent ambiguity and contextual nature of human languages. Finally, the proposed methodology is validated through sentiment classification.

Keywords Sentiment analysis · Opinion texts · Multiple languages · Text classification

1 Introduction

Sentiment analysis (also known as polarity analysis, opinion analysis, or feeling analysis) [1] focuses on understanding of people's sentiments and opinions about an interested subject (i.e. product, service, news, travel, and so on) [2]. The polarity

J. Polpinij (✉) · N. Srikanjanapert · C. Wongsin
Intellect Laboratory, Faculty of Informatics, Mahasarakham University,
Mahasarakham 44150, Thailand
e-mail: jantima.polpinij@gmail.com

N. Srikanjanapert
e-mail: nsrikanjanapert@gmail.com

C. Wongsin
e-mail: chetarin1995@gmail.com

P. Meesad et al. (eds.), *Recent Advances in Information and Communication Technology 2016*, Advances in Intelligent Systems and Computing 463,
DOI 10.1007/978-3-319-40415-8_4

of given text can be positive, negative, or neutral. In advance of sentiment analysis, it is called "beyond polarity" [3, 4], which also looks for people's feeling, for instance, at emotional states (such as angry, sad, and happy). Basically, sentiment analysis refers to Natural Language Processing (NLP), text analysis, and computational linguistics to identify and extract subjective information (or infer new knowledge) in sentiment or opinion texts. This is because sentiment analysis is driven on an engineering discipline that has used computers to do useful things with language [5] and extract subjective information (or infer new knowledge) in sentiment or opinion texts. This is because sentiment analysis is driven on an engineering discipline that has used computers to do useful things with language [5].

Today, customers' reviews can provide valuable information for companies, where an importance of information-containing customers' behavior has always been to seek out what other people think [6]. The importance of sentiment analysis can be summarized as follows [6, 7]. Sentiment analysis can help to understand how products and services of each company are perceived. The results of this study can yield clues about customers' satisfaction and expectations. Certainly, this knowledge can be used to determine their current and future needs and preferences. Furthermore, can help in understanding what product dimensions or attributes are important to each consumer group. It includes the discovery of new groups of consumers concerned about the same attributes or factors. Lastly Sentiment analysis can help to provide companies with a competitive market advantage and long-term stability, because it provides essential information to accomplish a successful competitor analysis. As searching of significant information from customers' opinions is of great importance, sentiment analysis is adopted to many business intelligence applications [6, 8, 9].

However, people can freely access to describe the items in some detail and evaluate them with different languages because of with the growth of the popularization of review sites, forums, blogs, and social media [10]. Therefore, it is possible that many companies can gather customer feedbacks with multiple languages, including special language (i.e. emoticon). For example, consider this sentence, "มันช่างดีจริงๆ ☺". This sentence is written with Thai and Emoticon. This becomes a very difficult task for sentiment analysis on the Web because understanding of this problem and its solution are still preliminary [2]. Therefore, this is an existing problem of sentiment analysis for today [10–13].

In the last decade, the research community in this field has actively proposed and improved methods to detect and classify the opinions and sentiments expressed in different types of text [12, 13]. In general, a technique of mapping subjectivity lexicon to other languages has been proposed for solving this problem [10, 14–18]. At present, it is a popular technique to apply for sentiment analysis. Although many researches have been presented that they work on multilingual sentiment analysis, these works actually concentrate on bilingual sentiment analysis (found in [10, 14–18]. This is because all researchers cannot understand in every natural language in the world. Therefore, they just presented the concept of multilingual sentiment analysis based on bilingual sentiment analysis.

As the problem mentioned above, it becomes a challenge in this study. This work aims to present a novel methodology, called Concept-based Sentiment Analysis (C-SA), for analyzing online opinion textual dataset with multiple-languages. A main mechanism of the CSA is Multilingual Sentiment WordNet (Msent-WordNet), which is used to prove and increase the results accuracy of multilingual sentiment analysis. By using the Msent-WordNet, all words in opinion texts having similar sense or meaning will be denoted and considered as a same concept. With the use of concept-level sentiment analysis, this will go beyond a mere word-level analysis of text and provide novel approaches to sentiment analysis that enables a more efficient solution from opinion text. Finally, the proposed methodology is validated through sentiment classification, where the opinion texts of feeling about attraction places in Thailand are collected from Facebook.

2 The Proposed Research Methodology

2.1 *Preliminary: The Msent-WordNet*

Msent-WordNet (**M**ultilingual **Sent**iment **WordNet**) is the controlled vocabulary thesaurus covering three languages (English, Thai, and Emotional1). It consists of sets of terms in a hierarchical structure that permits searching in the different languages [19]. The Msent-WordNet a set of sentiment concepts, $C = \{c_1, c_2, c_3, c_4, \ldots\}$, that is required in order to support the semantic analysis during text processing, where the inaccuracy of sentiment analysis can be due to the different languages between opinion texts. It is developed under the concept of WordNets [20].

A sentiment concept is a collection of words that are different forms and different languages. In this case, it concentrates on English, Thai, and Emoticon. Each word contains significant information, namely *morphological*, *syntactic*, and *semantic* information. Morphological information is word formation, while the syntactic information gives the grammatical classification of word. Semantic information is relevant to its synonyms and antonyms. It includes the words with similar sense, although they are different languages.

The MSent-WordNet is designed and developed in the modification format of XSD, where XSD is a standard of ontology language. In the MSent-WordNet, even though these sentiment words are from different languages, they are under the same concept. It is noted that the semantic information presents synonyms and antonyms in English can be obtained from WordNet, while the semantic information presents synonyms and antonyms in Thai have been obtained from LexiTron [21]. An example of the concept "*good*" stored in the Msent-WordNet is shown as Fig. 1.

An example of using the Msent-Wordnet can be illustrated following. Consider the following sentiment text, "Oh มันช่างดีจริงๆ ☺☺☺". It can be seen that this sentence

```
<?xml version = "1.0" encoding = "UTF-8"?>
<xsd: schema xmln = "http://w3.org/201/XMLScheme">
<xsd: sentiment_dictinary> MSent-WordNets
    ...
    <xsd: sentiment_word_ID> 0034
            <xsd: sentiment_word_concept> Good
                    <xsd: lanaguge> English
                            <xsd: morphology> single </xsd: morphology>
                            <xsd: syntactic> noun, N
                                    <xsd: semantics> goodness, righteousness,... </xsd: semantics>
                            </xsd: syntactic>
                            <xsd: syntactic> adjective, Adj
                                    <xsd: semantics> satisfying, trustworthy </xsd: semantics>
                            <xsd: syntactic> adverb, Adv
                                    <xsd: semantics> ... </xsd: semantics>
                            </xsd: syntactic>
                    </xsd: language>
                    <xsd: lanaguge> Thai, ไทย
                            <xsd: morphology> single </xsd: morphology>
                            <xsd: syntactic> adjective, adj, วิเศษณ์
                                    <xsd: semantics> ดี </xsd: semantics>
                            </xsd: syntactic>
                    </xsd: language>
                    ...
            </xsd: sentiment_word_concept>
    </xsd: sentiment_word_ID>
    ...
    ...
</xsd: sentiment_dictinary>
```

Fig. 1 An example of the Msent-Wordnet in the XSD format

contains three languages: English, Thai, and Emoticon. With the use of the Msent-Wordnet, it helps a system to understand that the words "ดี" and "☺" have the same meaning, because these words will be transformed into the same concept "*good*". Finally, the concept "*good*" will be used instead of "ดี" and "☺" during the sentiment analysis process.

To this end, concept-level sentiment analysis aims to go beyond a mere word-level analysis of text. Also, this may provide novel approaches to sentiment analysis that enable a more efficient solution from opinion text, where it can help to reduce the inherent ambiguity and contextual nature of human languages [22].

2.2 The Methodology of Concept-Based Sentiment Analysis for Opinion Texts with Multiple Language

In this section, it is to describe the methodology of concept-based sentiment analysis (C-SA) for opinion texts with multiple languages. The proposed methodology is a modification of lexicon-based approach. It consists of two main stages: language acquisition and sentiment analysis based classification. The overview of the C-SA methodology is shown as Fig. 2.

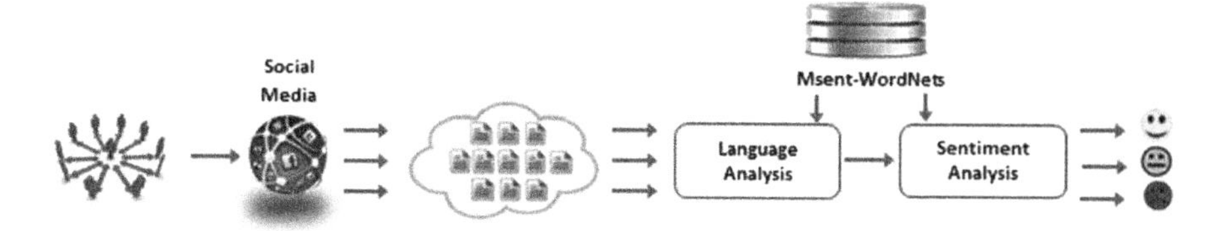

Fig. 2 The overview of conceptual methodology of CSA

Stage 1: Language Acquisition

This specific component is about language processing and comprehension, where natural language processing is a major source. In this stage, it consists of two processing steps: text tokenization and language analysis and transformation. They can be described as follows (Fig. 3).

A. Text Tokenization

It is the process of dividing a stream of opinion text up into words, phrases, symbols, or other meaningful elements called "*tokens*". Basically, most Natural Language Processing (NLP) applications require input text to be tokenized into individual terms or words before being processed further. In this study, the outcome becomes input for further processing such sentiment analysis. In general, text tokenization occurs at the word level, called lexicon analysis. Many tokenizers can rely on simple heuristics. Such English and emoticon, tokens can be separated by whitespace characters (i.e. a space or line break), or by punctuation characters. For Thai, while the main algorithm used to find words in text is *Longest Matching Algorithm* (*LMA*) [23]. The LMA is a well-known algorithm in Thai tokenization. This algorithm will scan an input sentence from left to right, and then select the longest match with a dictionary entry at each point. If the selected match cannot lead the algorithm to find the rest of words in the sentence, the algorithm will backtrack to find the next longest one and continue finding the rest, and so on.

B. Language Analysis and Transformation

This is a process of opinion text representation. After a word is searched and detected by using the Msent-Wordnet, the system will provide its concept and uses the concept as an index (surrogate) of a document. Simply speaking, we use the concepts as indexes of each document. It is noted that a document can have many

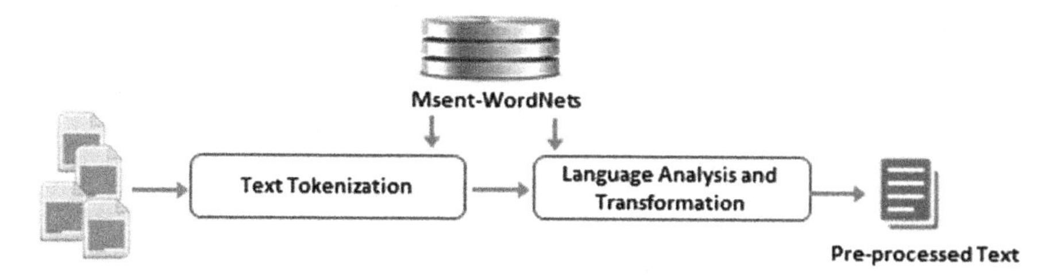

Fig. 3 The processes in the language acquisition stage

Table 1 Examples of sentiment words transformed into sentiment concepts

Opinion no.	Original opinion texts	Transforming to sentiment concepts by Msent-WordNet
1	ดีจริงๆ เลย ☺	good/good
2	It is a *good* product ☺ ☺ ☺	good/good/good/good

surrogates. With the use of the Msent-Wordnet, most indexes of each opinion text are the sentiment concepts. An example can be seen in Table 1.

Consider the examples in Table 1. By using the Msent-Wordnet, the original sentiment words will be transformed into the sentiment concept "*good*" after tokenizing text.

Stage 2: Sentiment Analysis based Classification

In general, sentiment analysis is to identify the orientation of opinion in a piece of text. In this work, the C-SA methodology is validated in the distinct setting of sentiment classification, which a popular task in sentiment analysis. The sentiment analysis based classification consists of two main processing steps: opinion text representation and sentiment classifier modeling. The overview of methodology can be shown as Fig. 4.

A. Opinion Text Representation

After tokenizing opinion text by using the Msent-Wordnet, the sentiment words will be represented specific words in a structured "*bag of words (BOW)*" [24]. Then, all sentiment words will be transferred into "*sentiment concepts*". For instance, if the words "*good*" and "ดี" are found in opinion texts, these words will be replaced with the sentiment concept "*good*". We do that in order to reduce an ambiguity problem in computational linguistics. After obtaining the concepts of each document, each document can be represented in a vector space, $d = \{c_{i,1}, c_{i,2}, \ldots, c_{i,t}\}$ called "*bag of concepts (BOC)*".

Afterwards, it is sentiment concept weighting. In general, *term frequency–inverse document frequency (tf-idf)* is the most common weighting method that has been widely used for determining the weight of a term in the vector space model [24]. The term weighting score indicates the number of occurrences of each feature within a document. This technique is the statistical measure used to evaluate how important a word is to a document in a collection. The term frequency tf_t, d of term t in document d is defined as the number of times that t occurs in d. Relevance does

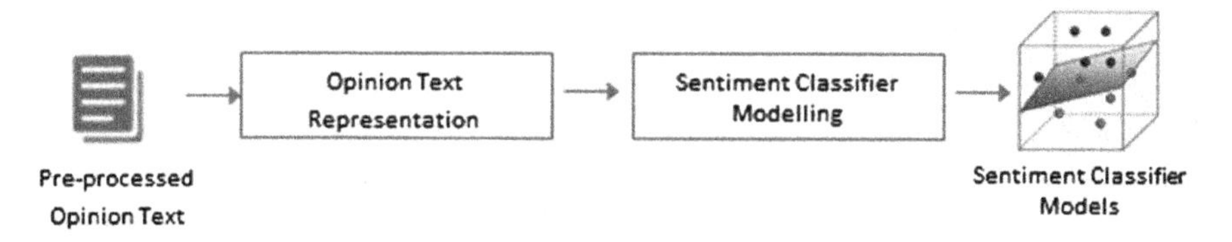

Fig. 4 The overview of sentiment analysis based classification

not proportionally increase with term frequency, while the score is *0* if none of the query terms are present in the document.

In this work, *tf-idf* is modified, where this work concentrates on '*concept*', not '*term*' or '*word*'. Therefore, *tf* is changed to *concept frequency* (*cf*). Then, each *cf* value should be greater than or equal to 1. The number of indexes exists in may be equal to the number of unique words in that entire opinion document. Finally, *tf-idf* is modified as *cf-idf.*

B. Sentiment Classifier Modelling

After the language acquisition component is done, it returns the outcome formatted as *BOC*. Later, the concepts containing in BOC are weighted by *cf-idf.* Finally, they is represented as, $d = \{c_{i,1}, c_{i,2}, \ldots, c_{i,t}\}$. Afterwards, the BOC is used for the processing of sentiment classifier modeling. To classify opinion texts into several groups, text classification is applied. This is because text classification is the process of grouping text documents into one or more predefined categories based on their content. By using "*concepts*" as document surrogate, another alternative name of our methodology can be called as the *concept-based sentiment classification.*

The majority approach to sentiment classification is the Naïve Bayes (NB) algorithm [25]. This algorithm is chosen because it is in the top 10 data mining algorithms identified by the IEEE International Conference on Data Mining (ICDM) [26, 27].

The main concept of NB is to use a set of training documents to estimate parameters and classes. In general context, the documents $D = \{d_1, d_2, \ldots, d_{|D|}\}$ to be classified are described by a vector of sentiment words $\omega = \{w_1, w_2, \ldots, w_i\}$. In this case, all sentiment words will be transformed into sentiment concepts, $\omega = \{c_1, c_2, \ldots, c_i\}$. Suppose $\nu = \{$ *negative*, *neutral*, *positive*$\}$ is a set of classes. A document will be assigned to a class by analyzing the *maximum a posterior classification*:

$$\nu = \underset{\nu \in class}{\arg\max}\, P(\nu|\omega) \tag{1}$$

Based on Bayes; theorem, it can be written as follows.

$$\nu = \underset{\nu \in class}{\arg\max}\, P(\nu) \times P(\omega|\nu) \tag{2}$$

The posterior probabilities $P(\omega|\nu) = P(c_1, c_2, \ldots, c_n|\nu)$ could be estimated directly from the training data, but are generally infeasible to estimate unless the available data is vast. So, the individual features (concepts) are conditionally independent of each other, given the classification, is introduced as:

$$P(c_1, c_2, \ldots, c_n|\nu) = \prod P(d_i|\nu) \tag{3}$$

With this strong assumption, formula (3) becomes the Naïve Bayesian Classifier:

$$v_{NB} = \arg\max_{v \in class} P(v) \prod \times P(d_i|v) \quad (4)$$

By this approach, we can develop the formal pattern with the ranging of candidate sentence probability estimation.

3 The Experimental Results

The conceptual methodology of C-SA is validated in the distinct setting of sentiment classification. This methodology is applied to classify a collection of reviews (opinion texts) into three groups: negative, neutral, and positive opinion. Then, the opinion texts are collected from Facebook and formatted as text file. These reviews present the feeling related to attraction places in Thailand. The collection consists of 1,000 reviews for Thai, and 1,000 reviews for English. Also, some of reviews contain some emoticons. We randomly select 70 % of texts for training the sentiment classifier models and using the rest for testing.

After the models of opinion text classifier are obtained, these models are typically evaluated by using performance measures from the standard information retrieval techniques [24]. Common performance measures for system evaluation are precision (P), recall (R), and F-measure (F). We test with 300 opinion texts related to attraction places in Thailand. We also compare the techniques of sentiment classification between the original methodology of text classification and the C-SA methodology with the Msent-Wordnet. The results can be shown in Table 2.

Consider in Table 2. It can be seen that the results of the C-SA methodology with the Msent-Wordnet are better than the result of the original methodology. This is because, when the concept is used as the surrogate of many words having the same meaning, it is to generate the semantics into these words. Therefore, it can reduce an ambiguous problem during sentiment analysis process. This can demonstrate that the proposed mythology can improve the accuracy of multilingual sentiment analysis based classification.

Table 2 The experimental results of automatic classification

Methodology	Recall (%)	Precision (%)	F-measure (%)
The tradition methodology	75.00	73	76.70
The proposed methodology	85.00	83	79.90

4 Conclusion

This work is to investigate a solution to solve the problem found in sentiment analysis, where an opinion text contains multiple-languages. Therefore, a novel methodology, called Concept-based Sentiment Analysis (C-SA), is proposed for analyzing online opinion texts with multiple-languages. A main mechanism of the C-SA is Multilingual Sentiment WordNet (Msent-WordNet), which is used to prove and increase the results accuracy of multilingual sentiment analysis. By using the Msent-WordNet, all words in opinion texts having similar sense or meaning will be denoted and considered as a same concept. With the use of concept-level sentiment analysis, this will go beyond a mere word-level analysis of text and provide novel approaches to sentiment analysis that enables a more efficient solution from opinion text. Finally, the proposed methodology is validated through sentiment classification, where the opinion texts of feeling about attraction places in Thailand are collected from Facebook. Finally, the conceptual methodology of C-SA is validated in the distinct setting of sentiment classification. The results of the C-SA methodology with the Msent-WordNet are better than the result of the original methodology. This is because, when the concept is used as the surrogate of many words having the same sense, it can reduce an ambiguous problem during sentiment analysis process. This can demonstrate that the proposed mythology can improve the accuracy of multilingual sentiment analysis.

References

1. Wilson, T., Wiebe, J., Hoffmann, P.: Recognizing contextual polarity: an exploration of features for phrase-level sentiment analysis. Assoc. Comput. Linguist. **35** (2009)
2. Liu, B.: Sentiment Analysis and Subjectivity. In: Handbook of Natural Language Processing. CRC Press, Taylor and Francis Group (2010)
3. Smith, P.: Sentiment analysis: beyond polarity. Thesis Proposal, School of Computer Science, University of Birmingham, UK (2011)
4. Reddy, A.S.S.: Polarity analysis through neutralization of non-polar words and segregation of polar words using training data. Int. J. Comput. Sci. Inf. Technol. **3**(5), 5176–5178 (2012)
5. Saif, H., Fernandez, M., He, Y., Alani, H.: SentiCircles for contextual and conceptual semantic sentiment analysis of Twitter. In: The Semantic Web: Trends and Challenges, pp. 83–98 (2014)
6. Funk, A., Li, Y., Saggion, H., Bontcheva, K., Leibold, C.: Opinion analysis for business intelligence applications. In: First international workshop on Ontology-Supported Business Intelligence (at ISWC). ACM, Karlsruhe (2008)
7. Plaza, L., de Albornoz, J.C.: Sentiment analysis in business intelligence: a survey. In: Customer Relationship Management and the Social and Semantic Web: Enabling Cliens Conexus, pp. 231–252 (2012)
8. Tsai, F.T., Lu, H.M., Hung, M.W.: The effects of news sentiment and coverage on credit rating analysis. In: The Pacific Asia Conference on Information Systems (PACIS) (2010)
9. Colleoni, E., Arvidsson, A., Hansen, L.K.. Marchesini, A.: monitoring corporate reputation in social media using real-time sentiment analysis. In: 15th International Conference on Corporate Reputation: Navigating the Reputation Economy, New Orleans, USA (2011)

10. Polpinij, J.: Multilingual sentiment classification on large textual data, In: IEEE 4th International Conference on Big Data and Cloud Computing (2014)
11. Denecke, K.: Using SentiWordNet for multilingual sentiment analysis, In: IEEE 24th International Conference on Data Engineering Workshop (ICDEW), pp. 507–512 (2008)
12. Balahur, A., Turchi, M.: Comparative experiments for multilingual sentiment analysis using machine translation. In: Proceedings of the First International Workshop on Sentiment Discovery from Affective Data (SDAD) (2012)
13. Balahur, A., Turchi, M.: Multilingual sentiment analysis using machine translation? In: Proceedings of the 3rd Workshop on Computational Approaches to Subjectivity and Sentiment Analysis, pp. 52–60 (2012)
14. Banea, C., Mihalcea, R., Wiebe, J.: A bootstrapping method for building subjectivitylexicons for languages with scarce resources. In: Proceedings of the Conference on Language Resources and Evaluations (LREC) (2008)
15. Banea, C., Mihalcea, R., Wiebe, J.: Multilingual subjectivity: are more languages better? In: Proceedings of the International Conference on Computational Linguistics (COLING), pp. 28–36 (2010)
16. Kim, J., Li, J.J., Lee, J.H.: Evaluating multilanguage-comparability of subjectivity analysis systems. In: Proceedings of the 48th Annual Meeting of the Association for Computational Linguistics (2010)
17. Steinberger, J., Lenkova, P., Ebrahim, M., Ehrman, M., Hurriyetoglu, A., Kabadjov, M., Steinberger, R., Tanev, H., Zavarella, V., Vazquez, S.: Creating sentiment dictionaries via triangulation. In: Proceedings of the 2nd Workshop on Computational Approaches to Subjectivity and Sentiment Analysis, Portland, Oregon (2011)
18. Tromp, E., Pechenizkiy, M.: SentiCorr: Multilingual sentiment analysis of personal correspondence. In: IEEE 11th International Conference on Data Mining Workshops (ICDMW) (2011)
19. Hu, X., Tang, J., Gao, H., Liu, H.: Unsupervised sentiment analysis with emotional signals. In: Proceedings of the 22nd International Conference on World Wide Web (2013)
20. Miller, G.A.: WordNet: a lexical database for English. Mag. Commun. ACM **38**(11), 39–41 (1995)
21. Mekpiroon, O., Tammarattananont, P., Apitiwongmanit, N., Buasroung, N., Pravalpruk, B., Supnithi, T.: Dictionary-based translation feature in open source LMS: a case study of Thai LMS: LearnSquare. In: The 8th IEEE International Conference on Advanced Learning Technologies (2008)
22. Coden, A., Gruhl, D., Lewis, N., Mendes, P. N., Nagarajan, M., Ramakrishnan, C., Welch, S.: Semantic lexicon expansion for concept-based aspect-aware sentiment analysis. In: Semantic Web Evaluation Challenge Communications in Computer and Information Science, pp. 34–40 (2014)
23. Meknavin, S., Charoenpornsawat, P., Kijsirikul, B.: Feature-based thai word segmentation. In: Proceedings of the Natural Language Processing Pacific Rim Symposium (NLPRS) (1997)
24. Baeza-Yates, R.A., Ribeiro-Neto, B.: Modern Information Retrieval. ACM Press, Addison-Wesley, New York (1999)
25. Nigam, K., McCallum, A., Thrun, S., Mitchell, T.: Text classification from labeled and unlabeled document using EM. Mach. Learn. **39**(2/3), 103–134 (2000)
26. Wu, X., Kumar, V., Quinlan, R.J., Ghosh, J., Yang, Q., Motoda, H., McLachlan, A., Ng, G.J., Liu, B., Yu, P.S., Zhou, Z.H., Steinbach, M., Hand, D.J., Steinberg, D.: Top 10 algorithms in data mining. Knowl. Inf. Syst. **14**, 1–37 (2008)
27. Aggarwal, C.C., Zhai, C.X.: A survey of text classification algorithms. In: Mining Text Data, pp. 163–222 (2012)

Automatic Advisor for Detecting Summarizable Chat Conversations in Online Instant Messages

Fajri Koto and Omar Abdillah

Abstract In this paper, we report the first work ever of detecting the summarizable chat conversation in order to improve the quality of summarization and system performance, especially in real time server-based system like online instant messaging. Summarizable chat conversation means that the document assessed could produce a meaningful summary for human. Our study intends to answer the question: what are the characteristics of a summarizable chat and how to distinguish it with non-summarizable chat conversation. To conduct the experiment, corpora of 536 chat conversations was constructed manually. Technically, we used 19 attributes and grouped them by feature sets of (1) chat attribute, (2) lexical, and (3) Rapid Automatic Keyword Extraction (RAKE). As result, our work reveals that the features can classify summarizable chat by 78.36 % as our highest accuracy, performed by feature selection with SVM.

Keywords Chat · Summarizable · Non-summarizable · Feature selection

1 Introduction

Text summarization has been actively investigated in recent years. It is caused by the overwhelming data available in the Internet, while information needs to obtain quickly along with technologies become more sophisticated. One of focused research, chat summarization also attracts the researcher interest due to the emerging of online instant messaging technologies that engender a big number of user and messages that are sent and received every day [1, 2]. As all of summarization approaches work by directly applying bunch of messages without considering whether these

F. Koto (✉) · O. Abdillah (✉)
Advanced Research Lab, Samsung R&D Institute Indonesia, Jakarta, Indonesia
e-mail: fajri.fajri@samsung.com
URL: http://www.samsung.com

O. Abdillah
e-mail: o.abdillah@samsung.com

P. Meesad et al. (eds.), *Recent Advances in Information and Communication Technology 2016*, Advances in Intelligent Systems and Computing 463,
DOI 10.1007/978-3-319-40415-8_5

Table 1 Example of summarizable and non-summarizable chat

Summarizable	Non-summarizable
A: Hi guys, lets have holiday somewhere, I need some fresh air	A: Hi dudes
	B: Yo
B: Where? I m free this weekend	A: How are you?
A: Florida beach on Sunday?	C: Tsup?
C: Nice idea	B: good
A: Horray, lets do it	C: hm? what?
B: coooool idea	A: nothing

messages have meaningful summary or not, the result of summarization is always produced. Whereas not all documents can produce a meaningful summary. Therefore, optimization is needed in order to provide better service to the user.

Summarizable chat conversation means that the document assessed could produce a meaningful summary for human. In other words, the purpose is to distinguish content-ful and content-less text in chat room discourse. As example in Table 1, the first conversation can be summarized as "*holiday to florida beach on Sunday*", while the second conversation is judged as non-summarizable conversation since there is no meaningful topic discussed in the conversation. We realize that the distinction between these conversations could be relative and depend on individual opinions. However, we consider this problem is possible to investigate by first constructing the relevant corpus of summarizable chat documents.

Our work is a part of building an automatic chat summarization for mobile application. Working on chat documents that contain many unstructured sentences is not a trivial matter. We argue that embedding summarization system into the user device will be such a dilemma. Though the storage and the CPU become cheaper and faster today, not all devices on user side have sufficient configuration to perform the summarization task. For example the summarization based on semantic similarity [3] will seize big resource in client device. Therefore, a server-based system for summarizer in online instant messaging becomes a better choice.

In Fig. 1 we illustrate our server-based summarization system for mobile instant messaging. Before the chat document of each clients going through the summarizer, each document will be examined by an advisor system to predict its summarizability. Here the advantage of advisor system is not only to improve the quality of the resulting summary, but also to optimize the usage of summarization engine that may take lot of resources.

The rest of this paper is structured as follows. Section 2 summarizes some related works of chat summarization. Section 3 provides the construction of our dataset. The analysis of dataset is also discussed. We then discuss feature of summarizable chat in Sect. 4. Experimental set-up and experiment results will be given in Sect. 5. Finally conclusion are drawn in Sect. 6.

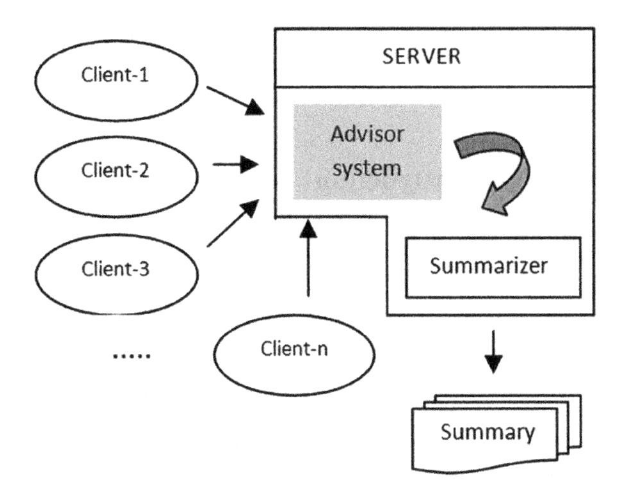

Fig. 1 Server-based summarization system with summarizable detection

2 Related Work

We realize that chat summarization is still in progressing stage among the researchers. There is only few number of works that have been published. As discussed in [2], it is caused by the difficulty in performing analysis of chat summarization. It includes: (1) chat characteristics that make it difficult to apply traditional NLP techniques, (2) Uncommon features such as frequent use of abbreviations, acronyms, deletion of subject pronouns, use of emoticons, abbreviation of nicknames, and stripping of vowels from words to reduce number of keystrokes [4]. Furthermore, working on multiple conversation also becomes a challenge in itself [5, 6].

In conducting this work, we only found some related study of chat summarization. [7] have worked on chat summarization by investigating summarizing chat logs in order to create summaries comparable to the human-made GNUe Traffic digests. [1] summarized real-time chat conversations which contain multiple users with frequent shifts in topic. Their approach consists of two phases: (1) leveraging topic modeling using web documents to find the primary topic of discussion in the chat, and (2) building a semantic word space to score sentences in performing summarization. In other study, [2] described their beginning stages of work on summarizing chat, which is motivated by their observations concerning the information overload of US Navy watchstanders. A similar work on summarization of media which share some similarities to chat has been also published. For example, [8] examined summarization of multi-party dialogues and [9] examined summarization of meeting recordings.

However, based on the related study, there is no any work that discusses about an advisor system that enable us to automatically distinguish summarizable chat conversation. Because it is the first investigation ever, until this stage there is no relevant corpora that can be used to perform quantitative evaluation. Therefore, we started the investigation by manually constructing the dataset with summarizable and non-summarizable label as discussed in next section.

3 Data Construction

The corpora was constructed by using seven WhatsApp groups chat in Indonesia language (Bahasa Indonesia). Most of the chat documents talk about daily life of the participants. We choose Bahasa Indonesia as language of our dataset in order to ease the labeling stage. In total they contain 51102 lines of chat and have 7 group member in average. Each chat in our raw data includes information of date and time, username and the contents.

In Fig. 2, our steps in constructing the dataset are described. First, we divided our chat documents into sections of 2 h conversation. As result, 3057 sections are produced. We then selected 1093 sections by performing stratified random sampling based on number of line. After that, two native speakers of Bahasa Indonesia conducted labeling manually. At the first stage of labeling, it produced a dataset with Kappa score equals to 0.46 [10]. According to [11], this number belongs to a moderate agreement. We then improved the dataset by conducting discussion between both annotators in which we decided an agreed new label or assigned label = 2 if two annotators still can achieve agreement.

The summary of the resulting corpora is given in Table 2. From 1093 annotated conversations, we did sampling to build a balanced dataset. Here, conversations containing more than 40 lines are excluded, caused by their tendency (98.6 %) as a summarizable documents. Therefore, we assume that the investigation of summarizable chat becomes a matter if it is conducted into chat documents containing few or some number of line (e.g. 1–10, 11–20, or less than 40). By the sampling, finally 536

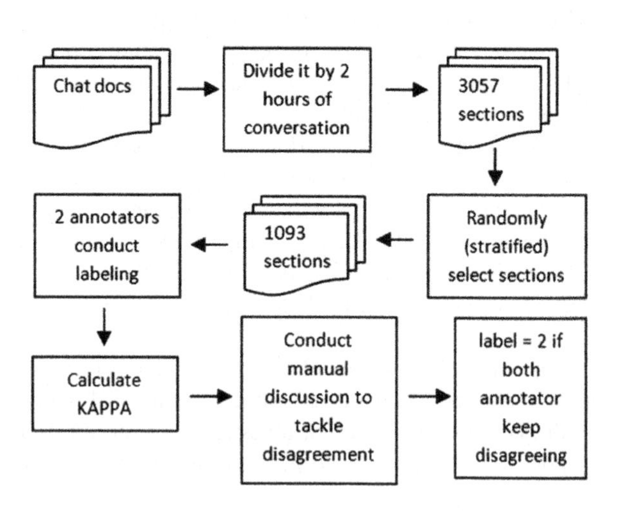

Fig. 2 Constructing the corpora

Table 2 Description of our raw data

#Line	Total	#Labeling	Label = 1		Label = 2		Label = 3
			Summarizable		Non-summarizable		Disagreement
			Total	Sampling	Total	Sampling	
1–10	1724	300	110	110	181	181	9
11–20	563	300	252	150	42	42	6
21–30	310	100	94	20	6	6	0
31–40	160	100	93	20	7	7	0
41–50	86	86	81	0	3	0	2
>50	207	207	205	0	1	0	1
Total	3057	1093	853	300	240	236	18

conversations are used, in which 300 belong to summarizable chat and 236 as non-summarizable chat.

4 Feature of Summarizable Chat Detection

Deciding a summarizable chat is often objective, it is strongly influenced by the message contents that are related to specific circumstances. However, we have derived set of features to approximate the difference based on our hypothesis. Nevertheless, these features do prove to be beneficial in the context of summarizable chat in Sect. 4. In total there are 19 features and all of them are summarized in Table 3. In this preliminary study we only analyze the text data and ignore the emoticon, pictures and video in the chat conversation.

We group features in Table 3 into three feature sets: (1) Chat attribute, (2) Lexical, and (3) Rapid Automatic Keyword Extraction (RAKE). The first and second group can be easily comprehended, while RAKE is used to verify our hypothesis of summarizable chats that tend to have more topic of conversation rather than non-summarizable. RAKE is an unsupervised, domain independent, and language independent method for extracting keywords from single document [12]. It can generate list of keywords from a document by performing 2 stages: (1) extraction of keywords candidate and followed with (2) scoring procedure.

5 Overview of RAKE

The input parameters for RAKE comprise a list of stop words (or stoplist), a set of phrase delimiters, and a set of word delimiters. RAKE uses stop words and phrase delimiters to partition the document text into candidate keywords, which are sequences of content words as they occur in the text.

Table 3 List of features for summarizable chat detection

Feature	Description
Chat attribute	
#user	Number of user participating in chat document
#chat	Number of chat section (line) in a document
ChatAvg	Average number of chat for each participant
Lexical	
#word	Number of words in a chat document
#unique	Number of unique words in a chat document
#non-stops	Number of non-stopword words in a chat document
#stops	Number of stopwords in a chat document
stopsRatio	Ratio between #stops and #word
nonStopsRatio	Ratio between #non-stops and #word
wordAvg	The average of non-stopword's frequency in a chat section
wordMax	Maximum value of non-stopword's frequency in a chat section
wordMin	Minimum value of non-stopword's frequency in a chat section
RAKE	
wordDegreeAvg	The average of words degree in the word co-occurrences graph
wordFreqAvg	The average of words frequency in the word co-occurrences graph
wordDegree	The ratio of wordDegreeAvg and wordFreqAvg
#keyword	number of extracted keywords generated by RAKE
keywordScoreAvg	The average of keyword score generated by RAKE
keywordScoreMax	Maximum value of keyword score generated by RAKE
keywordScoreMin	Minimum value of keyword score generated by RAKE

After the list of keywords is identified, RAKE works by evaluating several metrics for each unique word exist in the keywords list. Those metrics are: (1) word frequency ($freq(w)$), (2) word degree ($deg(w)$), and (3) ratio of degree to frequency ($deg(w)/freq(w)$). In summary, $deg(w)$ favors words that occur often and in longer candidate keywords. $deg(w)$ can be simply calculated by counting total number of words of keywords that contain word w. Finally, the score for each candidate keyword is computed as the sum of its member word scores.

Table 4 Classification result

Classifier	All (%)	Ft selection (%)	Selected feature
LR	77.24	77.05	#unique
NB	73.50	75.19	stopsRatio, #keyword
NN	**77.99**	78.18	#user, #keyword, keywordScoreMin
SVM	76.88	**78.36**	#keyword, keywordScoreMin

6 Experiment Result

In this work, the experiments were done by performing 5-fold cross validation in which 80 % of the corpus as training set as the remainder as testing set. We used an open source tools, Rapidminer [13] to classify the chat conversation by 4 different classifiers: (1) Linear Regression [14]; (2) Naive Bayes [15]; (3) Neural Network [16]; and (4) SVM [17] for the comparison. At the first, we used all features in classification. After that, feature selection with forward algorithm were conducted in order to obtain best feature combination for summarizability detection.

In Table 4 the performances of classifier by using all features are given. It shows that our features are able to distinguish summarizable chat by 77.99 % with Neural Network. By employing feature selection, our highest accuracy can reach 78.36 % performed by SVM, with using feature combination: (1) #keyword and (2) keywordScoreMin. According to this table, it also can be seen that some of selected features are RAKE group feature. Thus, It is in line with our hypothesis that shows the keyword or conversation topic is one of important parameter in detecting summarizable chat.

7 Conclusion and Future Work

As the first study on summarizable chat detection, this study reveals that summarizability of chat document can be observed. Our hypothesis defined that summarizable chat tends to have more number of conversation topic than non-summarizable chat. By employing three feature sets: (1) chat attributes, (2) lexical, and (3) RAKE (a keyword extraction procedure), we can distinguish summarizable chat by 78.36 % as the highest accuracy performed by feature selection with SVM classifier. The selected features are (1) #keyword and (2) keywordScoreMin and it directly verifies our hypothesis. As the future work, research on this field still has some challenges to improve the quality of the system. For example semantic approach can be implemented to further understand the characteristic of summarizable chat.

References

1. Sood, A., Mohamed, T.P., Varma, V.: Topic-focused summarization of chat conversations. In: Advances in Information Retrieval, pp. 800–803 (2013)
2. Uthus, D.C., Aha, D.W.: Plans toward automated chat summarization. In: Proceedings of the Workshop on Automatic Summarization for Different Genres, Media, and Languages. ACL (2011)
3. Koto, F., Sakriani S., Neubig, G., Toda, T., Adriani, M., Nakamura, S.: The use of semantic and acoustic features for open-domain TED talk summarization. In: Proceedings of The 6th Asia Pacific Signal and Information Processing Association (APSIPA). Siem Reap, Cambodia (2014)
4. Werry, C.C.: In: Linguistic and Interactional Features of Internet Relay Chat, pp. 47–64 (1996)
5. Hering, S.C.: Interactional coherence in CMC. In: Proceedings of the Thirty-Second Annual Hawaii International Conference on System Sciences (1999)
6. Hering, S.C.: Computer-mediated conversation: introduction and overview. In: Language@ Internet (2010)
7. Zhou, L., Hovy, E.: Digesting virtual "geek" culture: the summarization of technical Internet Relay Chats. In: Proceedings of the 43rd Annual Meeting on Association for Computational Linguistics, pp. 298–305. ACL (2005)
8. Zechner, K.: Automatic summarization of open-domain multiparty dialogues in diverse genres. Comput. Linguist. **28**(4), 447–485 (2002)
9. Murray, G., Renals, S., Carletta, J., Moore, J.: Evaluating automatic summaries of meeting recordings. In: Proceedings of the ACL Workshop on Intrinsic and Extrinsic Evaluation Measures for Machine Translation and/or Summarization, pp. 33–40. ACL (2005)
10. Cohen, J.: A coefficient of agreement for nominal scales. Educ. Psychol. Meas. **20**(1), 37–46 (1960)
11. Altman, D.G.: Practical Statistics for Medical research, vol. 20(1). Chapman Hall/CRC Press, London (1990)
12. Berry, M.W., Kogan, J.: "Text Mining": Applications and Theory. Wiley, West Sussex, PO19 8SQ, UK (2010)
13. Akthar, F., Hahne, C.: RapidMiner 5 Operator Reference. In: Rapid-I GmbH (2012)
14. Montgomery, D.C., Peck, E.A., Vining, G.G.: Introduction to Linear Regression Analysis, vol. 821. Wiley, West Sussex, PO19 8SQ, UK (2012)
15. Lewis, D.D.: Naive (Bayes) at forty: the independence assumption in information retrieval. In: Educational and Psychological Machine learning: ECML-98, pp. 4–15. Springer, Berlin, Heidelberg (1998)
16. Fu, L.M.: Neural Network in Computer Intelligence. MIT-Press, McGraw-Hill International Edition (1994)
17. Cristianini, N., Shawe-Taylor, J.: An Introduction to Support Vector Machines and Other Kernel-Based Learning Methods. Cambridge university press (2000)

Optical Character Recognition for Nepali, English Character and Simple Sketch Using Neural Network

Subarna Shakya, Abinash Basnet, Suman Sharma and Amar Bdr Gurung

Abstract Optical Character Recognition (OCR) is the process of text extraction from of images of typewritten or handwritten text. It deals with the recognition of optically processed characters, with the advent of digital optical scanners a lot of paper based books, textbooks, magazines, articles and documents are being transformed into an electronic version that can be manipulated by a computer. Unlike English character recognition, Nepali languages are complicated in terms of structure and computations. Nepali language are derived from Devanagari Script; written from left to right fashion having common features of containing straight line on top 'Shiro Rekha'. The OCR systems developed for the Nepali language carry a very poor recognition rate due to error in character segmentation, ambiguity with similar character, unique character representation style. In this paper we proposed an OCR for Nepali text in Devanagari script, using multi-layer feed forward back propagation Artificial Neural Network (ANN), which improved its efficiency and accuracy. Adaptive learning rate with Gradient descent algorithm is implemented in Neural net with 2 hidden layers used with input and output and MMSE is the performance criteria. Various classifiers for training characters are created and stored. De-noised test sheet is carefully segmented and inputted in trained neural net resulted higher accuracy. Also we have included recognizing simple sketch like as tree, home, and ball.

Keywords OCR · Neural net · Nepal font · Image processing

S. Shakya (✉) · A. Basnet · S. Sharma · A.B. Gurung
Department of Electronics and Computer Engineering,
Central Campus Institute of Engineering, Tribhuvan University,
Kirtipur, Nepal
e-mail: drss@ioe.edu.np

A. Basnet
e-mail: abinash@ioe.edu.np

S. Sharma
e-mail: 069msice619@ioe.edu.np

A.B. Gurung
e-mail: amargurung@ioe.edu.np

P. Meesad et al. (eds.), *Recent Advances in Information and Communication Technology 2016*, Advances in Intelligent Systems and Computing 463,
DOI 10.1007/978-3-319-40415-8_6

1 Introduction

OCR (Optical Character Recognition) also called Optical Character Reader is a system that provides a full alphanumeric recognition of printed or handwritten characters at electronic speed by simply scanning. Recognition is the mapping of a low-level vector to a higher-level concept, For example mapping bitmaps to characters. Learning is to find out which low-level vectors correspond to high-level concepts.

Intelligent Character Recognition (ICR) has been used to describe the process of interpreting image data, in particular alphanumeric text. Images of handwritten or printed characters are turned into ASCII data (machine-readable characters). Usually, OCR uses a modular architecture that is open source, scalable, and workflow controlled. It includes forms definition, scanning, image pre-processing, and recognition capabilities.

Artificial Neural Network (ANN) is nonlinear parallel distributed highly connected mathematical model or computational model network having capability of adaptively, self-organization, fault tolerance, evidential response and closely resemble with physical nervous system. ANN system can perceive and recognize a character based on its topological features such as shape, symmetry, closed or open areas, and number of pixels. The advantage of such a system is that it can be trained on samples and then can be used to recognize characters having a similar (not exact) feature set. The ANN used in this system gets its inputs in the form of Feature Vectors i.e. every feature or property is separated and assigned a numerical value [1]. The training and testing process diagram is shown in Fig. 1.

Input: Samples are read to the system through a scanner.

Preprocessing: Preprocessing converts the image into a form suitable for subsequent processing and feature extraction.

Segmentation: The most basic step in Character Recognition is to segment the input image into object from noisy background. This step separates out sentences from text and subsequently words and letters from sentences also.

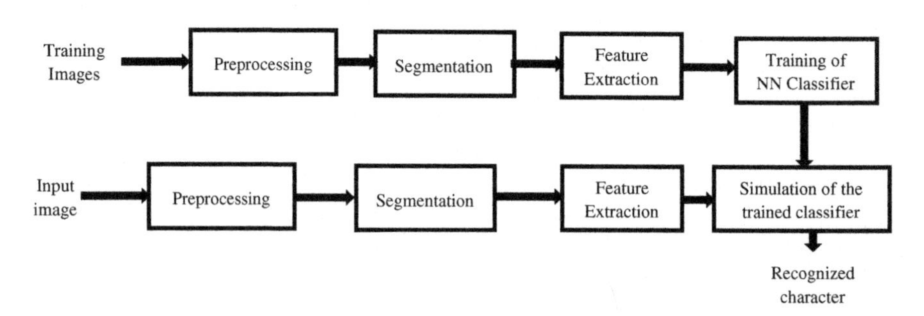

Fig. 1 Training and testing process for generic character recognition system

Feature extraction: Extraction of features i.e. geometry, of a character forms a vital part of the recognition process. Feature extraction collects the required information details of any characters.

Classification: During classification, a character is placed in the appropriate class to which it belongs. In training, the back propagation training algorithm subtracts the training output from the target (desired answer) to obtain the error signal. It then goes back to adjust the weights and biases in the input and hidden layers to reduce the error. Feed forward means there are no paths where signals travel backwards or sideways.

Post Processing: Evaluate MMSE with Combines the train and test classifier and does for all trained classifiers. If MMSE is less then defined value then directly goes for next character recognition which reduces execution time.

2 Background and Literature Review

In Tamil Character Recognition by using Kohonen SOM technique to classifies handwritten and also printed Tamil characters. But it was not for joined letters and had less segmentation accuracy [2]. Recognize printed and handwritten characters by projecting them on different sized grids results showed that the precision of the character recognition depends on the resolution of the character projection [3]. The smaller letters were better recognized with the network with smaller resolution. Regardless the difference of the orientation, size and place of the characters, the network still had a 60 % precision [3]. Simple pattern recognition can be done using artificial neural network to simulate character recognition. A simple feed-forward neural network model has been trained with different set of noisy data. The back-propagation method is used for learning in neural network. The experiment result shows recognition rate is 70 % for noisy data to up to 99 % [4].

Combining with several other modes of minimizing the searching space and helping the recognition with dictionary methods, neural networks can be a promising solution. In general, documents contain text, graphics, and images. The procedure of reading the text component in such a document can be divided into three steps: First, document layout analysis in which the text component of the document is extracted. Second, extraction of the characters from text component of the document, and finally recognized the segmented characters [5] (Fig. 2).

Line Segmentation, during letter segmentations, frequent problem occurs due to abnormally written characters therefore segmentation should be so precise. In line, the corresponding line axes are extracted through a skeletonization algorithm and the conflicts between adjacent cutting lines are solved by some heuristics [6]. In line segmentation our aim is to separate out the line of text from the image. For this global horizontal projection profile method is used which constructs a histogram of all the black pixels in every row as shown in Fig. 3. Based on the peak/valley points of the histogram, individual lines are separated. Word Segmentation, after line segmentation the boundary of the line (i.e. the top and bottom of the line) is known.

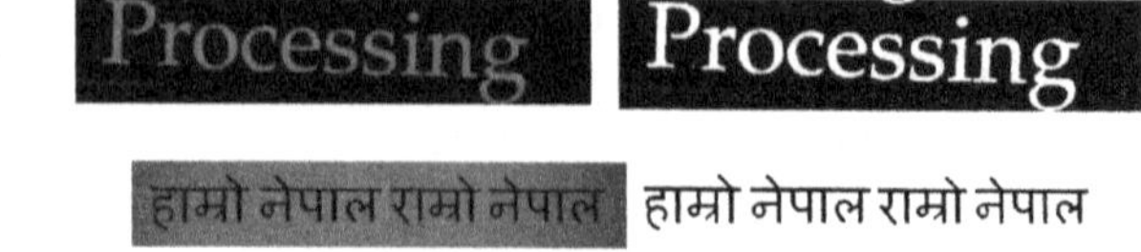

Fig. 2 Image before binarization (*left*); Image after binarization (*right*)

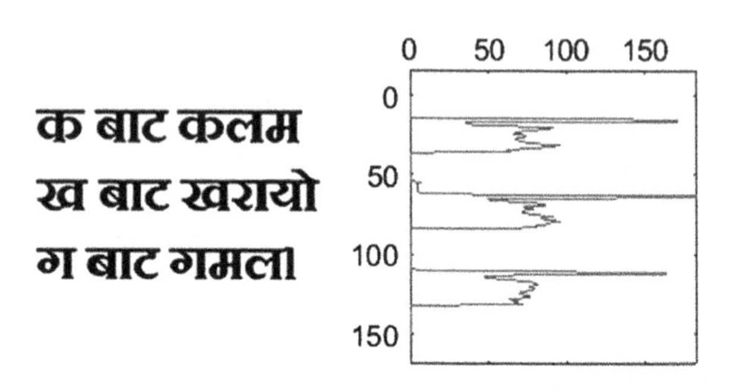

Fig. 3 Horizontal line profiles of a document for line segmentation

Binarization: Printed documents generally are black text on white background. Process of converting colored or gray scale images to bi-level image is often known as binarization or thresholding. Binarization of image on both English and Nepali is shown in Fig. 2. The pixel values of the binary image are stored in an array. All the pixel values in the array are compared with their horizontally adjacent pixel values, row by row, for the presence of collinear points (i.e., a line). It is done by detecting the continuity of either the white or black pixels accordingly. Once the continuity is detected, the starting and end coordinates are displayed as an intermediate result [7].

Segmentation phase is a very crucial stage since this is where most of the errors occur. Even in good quality documents, sometimes adjacent characters touch each other due to inappropriate scanning resolution or the design of characters. This can create problems in segmentation. Incorrect segmentation leads to incorrect recognition. Its phase includes line, word and character segmentation. It occurs in three steps for OCR: line segmentation, word segmentation and character segmentation.

Line Segmentation, during letter segmentations, frequent problem occurs due to abnormally written characters (which misguide the system during recognition) therefore segmentation should be so precise. In line segmentation our aim is to separate out the line of text from the image. For this global horizontal projection profile method is used which constructs a histogram of all the black pixels in every row as shown in Fig. 3. Based on the peak/valley points of the histogram, individual lines are separated. Word Segmentation, after line segmentation the boundary of the line (i.e. the top and bottom of the line) is known. Character recognized techniques on the basis of projection profile (including horizontal projection profile and stripe) in the experiment are best technique for single-column for sorting and distinguishing from document [8].

3 Proposed System

Neural network learning is based on learning from examples and their respective classes. And in supervised learning, main goal is to build a classification system from a set of patterns available. Because of the variety of patterns and the difficulties in expressing empirical rules, character recognition is very often based on training a system with patterns. Neural networks are especially suitable for this recognition purpose.

Following steps have been followed in the training of characters system:

Preprocessing: First scanned colored RGB image is converted to gray scale and then gray scale image is converted to binary. Preprocessing has done to improve the accuracy of the recognition algorithm. Main steps in preprocessing are salt and pepper noise removal, binarization, and skew correction. Then boundary of each character is detected. From histogram analysis 'shiro rekha' is detected as it has highest number of lower values of intensity pixels in text as shown on right portion of Fig. 3.

Word segmentation is done in the same way as line segmentation but in place of horizontal profiling, vertical projection profiling is done as shown in Figs. 3 and 4.

Morphological operation: It is used to create morphological structuring of square element completed with erosion and then dilates to address each character. It was useful to de-noised image as well.

The main sources of noise in the input image are as follows:

- Noise due to the quality of paper on which the printing is done.
- Noise induced due to printing on both sides of paper or the quality of printing.
- Noise added due to the scanner source brightness and sensors.

We found the boundary of the image which was done by finding blank spaces at left/right/top/bottom. For that we measured properties of image regions i.e. 'Shape', 'Area', 'Centroid', 'Filled Area' and 'Major Axis Length'. And result is reshaped to 5×7 character representations in single vector from binary image. Centroid of Processed image works as Feature to recognize.

Training of Classification: In Neural net we used feed forward back propagation with 2 hidden layers are trained as shown in Fig. 5. Parameters are used as:

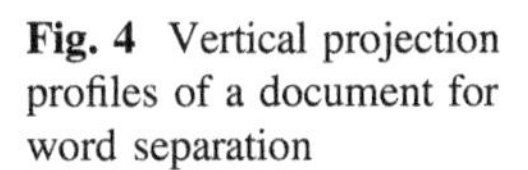

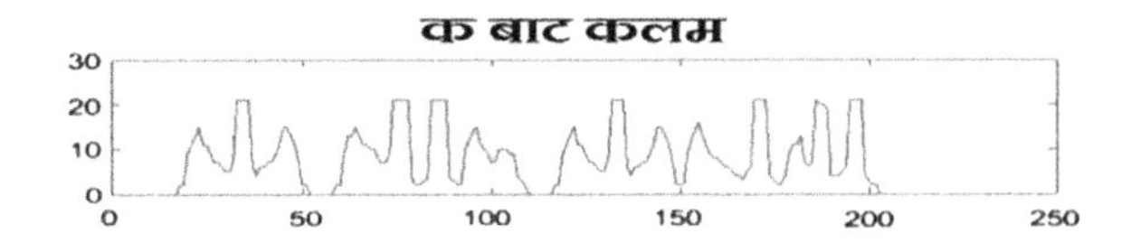

Fig. 4 Vertical projection profiles of a document for word separation

Accuracy goal = 0.01;
Epochs = 5000;
Machine train parameter = 0.95;
Input layers = output layer = 1; Hidden Layers = 2;

Train is done to the neural network such that with input data and target data clustered and returns the network after training it.

Feed forward back propagation algorithm is a method that depends on the gradient value of the moment. The learning starts when all of the training data was showed to the network at least once. For every network learning algorithm, consists of the modification of the weights. We used the gradient of the criteria field to determine the best weight/modification to minimize the mean square error.

Comparison with trained and test classifier and if trained classifier have less MMSE with test. Here we have lastly simulated with SIM: Evaluate network outputs given inputs. Training data sets are shown in Figs. 6, 7 and 8.

After extracting each line from a do-noised whole sheet image, start and end of each character word can be found from vertical projection profile. Here our logic is end of character ends with a maximum or symmetry with maximum or some additive length tailed from maximum defined by some higher pixel. This separated character was taken as test samples. Then each test character is analyzed with stored classifier and calculated MSE.

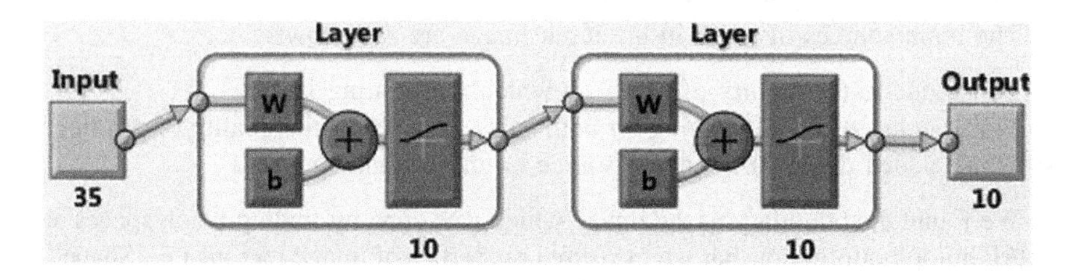

Fig. 5 Structure of used NN

Fig. 6 Nepali alphabets

क ख ग घ ङ च छ ज झ ञ
ट ठ ड ढ ण त थ द ध न प
फ ब भ म य र ल व श ष स
ह क्ष त्र ज्ञ

Fig. 7 Nepali numeric character

१ २ ३ ४ ५ ६ ७ ८ ९ ०

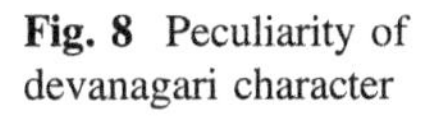

Fig. 8 Peculiarity of devanagari character

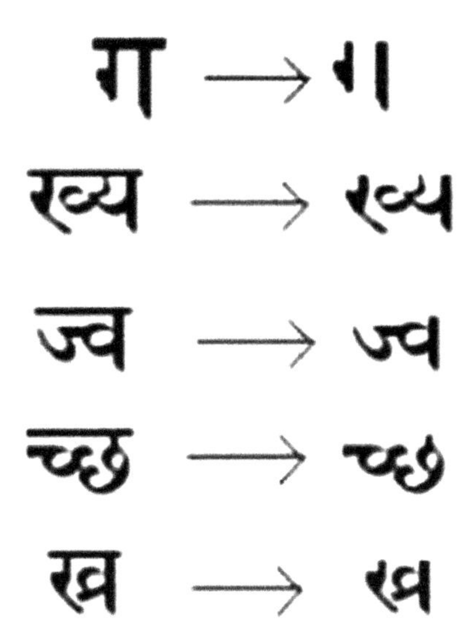

4 Results and Discussions

All the character set available in Nepali fonts are typed and written in paint are converted as train images. These train images are stored and processed with gradient descent with adaptive learning rate algorithm in neural net with 2 hidden layers used with input and output. And segmented test image are feed as input in the net resulted as in Fig. 9.

Along with Nepali alphanumeric characters, this work is able to recognize English alphanumeric also. Actually, Nepali character recognition is extended form of English with extra processing in training and testing data set hence train data sets are separately stored and processed while for testing we check for if maximum characters are likelihood with English or Nepali.

A Character is chosen from trained set as recognized character who's MMSE to train classifier is minimum and Corresponding character, is displayed. In addition in doc mode, corresponding train character is written and displayed in trained language, i.e. if Nepali then written and display in Nepali as shown in Figs. 10 and 11.

If handwritten character is rotated with some angle then MMSE increases and results less recognition but still all characters are trained with multiple probable sets hence result are better. For few characters, that are more distinct then other, are

Fig. 9 Simple Nepali handwritten character recognition

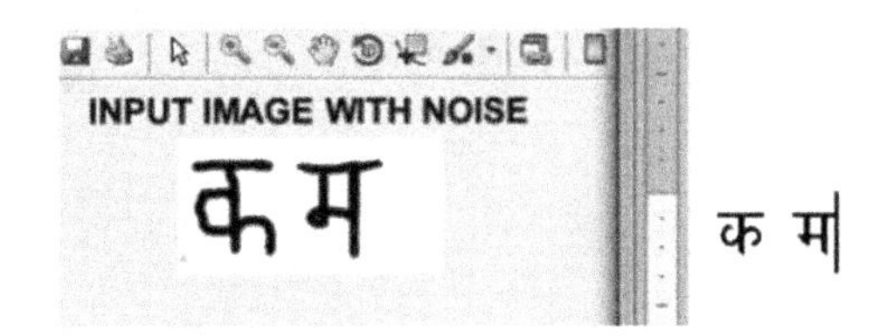

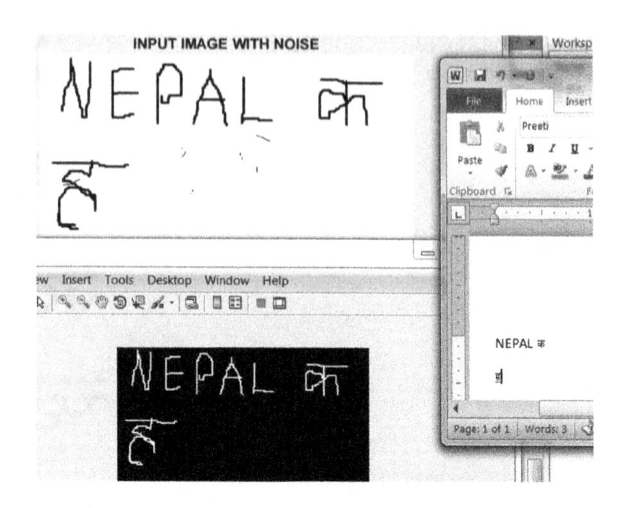

Fig. 10 English and Nepali character recognition in doc mode

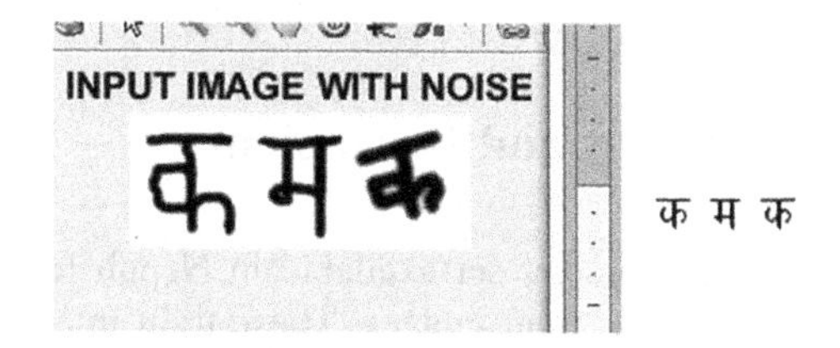

Fig. 11 Rotation invariant

rotation invariant. Simple sketches are also successfully trained in network to make interactive through object recognition e.g. tree, house, ball, and etc. as shown in Fig. 12. Thus we concluded that the proposed system gives fairly good results on the test samples that were presented to it. Result shows that increase in different train sample increases recognition accuracy and we got about 90 % of accuracy in most of Nepali characters.

The experiments have illustrated that the artificial neural network concept can be applied successfully to solve the Nepali Optical Character Recognition Problem. It is also concluded that the output gives better results than present available systems. Result also showed that training with Nepali character only and probabilistic approach in case of testing dilemma gave optimally with Nepali characters recognition. The recognition rate of OCR system using ANN with the image document of Nepali script is quite high. Our proposed system recognized the simple sketch form of structure like as home, tree etc. Character segmentation method

Fig. 12 Sketch recognition

which is incorporated in this paper can handle large variety of touching characters that occur often in images obtained from inferior-quality documents with some modification. In future dictionary words implantation can be use to improve the performance of OCR system. Furthermore Multi factorial Fuzzy System can be used for segmenting the characters in hand written documents.

We have used ANN rather than because each support vector machine would recognize exactly one digit or character only, and fail to recognize all others. Since each handwritten digit can-not be meant to hold more information than just its class, it makes no sense to try to solve this with an artificial neural network.

References

1. Gunasekaram, M., Ganeshmoorthy, S.: OCR recognition system using feed forward and back propagation neural network. In: Second National Conference on Signal Processing, Communications and VLSI, Coimbatore (2010)
2. Banumathi, P., Nasira, G.M.: Handwritten Tamil character recognition using artificial neural networks. In: International Conference on Process Automation, Control and Computing (PACC), pp. 1–5. Coimbatore (2011)
3. Arnold, R., Miklós, P.: Character recognition using neural networks. In: 11th IEEE International Symposium Computational Intelligence and Informatics (CINTI) (2010)
4. Mani, N., Srinivasan, B.: Application of artificial neural network model for optical character recognition. In: IEEE International Conference on Systems, Man, and Cybernetics, 1997. Computational Cybernetics and Simulation, pp. 2517–2520, Orlando (1997)
5. Alpaydin, E.: Optical character recognition using artificial neural networks. In: First IEEE International Conference on Artificial Neural Networks IET, pp. 191–195 (1989)
6. Sanchez, A., Suarez, P.D., Mello, C.A.B., Oliveira, A.L.I., Alves, V.M.O.: Text line segmentation in images of hand written historical documents. In: First Workshops on Image Processing Theory, Tools and Applications, pp. 1–6 (2008)
7. Manikandan, V., Venkatachalam, V., Kirthiga, M., Harini, K., Devarajan, N.: An enhanced algorithm for character segmentation in document image processing. In: IEEE International Conference on Computational Intelligence and Computing Research (ICCIC), pp. 1–5 (2010)
8. Surinta, O.: Optimization of line segmentation techniques for Thai handwritten documents. In: Eighth International Symposium on Natural Language Processing, pp. 180–183 (2009)

Segmentation of Overlapping Isan Dhamma Character on Palm Leaf Manuscript's with Neural Network

Mahasak Ketcham, Worawut Yimyam and Narumol Chumuang

Abstract The segmentation is the first important step for optical character recognition (OCR) system. It separate the image text documents into line, characters and word. The accuracy of the recognition system mainly rely on the algorithm in segmentation. The challenge of segmentation is overlapping therefore this paper presents a new high performance algorithm in the segmentation on Palm leaf manuscript's Isan Dhamma overlap characters.

Keywords Segmentation · Overlapping · Isan Dhamma · Manuscript · Neural network

1 Introduction

The method to segmentation can separate into three categories such as to apply the rule by using a dictionary and the use of corpus. The word segmentation with rule base is text wrapping by checking the rules of orthography of the nature of the compound word, the character spacing and carriage. Provide a basis for determining the scope of words. This approach has limitations in the accuracy of the work is

M. Ketcham (✉)
Faculty of Information Technology, Department of Information Technology Management, King Mongkut's University of Technology North Bangkok, Bangkok, Thailand
e-mail: mahasak.k@it.kmutnb

W. Yimyam (✉)
Faculty of Management Science, Department of Computer Business, Phetchaburi Rajabhat University, Phetchaburi, Thailand
e-mail: worawut_yimyam@hotmail.com

N. Chumuang (✉)
Faculty of Science and Technology, Muban Chombueng Rajabhat University, Chom Bueng, Thailand
e-mail: lecho20@hotmail.com

P. Meesad et al. (eds.), *Recent Advances in Information and Communication Technology 2016*, Advances in Intelligent Systems and Computing 463,
DOI 10.1007/978-3-319-40415-8_7

wrapping a syllable but the accuracy of the cut is relatively low. The advantage of this method is a fast and efficient and use less resources [1, 2]. The segmentation by wrapping the string dictionary is compared to a word in the dictionary. It will be retained in the dictionary. This method has a higher accuracy in cutting the use of rules but will take over [3–5]. The segmentation of the corpus is wrapping over the statistical methods used in processing language. The data warehouse is a linguistic knowledge base stores the frequencies used in the wrapping. The wrapping using data warehouse is divided into two methods: Probabilistic word segmentation and Feature-based word segmentation [6, 7]. The segmentation method through the probability is text wrapping using word n-gram model to find a style of segment the words and the word sequences are the most likely. By the way this will require the use of a data warehouse wrapping and supervision of the group to keep it. This way the result is an alternative form of the cut is probably the most. The process in the segmentation by using features extraction the words. This will resolve the error by wrapping. Relies on the probability of limited of the group is a model in the wrapping. This method is word based on the characteristics of the hybrid approach. The generally technique used to segmentation are longest word pattern matching, shortest word pattern matching, word usage frequency and word usage frequency.

Handwriting in Thai like as English sentences which they overlap between the characters. Therefore, processing language called word segmentation and word separation, thus dividing the each word in the sentence apart for the purpose of use in areas such as formatting a document in word processing, Thai spelling check, syntax analysis, machine translation, document indexing, thesaurus, natural language processing, speech synthesis, syntactic rules analysis and so on. The word in Thai language may consist of the accompanying vowel. That's mean Thai words are different from the other vowel several factors and some consonants also can act as either a vowel spelling too. Especially for handwritten Isan Dhamma characters are overlap between character and character see in Fig. 1. However, the segmentation's problem shown the problem in segmentation the overlapping of Isan Dhamma characters effectively as a result the accuracy in recognizing poor see in Fig. 2.

This paper presents a new segmentation of Overlapping Isan Dhamma Character on Palm Leaf Manuscript's algorithm. The results have shown this new algorithm can wrapping the overlapping more effectively. The details of the process is described in the next subsection.

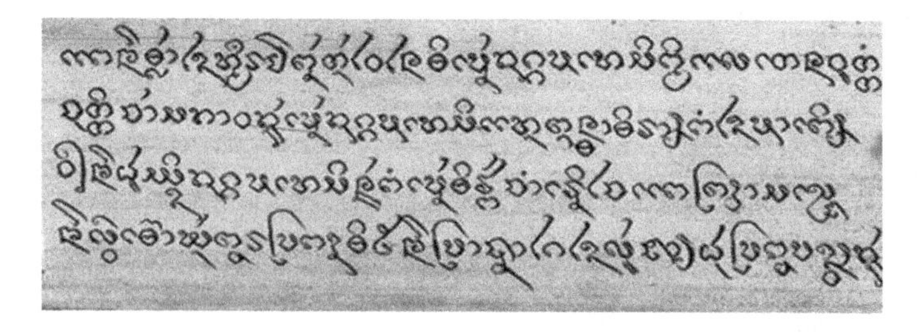

Fig. 1 Example of Isan Dhamma characters on palm leaf manuscripts

Fig. 2 Segmentation on the overlapping characters

The organize paper is (1) Introduction (2) Related literature (3) Propose Segmentation Algorithm (4) Experimental and results (5) Conclusion.

2 Related Literature

The typical character segmentation will consist of two main processes are the fundamental image processing and the character segmentation. The initial image processing, there are many ways for example enhance contrast, noise reduction and binarization etc. This method improve the image which are used in the research. Most of the district were used to match each category. Basically the text segmentation consists of two processes: line segmentation and character segmentation [8–10].

Surinta and Chamchong [8] proposed image segmentation of historical handwriting from palm leaf manuscripts by applying Gaussian Filter for noise removed after convert RGB to a gray-scale image. They used Otsu's algorithm to calculate the optimize threshold for convert a gray scale image into a binary image by isolating the background out of character and projection profile to cut out the text line separation.

Ntogas Nikolaos et al. [11] binarized algorithm is proposed for historical handwriting document. The separation character out of background by using threshold basically and filtering in image processing. Threshold was calculated with Otsu'a algorithm shown the best results for the binary image with the appearance of a thin strokes.

In addition [8–11] still [12] presented by Pradeep et al. was used to separate the character from background by using a threshold Otsu's algorithm as well. Then they found a way to extract text from the background by using the threshold with Otsu's algorithm is a method that has been popular. In a separate character by projection profile found some bugs. The reason of this method does not suit the character handwriting without spaces between lines or characters [13]. But this paper presents a way to do both vertical and horizontal histogram to find pattern of Isan Dhamma characters then neural networks used to learn and recognize with made the point more clearly in the segmentation characters.

3 Proposed Segmentation Algorithm

The segmentation processes handwritten character images from a document written by Isan Dhamma on palm leaves manuscripts consist of two main steps are the steps for pre-processing and segmentation processes separate characters Fig. 3 was shown.

3.1 Pre-processing Process

The first reason we have to pre-processing is the overlapping character images used in this paper is RGB. These images are different of dark and brightness. Then in this step focus to improve the appropriate quality and it can improves the accuracy rate of our segmentation. The process consists of small four steps.

- *Crop Image.* The image will be trimming it to a specific character. Then it is used to calculate the average value of the pixel. This average is used as the basis for grouping images. The similar were classified into the same group. Can be divided into five groups.

Group 1 The image has smeared scattered throughout and it is depicted at least brightness

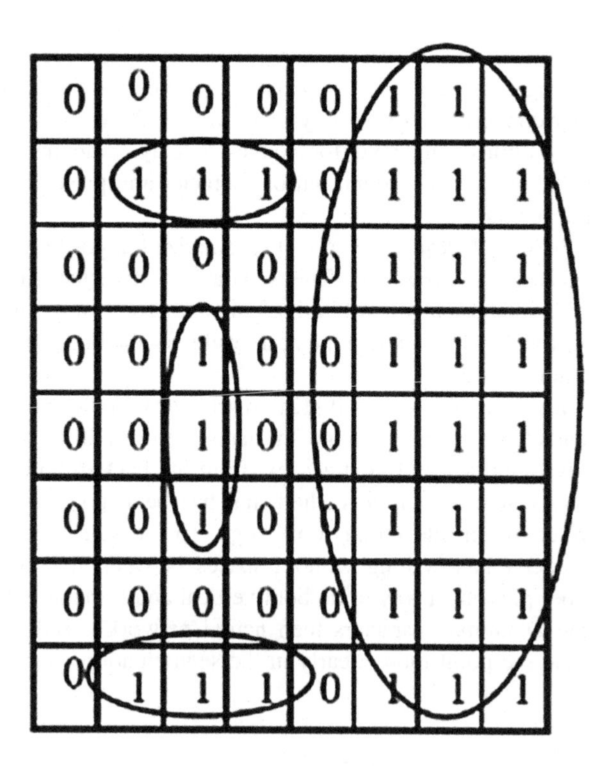

Fig. 3 Binary image with 4-connected components

Group 2 This group was small font size and bold. But with the brightness uniformity
Group 3 Character in this group are easy to read. The font size and thickness
Group 4 The image that contains mixed both of characters that are bold and thin
Group 5 This group consists of characters with thinness strokes and unclear

- *Binarization.* This step is convert input image to binary with (1),

$$g(x,y) = \begin{cases} 1 : f(x,y) \neq 0 \\ 0 : f(x,y) = 0 \end{cases} \tag{1}$$

  where $g(x, y)$ are the coordinate pixel (x, y) of binary image and $f(x, y)$ are the coordinate pixel (x, y) of original image. The value of pixel = 1 is background (white) or 0 is character (black).
- *Adjustment Contrast.* The image is crop to a specific text ago. It will be updated as needed to break them. Because each group has the appearance of darkness and brightness of the image is different. Then each image continues to have a complete detail as possible. In this paper, the method to adjust is chosen to improve the contrast of the image by the image with contrast stretching [14].
- *Noise Reduction.* The image is converted to binary it has noise going on. The eliminate interference occurs. This paper was using the median filter to filter out noise from the image data.

3.2 Connected Component Labeling

The connection of components is to define the character of the elements sequentially. In this paper a set of connected elements 4-connected as shown in Fig. 3.

In this process the character is isolated as a single character. The non-alphanumeric characters (such as borders, noise) mixed together and we separated the single character. The visual character of the single step will be taken to determine the average height and width of the character. The single character will be taken into consideration again. If it is a special character had to be taken to determine the average height and width of the character as well. If that single character written feedback will be used to de splitters Ignatius times and the non-letter characters are removed.

3.3 Find Character

This process are split overlap character to single character.

Fig. 4 The skeleton of character

Step 1 the character analyzes for skeleton image [15–17] see in Fig. 4
Step 2 we calculate pure width and pure height of character follows pseudo in below (Fig. 5):

```
//vertical segmentation
        let A = number of the columns have black pixel
          for v = 1 to A
            column[i]= count the number of black pixel
            if column[i] = 0 then split
```

```
//horizontal segmentation
        let B = number of the rows have black pixel
          for v = 1 to B
             row[j]= count the number of black pixel
            if row[j] = 0 then split
          end
```

Step 3 the overlap character image divided into rule base as follows:
Rule 1: If $w < Aw$ then split vertical projection to single character, where w is the width of character,
Aw is the average of the single character width see in Fig. 6.
Rule 2: If $h < Ah$ the then split horizontal projection to single character, where h is the height of image,
Ah is the average of the single character height see in Fig. 7.
Rule 3: Else split special character. The character image in this group is not in the above rule. This special character used neural network to training the pattern of them shown in Fig. 8.

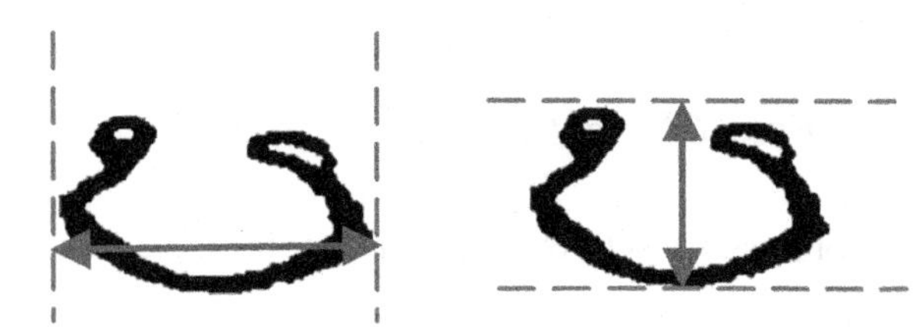

Fig. 5 Calculating pure width and pure height of character

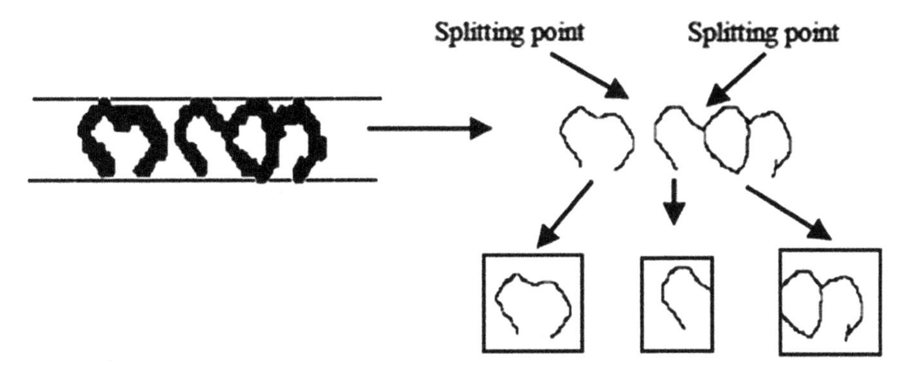

Fig. 6 The vertical projection of the character

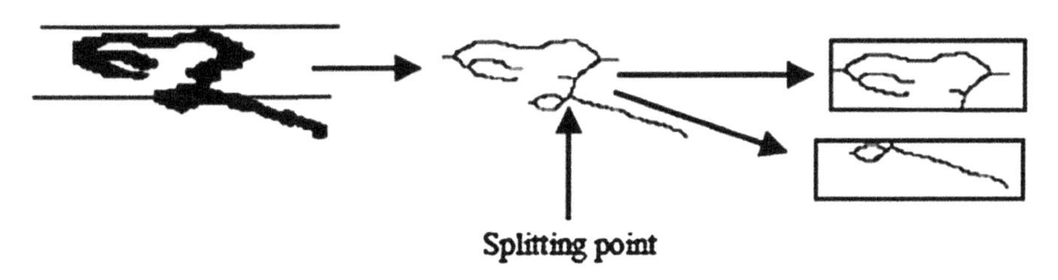

Fig. 7 The horizontal projection of the character

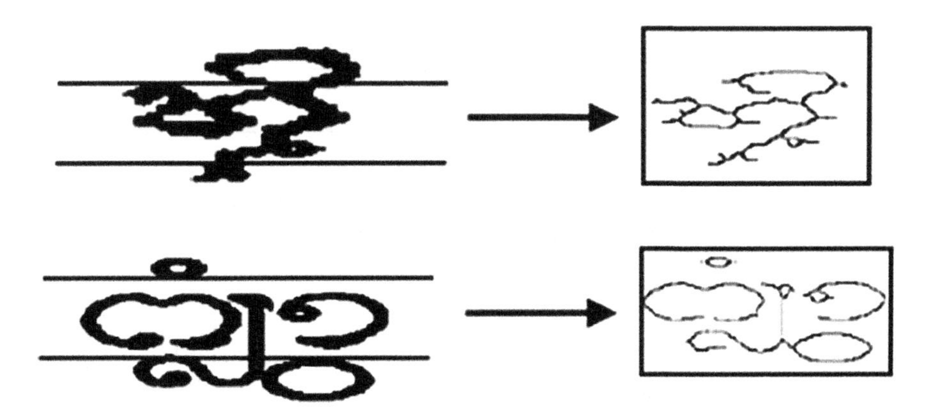

Fig. 8 Example the special character of Isan Dhamm character

3.4 Training Pattern with Multilayer Perceptron

MLP is the feed-forward artificial neural network model that used in widely pattern recognition research. The MLP utilizes a supervised learning technique. The algorithm consists of many hidden layer can adaptive the weight and bias to appropriate output. This paper used MLP for training the pattern of the character for text segmentation.

After we split the characters was described in above. The vertical and horizontal projection came to input into neural network see example in Fig. 9.

The first layer is the input layer which accepts input directly from an external network. The process of back-propagation can describe with pseudo code follow as:

```
Initialize all weights and biases in network;
while terminating condition is not satisfied {
for each training sample P in sample {
   //Propagate the inputs forward:
    for each hidden or output layer unit j {
      I_r = ∑_q w_qr H_q + θ_r; // where I_r is the net input to unit to compute the net input of unit r
                                 // with respect to the previous layer q

      H_s = 1/(1+e^-1) ; } // compute the output of each unit r

   //Backpropagate the errors:
    for each unit r in the output layer
      Err_r H_r (1-H_r)(T_r-H_r); // T_r is the true value output for compute the error

    for each unit r in the hidden layers from the last to the first hidden layer
      Err_r = H_r(1-H_r)∑_s Err_s w_rs:// compute the error with respect to the next higher
layer s

    for each weight w_qr in network {
      Δw_qr = (l)Err_r H_q; // where l is the learning rate  for weight increment
      w_qr = w_qr + Δw_qr ; } // where Δw_qr is the change in weight w_qr  for weight update

    for each bias θ_r in network {
      Δθ_r = (l)Err_r ; //where Δθ_r is the change in bias θ_r for bias increment
      θ_r = θ_r + Δθ_r; } // bias update
    }}
```

After this step, we find the relation of the network N as the input to the output of each layer in the next layer which have the relationship as following in (5),

$$a^{m+1} = f^{m+1}\left(w^{m+1}a^{m} + b^{m+1}\right) \tag{5}$$

where a^m, f^n, w^m and b^m are output, transfer function, weight and bias of layer m and $m = 0, 1, 2, \ldots, N - l$ (l is layer).

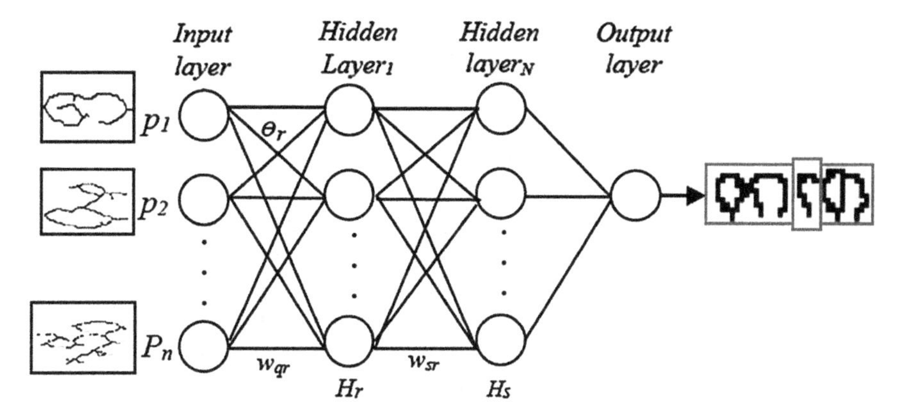

Fig. 9 The process of multilayer feed-forward network

4 Experimental and Result

The vertical and horizontal projection are trained with back-propagation algorithm in MLP. We used 116 images were scanned to input with resolution 300 pixels per inches.Type of input file is ".jpg". The system has accuracy in splitting character 91.00 % by the example single characters and special characters can detected shown in Fig. 10.

After training process, the trained model of MLP are evaluated on Isan Dhamma characters on palm leaf manuscript. The image 186 files are divided into 5 groups. The recognition rates of these images were illustrated in Table 1.

The recognition rates are higher than 89 % in all groups. After the overlapped characters are trained with MLP. The results table I shown the efficient segmentation by the sample of results in detection is shown in Fig. 11.

Fig. 10 Example single character and special characters can detected

Table 1 Recognition rate

Group	The accuracy of the character segmentation		Accuracy (%)
	True	*False*	
1	184	21	89.76
2	4,923	443	91.74
3	4,132	302	93.19
4	14,756	645	95.81
5	9,817	495	95.20
Average			**93.14**

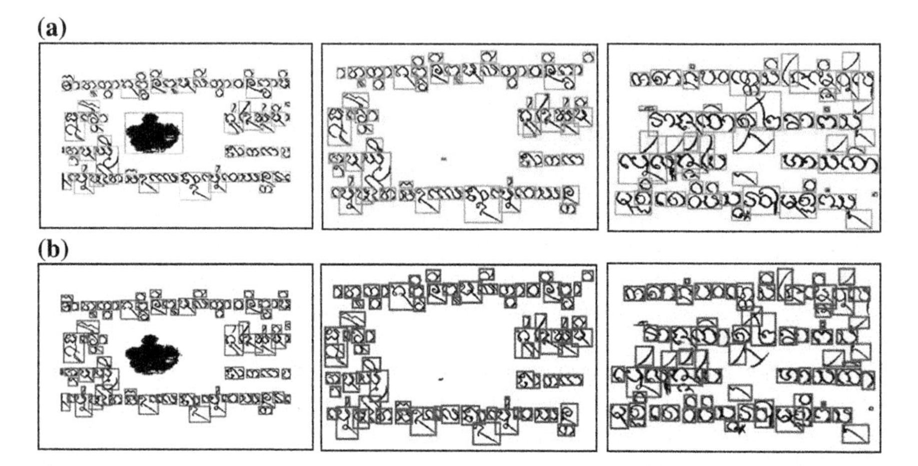

Fig. 11 Examples of the segmentations used projection. **a** The characters detection without MLP training. **b** The characters detection with MLP training

5 Conclusions

This paper presented the segmentation of overlapping Isan Dhamma character on palm leaf manuscript's with neural network. Our method increase more accuracy then we focus to improve the efficient segmentation algorithm. The projections of characters are the features used to pattern frame training by MLP. With our algorithm, the more accuracy in the characters overlapped segmentations. The experimental results shown the recognition rates are well. The average accuracy is 93.14 %. In future work, the accuracies of these characters may be improved by using dictionaries in post processing. And other technique may be applied in the recognition process such as support vector machine, sequential minimal optimization and others. Moreover, the parameters tuning may improve the recognition rate of learning.

Acknowledgements This work was supported by King Mongkut's University of Technology North Bangkok. Contract no. KMUTNB-GEN-58-37.

References

1. Martin, S., Liermann, J., Ney, H.: Algorithms for bigram and trigram word clustering. Speech Commun. **24**, 19–37 (1998)
2. Rainer, L.,Wolfgang, E.: Automatic text segmentation and text recognition for video indexing. In: ACM/Springer Multimedia Systems, vol. 8, pp. 69–81. Springer-Verlag (2000)
3. Christian, W., Jean, M.J.: Object count/area graphs for the evaluation of object detection and segmentation algorithms. Int. J. Doc. Anal. Recogn. (IJDAR) 280–296 (2005)
4. John, C.: A computational approach to edge detection. IEEE Trans. Pattern Anal. Mach. Intell. **PAMI-8**(6), 679–698 (1986)
5. Kim, W., Kim, C.: A new approach for overlay text detection and extraction from complex video scene. IEEE Trans. Image Process. **18**(2), 401–411 (2009)
6. Med, A.H.: Modified algorithm marker-controlled watershed transform for image segmentation based on curvelet threshold. Middle-East J. Sci. Res. **20**(3), 323–327 (2014)
7. Chyzhyk, D., Borja, A., Josu, M.: Active learning with bootstrapped dendritic classifier applied to medical image segmentation. Elsevier Pattern Recogn. Lett. **34**(14), 1602–1608 (2013)
8. Surinta, O., Chamchong, R.: Image segmentation of historical handwriting from palm leaf manuscripts. IFIP Int. Fed. Inf. Process. **288**, 182–189 (2008)
9. Rejean, P., Ecole, P., Montreal, Q., Sargur, N.S.: Online and off-line handwriting recognition: a comprehensive survey. IEEE Trans. Pattern Anal. Mach. Intell. 63–84 (2000)
10. Berrin, Y., Peter, A.S.: Segmentation of off-line cursive handwriting using linear programming. Elsevier Pattern Recogn. **31**(12), 1825–1833 (1998)
11. Ntogas, N., Ventzas, D.: A binarization algorithm for historical manuscripts. In: The 12th WSEAS International Conference on Communications, pp. 41–51 (2008)
12. Pradeep, K.J., Srinivasan, E., Himavathi, S.: Diagonal based feature extraction for handwritten alphabets recognition system using neural network. Int. J. Comput. Sci. Inf. Technol. (IJCSIT) (2011)
13. Zaidi, R., Khansa, Z., Mohd, Y.I.I., Emran, M.T., Mohd, N.M.N., Rosli, S., Mohd, Y., Yusof, Z.M., Mashkuri, Y.: Off-line handwriting text line segmentation: review. IJCSNS Int. J. Comput. Sci. Netw. Secur. 12–20 (2008)
14. Rafael, C.G., Richard, E.W., Steven, L.E.: Digital Image Processing using Matlab. Upper Saddle River, New Jersey, Pearson Prentice Hall (2004)
15. Naruemol, C., Mahasak, K.: Intelligent handwriting thai signature recognition system based on artificial neuron network. In: TENCON 2014—2014 IEEE Region 10 Conference Bangkok, pp. 1–6 (2014)
16. Precha, K., Narumol, C., Mahasak, K.: Isan Dhamma handwritten characters recognition system by using functional trees classifier. In: 11th International Conference on Signal-Image Technology & Internet-Based Systems (SITIS), pp. 606–612 (2015)
17. Narit, H., Narumol, C., Mahasak, K.: Thai handwritten verification system on documents for the investigation. In: 11th International Conference on Signal-Image Technology & Internet-Based Systems (SITIS), pp. 617–622 (2015)

Comparative Results of Attribute Reduction Techniques for Thai Handwritten Recognition with Support Vector Machines

Tanasanee Phienthrakul and Massaya Samnienggam

Abstract Data reduction is an important step in machine learning and big data analysis. The handwritten recognition is a problem that uses a lot of data to get the good results. Thus, the attribute reduction can be applied to improve the accuracy of classification and reduces the learning time. In this paper, the attribute reduction techniques are studies. These techniques are applied to the Thai handwritten recognition problems. Support vector machines (SVMs) are used to verify the results of 4 attribute reduction techniques, i.e., principle component analysis (PCA), local discriminant analysis (LDA), locality preserving projection (LPP), and neighborhood preserving embedding (NPE). All of these 4 techniques will transform the original attributes to a new space with the different methods. The results show that LDA is a suitable data reduction technique for classifying the handwritten character with SVM. Only 10 % of features can give the accuracy about 47.68 % for 89 classes of the characters. This technique may give a better result when the suitable feature extraction techniques are applied.

Keywords Attribute reduction · Handwritten recognition · Support vector machines · Principal components analysis · Local discriminant analysis · Locality preserving projection · Neighborhood preserving embedding

1 Introduction

The optical character recognition (OCR) is a way to help communicate with the computer easier. The OCR uses artificial intelligence methods to convert visual information into a text document. The OCR can be used to make electronic book (E-Book), verify the handwriting, input data via touch screen, etc. The OCR can be divided into two main types of printed character recognition and handwritten

T. Phienthrakul (✉) · M. Samnienggam
Faculty of Engineering, Department of Computer Engineering,
Mahidol University, Nakornpathom, Thailand
e-mail: tanasanee.phi@mahidol.ac.th

P. Meesad et al. (eds.), *Recent Advances in Information and Communication Technology 2016*, Advances in Intelligent Systems and Computing 463,
DOI 10.1007/978-3-319-40415-8_8

recognition. Due to the differences in individual handwritten, the handwritten recognition is somewhat more complicated in recognition. Even the same person's handwritten is also different if the font size, stationery, and other environmental changes. Currently there are efforts to bring the different methods used for handwritten recognition in multiple languages. For Thai handwritten recognition, there continues to develop continuously.

Thai handwritten recognition in the past have been used various methods of artificial intelligence, such as Ant-Miner Algorithm [1] which provides the accuracy of the information learned on the testing data is 82.7 %. Moreover, it also includes the Artificial Neural Network [2] which provides the accuracy of the information learned on the testing data is 85 %. The Thai handwritten recognition also needs to improve a lot when we compare it with the recognition of the other languages, such as the English language that have a right to recognition at 99.87 %.

When we are considering the handwritten recognition in the languages such as English [3], Tamil [4, 5], Japanese characters (Hiragana) [6], or the number recognition of Kannada (the language commonly used in southern India) [7], we found the recognition with support vector machine is a learning algorithm that provides a good classification than others. Therefore, the SVMs are used for Thai handwritten recognition, it likely to yield good recognition.

In addition, the selection of key features of the image also affects the accuracy of OCR. Since most research cannot be used all pixel in the image as a feature to learn because when the pixels of character image data to be converted to the form of vector data, the recognition has many feature or more dimension that take the time to learn and classification as well. Moreover, some of the pixels in the image do not affect recognition. Although choosing the appropriate attributes will spend less time in recognition but the difficulty is not knowing what is the features appropriate in the recognition.

Data Reduction is an interesting technique used before the data used in learning. The data reduction has several methods, such as ReliefF, Principal Component Analysis (PCA), Multidimensional Scaling (MDS), and Local Discriminant Analysis (LDA). Those methods are the ways to reduce the data widely known and they are used in the data analysis and recognition. There are also having other data reduction techniques such as Isometric Mapping (Isomap) [8], Locality Preserving Projection (LPP) [9], Local Discriminant Embedding (LDE) [10], and Discriminant Neighborhood Embedding (NEC) [11]. Those methods above should be applied to the Thai handwritten recognition.

This paper introduced the data reduction techniques to reduce the Thai handwritten data before bring to learned with Support Vector Machines (SVMs) for finding the reduction technique that is suitable with SVMs. The reduction techniques are four techniques include Principal Components Analysis (PCA), Local Discriminants Analysis (LDA), Locality Preserving Projection (LPP), and Neighborhood Preserving Embedding (NPE).

2 Support Vector Machines

In this paper, we bring recognition with SVMs to train and test the data sets to know the accuracy on the testing data. SVMs were introduced by Vladimir Vapnik and colleagues [12] which are a learning method used for binary classification. The idea of this method is to find a hyperplane which separates the d-dimensional data into its two classes. Since example data is often not linearly separable, SVMs introduce the notion of a "kernel induced feature space" which casts the data into a higher dimensional space where the data is separable. SVMs are intuitive, theoretically well-founded, and have shown to be practically successful. SVMs have also been extended to solve regression tasks. SVMs can be described based on mathematical equations as follows.

Given l training example $\{x_i, y_i\}, i = 1, \ldots, l$, where each example has d inputs $(x_i \in R^d)$, and a class label with one of two values $(y_i \in \{-1, 1\})$. All hyperplanes in R^d are parameterized by a vector (w), and a constant (b), expressed in the Eq. (1).

$$w \cdot x + b = 0 \tag{1}$$

Given such a hyperplane (w, b) that separates the data, this give the function (2)

$$f(x) = sign(w \cdot x + b) \tag{2}$$

This correctly classifies the training data (note the other testing data has not seen yet). However, a given hyperplane represented by (w, b) is equally expressed by all pair $\{\lambda w, \lambda b\}$ for $\lambda \in R^+$. So, the canonical hyperplane was defined to be that which separates the data from the hyperplane by a distance of at least one example on both sides has a distance of exactly 1. Thus, for a given hyperplane, the scaling (the λ) is implicitly yet. Consider those that satisfy:

$$x_i \cdot w + b \geq +1 \text{ when } y_i = +1 \tag{3}$$

$$x_i \cdot w + b \leq -1 \text{ when } y_i = -1 \tag{4}$$

Or more compactly

$$y_i(x_i \cdot w + b) \geq 1 \quad \forall i \tag{5}$$

All such hyperplane have a function distance more than one. For a given hyperplane (w, b), all pairs $\{\lambda w, \lambda b\}$ define the same hyperplane, but each has a different functional distance to a given data point. So, we must normalize by the magnitude of w to obtain the geometric distance from the hyperplane to a data point. This distance is simply:

$$d((w,b),x_i) = \frac{y_i(x_i \cdot w + b)}{\|w\|} \geq \frac{1}{\|w\|} \quad (6)$$

We want the hyperplane that maximizes the geometric distance to the closest data points. See Fig. 1.

The main method of minimizing w is with Lagrange multipliers [12, 13]. The problem is eventually transformed into:

$$\text{Minimize:} \quad W(\alpha) = -\sum_{i=1}^{l}\sum_{j=1}^{l} y_i y_j \alpha_i \alpha_j (x_i \cdot x_j)$$

$$\text{Subject to:} \quad \sum_{i=1}^{l} y_i \alpha_i = 0$$

$$0 \leq \alpha_i \leq C \quad (\forall i),$$

where α is the vector of l non-negative Lagrange multipliers to be determined, and C is a constant. The matrix $(H)_{ij} = y_i y_j (x_i \cdot x_j)$ is defined, and introduce more compact notation:

$$\text{Minimize:} \quad W(\alpha) = -\alpha^T 1 + \frac{1}{2}\alpha^T H \alpha \quad (7)$$

$$\text{Subject to:} \quad \alpha^T y = 0 \quad (8)$$

$$0 \leq \alpha \leq C1 \quad (9)$$

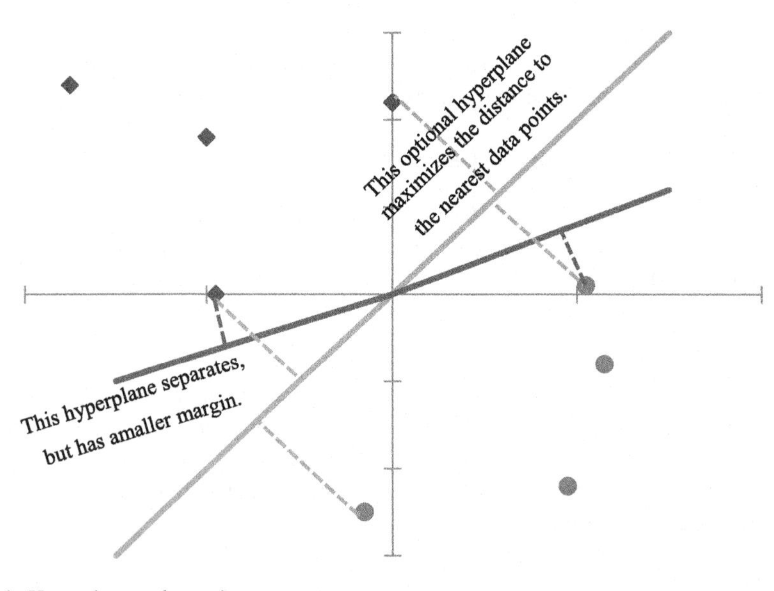

Fig. 1 Hyperplane and margin

In addition, from the derivation of these equations, the optimal hyperplane can be written as:

$$w = \sum_i \alpha_i y_i x_i \tag{10}$$

The vector w is a linear combination of the training example and it can also be shown that

$$\alpha_i(y_i(w \cdot x_i + b) - 1) = 0 \;\; (\forall i),$$

when the functional distance of an example is strictly greater than 1 (when $y_i(w \cdot x_i + b) > 1$), then $\alpha_i = 0$. So only the closet data points contribute to w. These training examples for which $\alpha_i > 0$ are termed support vector. They are the only ones needed in defining the optimal hyperplane. The support vectors are the "borderline cases" in the decision function. That α_i can be thought of as a "difficulty rating" for the example x_i—how important that example was in determining the hyperplane.

Assuming we have the optimal α (from which we construct w), we must still determine b to fully specify the hyperplane. To do this, take any "positive" and "negative" support vector, x^+ and x^-, for which we know

$$(w \cdot x^+ + b) = +1$$

$$(w \cdot x^- + b) = -1$$

Solving these equation gives:

$$b = -\frac{1}{2}(w \cdot x^+ + w \cdot x^-) \tag{11}$$

when $C = \infty$, the optimal hyperplane will be the one that separates the data (assuming one exist). For finite C, this change the problem to finding a "soft-margin" classifier, which allow some of the data to be misclassified. If C as a tunable parameter, higher C corresponds to more importance on classifying all the training data correctly, and lower C results in a "more flexible" hyperplane that tries to minimize the margin error for each example. Finite values of C are useful in situations where the data is not easily separable.

3 Attribute Reduction Techniques

To study the efficiency of attribute reduction techniques, 4 techniques are selected. The brief descriptions of these techniques are reviewed.

3.1 Principal Components Analysis (PCA)

PCA [14] is a statistical technique for reducing the complexity of the data set which was first proposed by Pearson [15] and Hotelling [16] while the most reference is Jolliffe [17]. The purpose of this method is to reduce the dimensions of multi-variables while it is still maintaining the relationship of data as much as possible. The PCA is an unsupervised learning which input data without any reference to the corresponding target data.

PCA constructs a low-dimensional representation of the data that describes as much of the variance in the data as possible. This method will find a linear basis of reduced dimensionality for the data, which the amount of variance in the data is maximal. PCA attempts to find a linear transformation T that maximizes $T^T cov_{X-\overline{X}} T$, where $cov_{X-\overline{X}}$ is the covariance matrix of the zero mean data X. We can see this linear mapping is formed by the d principal eigenvectors of the covariance matrix of the zero-mean data. So, PCA solves the Eigen problem.

$$\underset{X-\overline{X}}{cov} v = \lambda v \tag{12}$$

The eigenproblem is solved for the d principal Eigen values λ. The corresponding eigenvectors form the columns of the linear transformation matrix T. The low-dimensional data representations y_i of the data points x_i are computed by mapping them on to the linear basis T, such as $Y = (X - \overline{X})T$.

The problem of PCA is the size of the covariance matrix is proportional to the dimensionality of the data points. Thus the computation of the eigenvectors might be infeasible for very high-dimensional data (under the assumption that $n > D$).

3.2 Local Discriminant Analysis (LDA)

LDA [14] tries to maximize the linear reparability between data points belonging to the different classes. LDA is a supervised technique which finds a linear mapping M that maximizes the linear class separability in the low-dimensional representation of the data. LDA use the within-class scatter S_w and the between-class scatter S_b as criteria to formulate linear class separability, which are defined as

$$S_w = \sum\nolimits_c p_c \underset{X^c - \overline{X}^c}{cov} \tag{13}$$

$$S_b = \sum\nolimits_c \underset{\overline{X}^c}{cov} = \underset{X-\overline{X}}{cov} - S_w \tag{14}$$

where p_c is the class prior of class label c, $\underset{X^c - \overline{X}^c}{cov}$ is the covariance matrix of the zero mean data points x_i assigned to class $c \in C$ (where C is the set of possible classes), $\underset{\overline{X}^c}{cov}$ is the covariance matrix of the cluster means, and $\underset{X-\overline{X}}{cov}$ is the covariance matrix of the zero mean data X.

LDA used Fisher criterion to optimizes the ratio between the within-class scatter S_w and the between-class scatter S_b in the low-dimensional representation of the data, by finding a linear mapping M.

$$\frac{T^T S_b T}{T^T S_w T} \tag{15}$$

This maximization can be performed by solving the generalized eigen problem

$$S_b v = \lambda S_w v \tag{16}$$

The eigenproblem is solved for the d largest eigenvalues (under the requirement that $d < |C|$). The eigenvectors v form the columns of the linear transformation matrix T. The low-dimensional data representation Y of the data points in X can be computed by mapping them onto the linear basis T, such as $Y = (X - \overline{X})T$.

3.3 *Locality Preserving Projection (LPP)*

LPP [14] is a technique that aims at combining the benefits of linear techniques and local non-linear techniques for dimensionality reduction by finding a linear mapping that minimizes the cost function of Laplacian Eigen maps. LPP constructs a nearest neighbor graph in every data points x_i is connected to its k nearest neighbors x_{ij}. LPP computes weights of the edges in the graph using equation below.

$$w_{ij} = e^{-\frac{\|x_i - x_j^2\|}{2\sigma^2}}, \tag{17}$$

where σ indicates the variance of the Gaussian. Then LPP solves the generalized eigenproblem

$$(X-\overline{X})^T L (X-\overline{X}) v = \lambda (X-\overline{X})^T M (X-\overline{X}) v, \tag{18}$$

where L is the graph Laplacian, and M is the degree matrix of the graph. The eigenvectors v_i corresponding to the d smallest nonzero eigenvalues form the columns of the linear mapping T that minimizes the Laplacian Eigen map cost function (see Eq. 19). The low-dimensional data representation Y is thus given by $Y = (X - \overline{X})T$

$$\varnothing(Y) = \sum_{ij} (y_i - y_j)^2 w_{ij}. \tag{19}$$

3.4 Neighborhood Preserving Embedding (NPE)

NPE [14] minimizes the cost of a local nonlinear technique for dimensionality reduction under the constraint that the mapping from the high-dimensional to the low-dimensional data representation is linear. NPE is the linear approximation to Local Linear Embedding (LLE). NPE defines a neighborhood graph on the dataset X, and computes the reconstruction weights W_i. The cost function of LLE is optimized by solving the following generalized eigenproblem for the d smallest nonzero eigenvalues

$$(X-\overline{X})^T(I-W)^T(I-W)(X-\overline{X})v=\lambda(X-\overline{X})^T(X-\overline{X})V, \tag{20}$$

where I represents the $n \times n$ identity matrix. The low-dimensional data representation is computed by mapping X onto the obtained mapping T, such as $Y=(X-\overline{X})T$.

4 Comparative Results of the Attribute Reduction Techniques

This paper use Principal Components Analysis (PCA), Local Discriminant Analysis (LDA), Locality Preserving Projection (LPP), and Neighborhood Preserving Embedding (NPE) as tools for data reduction. The document forms were designed for collecting handwritten, then handwritten sample were collected. Each character was written many times by several people and a character were written at least 200 times. These handwritten samples were scanned to keep on digital files (BMP Image) that have the size of 64 × 64 dimensions. Next, we made data preparation to provide data in a format that can be learned. The handwritten sample in digital format files was cut into partly based on a character, then it will be converted to a vector or matrix or other appropriate learning. So, the handwritten character is in text files format (.txt).

Since we plan to use 10 folds cross-validation with SVMs, the data were divided into 10 parts. In the next step, the attribute reduction is used to reduce the training data set. We set the numbers of attributes are 10, 20, 30, 40, 50, 60 % of its total attributes. Then we will get the set of data that have low-dimensions with number of dimension as defined. The result is the accuracy in the classification of the handwritten characters (Fig. 2).

We compared the recognition rates of each data reduction techniques. Then the results were analyzed to find the data reduction technique that suitable to classify

Fig. 2 Examples of handwritten characters

Thai handwritten character with SVMs. The results are the accuracies of recognition with SVMs based on the 4 attribute reduction techniques: Principal Components Analysis (PCA), Local Discriminant Analysis (LDA), Locality Preserving Projection (LPP), and Neighborhood Preserving Embedding (NPE). The sample data has 4977 character images with 78 classes. Each image was resize to 64 × 64 pixels and these pixels are used as the attributes of classification. The results are shown as the table below.

The data reduction techniques that we used to reduce the data before Thai handwritten recognition with SVMs: Principal Components Analysis (PCA), Local Discriminant Analysis (LDA), Locality Preserving Projection (LPP), and Neighborhood Preserving Embedding (NPE). We used "svm-train.exe" to train the data and "svm-predict.exe" to test the data. The result is a percent of accuracy. Another problem, memory of use in running the programs is not enough. In this work, we use computer with 8 GB of ram so we just can reduce the data at 10 to 60 % of number of the data dimension. The programs take the time to run program quite a long time.

As the result in Table 1, the accuracy each of the data reduction techniques are likely to decrease which differs from the definition that is more training data then more accuracy as well. We can probably assume that the quality of Thai handwritten is not good enough, the data format is not suitable for the technique, and the features that are chosen are not good enough to classify the data.

Since the problem above, we have the accuracy based on the data reduction techniques that have 10–60 % of number of the data dimension. We found the LDA is a data reduction technique that is suitable with SVMs the best based on the accuracy. However, the accuracy before data reduction cannot be compared to the accuracy after data reduction because of the problem on memory.

Table 1 The accuracy of handwritten recognition based on the data reduction techniques

Reduction techniques	Number of used attributes					
	10 %	20 %	30 %	40 %	50 %	60 %
PCA	1.2698	1.2698	1.2698	1.2698	1.2698	1.2698
LDA	**47.6793**	46.1744	45.5676	45.4511	46.0241	46.6607
LPP	40.5586	38.0671	37.6472	36.9439	36.5120	35.7163
NPE	45.5883	42.0605	39.9819	38.4690	37.2112	36.1403

5 Conclusion

This paper compared 4 attribute reduction techniques, i.e., Principal Components Analysis (PCA), Local Discriminant Analysis (LDA), Locality Preserving Projection (LPP), and Neighborhood Preserving Embedding (NPE) on the handwritten recognition problems. The experimental results showed that only 10 % of attributes that are selected with LDA yields the good performance when it is compared to other attribute reduction techniques. Although the accuracies are below 50 % chance, it causes by we used the simple attributes and a simple method for classification on 78 classes. The original set of attributes that are used for testing in this paper is the pixel values which are the simplest attributes of images. Thus, if the suitable features are extracted from the handwritten images, the accuracies of classification may be improved.

Acknowledgement This paper was supported by the young researcher fund, Mahidol University and Faculty of Engineering, Mahidol University.

References

1. Phokharatkul, P., Sankhuangaw, K., Somkuarnpanit, S., Phaiboon, S., Kimpan, C.: Off-line Handwritten Thai Character Recognition using Ant-miner Algorithm. World Academy of Science, pp. 195–197 (1981)
2. Surinta, O., Jareanpon, C.: Comparison of image analysis for Thai handwritten character recognition. Intell. Inf. Proc. III, IFIP Int. Fed. Inf. Proc. **228**(2007), 373–382 (2007)
3. Hiremath, S., Shuhangi, D.: Handwritten english character and digit recognition using multiclass SVM classifier and using structural micro feature. Int. J. Recent Trends Eng. **2**, 193–195 (2009)
4. Milgram, J., Cheriet, M., Sabourin, R.: One Against One or One Against All which One is Better for Handwriting Recognition with SVM?. Canada, Montreal (2006)
5. Venkatesh, J., Sureshkumer, C.: Handwritten tamil character recognition using SVM. Int. J. Comput. Netw. Secur. **1**, 29–33 (2009)
6. Miyao, H., Matayama, M., Nakano, Y., Hananoi, T.: Off-line handwritten character recognition by SVM based on the virtual examples synthesized from on-line characters. In: Proceedings of the 8th International Conference on Document Analysis and Recognition, vol. 1, pp. 494–498 (2005)
7. Rajashekararadhya, S., Ranjan, P.: Support vector machine based handwritten numeral recognition of Kannada script. In: International Advance Computing Conference pp. 381–386. Patiala (2009)
8. Tenenbaum, J.B., Silva, V., Langford, J.C.: A global geometric framework for nonlinear dimensionality reduction. Science **290**(5500), 2319–2323 (2000)
9. He, X., Niyogi, P.: Locality Preserving Projections. Advances in Neural Information Processing Systems (NIPS), 16, Vancouver, Canada (2003)
10. Chen, H.T., Chang, H.W., Liu, T.L.: Local discriminant embedding and its variants. Comput. Vis. Pattern Recognit. **2**, 846–853 (2005)
11. Zhang, W., Xuea, X., Suna, Z., Lua, H., Guoa, Y.F.: Optimal dimensionality of metric space for classification. In: Proceedings of the 24th International Conference on Machine Learning, pp. 1135–1142. Corvalis, Orego (2007)

12. Cortes, C., Vapnik, V.: Support-vector network. Mach. Learn. **20**, 273–297 (1995)
13. Burges, C.J.C.: A tutorial on support vector machines for pattern recognition. Data Min. Knowl. Disc. **2**, 121–167 (1998)
14. Maaten, L.: Matlab Toolbox for Dimensionality Reduction. http://lvdmaaten.github.io/drtoolbox
15. Karl, P.: On lines and planes of closest fit to system of points in space. Philos. Mag., Ser. 6, **2**(11), 559–572 (1901)
16. Harold, H.: Analysis of a complex of statistical variables into principal components. J. Educ. Psychol. **24**(6&7), 417–441 & 498–520 (1933)
17. Jolliffe, I.T.: Principal Component Analysis. Springer Series in Statistics, 2nd edn. Springer-Verlag New York, New York (2002)

Human Action Invarianceness for Invarianceness Using Integration Moment for Human Action Recognition in Video

Nilam Nur Amir Sjarif, Siti Mariyam Shamsuddin, Siti Zaiton Mohd Hashim, Aida Ali and Zanariah Zainudin

Abstract The uniqueness of the human action shape or silhouette can be used for the human action recognition. Acquiring the features of human silhouette to obtained the concept of human action invarianceness have led to an important research in video surveillance domain. This paper discusses the investigation of this concept by extracting individual human action features using integration moment invariant. Experiment result have shown that human action invarianceness are improved with better recognition accuracy. This has verified that the integration method of moment invariant is worth explored in recognition of human action in video surveillance.

Keywords Human action recognition · Human action invarianceness · Integration moment invariant

N.N.A. Sjarif (✉)
UTM Advanced Informatic School, Universiti Teknologi Malaysia,
Kuala Lumpur, Malaysia
e-mail: nilamnini@gmail.com

S.M. Shamsuddin (✉) · S.Z.M. Hashim · A. Ali · Z. Zainudin
UTM Big Data Centre, Universiti Teknologi Malaysia, Skudai, Johor, Malaysia
e-mail: mariyam@utm.my

S.Z.M. Hashim
e-mail: sitizaiton@utm.my

A. Ali
e-mail: aida@utm.my

Z. Zainudin
e-mail: zanariah86@gmail.com

S.M. Shamsuddin · S.Z.M. Hashim · A. Ali
Faculty of Computing, Universiti Teknologi Malaysia, Skudai, Johor, Malaysia

P. Meesad et al. (eds.), *Recent Advances in Information and Communication Technology 2016*, Advances in Intelligent Systems and Computing 463,
DOI 10.1007/978-3-319-40415-8_9

1 Introduction

Human action recognition (HAR) has become essential parts of many computer vision applications, and it involves a series of processes such as image data acquisition, feature extraction and representation and classification for recognition [1, 2]. The action classification is crucial in recognizing the human action because the extracted features of the motion must be classified accordingly. The increasing number of action classification in a video will lead to more challenges due to the higher overlapping between classes. Random movements cause scattered data, and consequently, these problems cause the features are not labeling action class correctly and representation of data features are not standardized properly [3–5]. This shows that inappropriate feature extraction and representation may directly cause a low accuracy in classification of human action. Based on the literature, many studies have been done on global approach for feature extraction. The Global representations are powerful since much of the information is encoded, as they focus on global information. The conventional global based moment invariant that was put forward by [6], Geometric Moment Invariant (GMI), was widely used for feature extraction and classification in human action recognition [6]. This method has a unique characteristic in identifying an image due to its invariant to orientation, size and position of the shape image. However, in terms of the invarianceness characteristic of human still have some weaknesses. The most important characteristic of invariance of the human action is the stability under different views but distinctive from the different classes.

Consequently, this method leads some issues, especially in terms of the complexity of data feature representation during the process of feature extraction in term of intra-class and inter-class variance. The weaknesses of GMI are includes:

i. Reliant and incomplete invariant under translation, rotation and scaling [7]
ii. Lost scale invariant in discrete condition [8]
iii. Applied only a small subset of moment invariant [9]
iv. Produce errors if the transformations are subjected to unequal scaling data transformation [10–18]
v. Data position of pixel is far away from centre coordinate [19]
vi. Problem in region, boundary and discrete condition [20]

Therefore, more standardize uniform representation of data distributions are needed for recognition of human action. In this paper, a human action invarianceness (HAI) is present using feature extraction techniques and representation methods based on the integration of moment invariant called Higher United Moment Invariant (HUMI); the function is introduced in order to replace the conventional Geometrical Moment Invariant (GMI). A good feature extraction approach should be able to generalize over variations within class (intra-class) and distinguish between actions of different classes (inter-class). Therefore, the variation of the features of human action can minimize variation for intra-class and maximize variations for inter-class for human action. The human features representation is then improved with the invariant discretization method in order to raise the performance of the system and standardize the amount of accessible information to a manageable size of feature vector representation without losing any valuable information.

The paper is systematized as follows. Human action is explained in Sect. 2. Followed with the Human action invarianceness through intra-class and inter-class concept in Sect. 3. Section 4 describes the proposed integration of moment process in this work. The experiment result is discussed in Sect. 5. And finally, the conclusion is drawn in Sect. 6.

2 Human Action

Human Action is considered as the classification of time varying feature data, such as matching unknown sequence with a group of labeled sequences representing typical actions. Human silhouette data contains a different number of frames in different actions and located at different places. It is difficult to determine the invarianceness distance as close as possible when the person is in motion. Moreover, the center of the silhouette moves due to actions and different view sets. The multitude of action features can either be based on shapes or motion; in this paper, the focus of features extraction is based on the shapes features. Generally, the image is basically based on the binary image. Black pixel represents the background and white pixel represents the foreground of an image, as shown in Fig. 1.

Hence, the features that describe the human body silhouette could capture the shape characteristics of the human subject, either region or boundary representation of an action, over time by using the proposed approach. These features then form a main source of information, especially with regards to the interpretation and understanding of human action based on image sequence. Therefore, the invarianceness concept is important when trying to recognize human action that has many variations of the same action under certain transformations, such as translation, scaling and rotation.

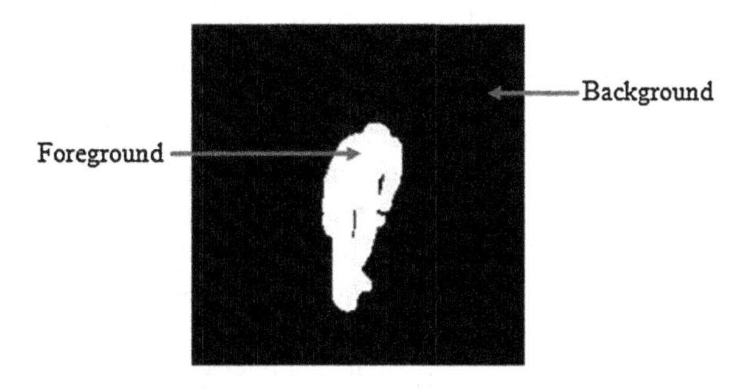

Fig. 1 Background and foreground pixel of the image frame

3 Human Action Invarianceness

The invarianceness of the human action is very important as most action features based on silhouettes are dependent on the movement of the silhouette in the image space. Human action invarianceness (HAI) is defined as the *preservation of an image regardless of its transformations, where it gives small similarity error for intra-class of action (same action) and large similarity error for inter-class of action (different actions).*

3.1 Human Action Invarianceness Procedure

HAI procedure consist of three processes: extracting global features from moment representation, similarity measurement of the variance between features, and intra-class and inter-class analysis. The incorporation of features is considered essential in order to handle the task of recognition of human actions performed by different people, especially in intra-inter class variations [21]. These representation features are form as invariant feature vectors of the image frame resulting from the feature extraction task.

3.2 Feature Extraction

Feature extraction is a fundamental source of information regarding the interpretation of specific of human action, generally viewed as the core and time consuming part in any action recognition systems [22]. The process of feature extraction in human action involves transforming the input data that describe the shape of a segmented silhouette of a moving person into the set of represented features of action poses. The quality of the extracted features is demonstrated by the success of the relatively simple classification scheme that could be achieved basically by reducing the form instead of full representation to the relevant information [23]. Low performance in terms of accuracy is due to the various features that represent the same action, and this makes the recognition process difficult and complex. Similar characteristics of action behavior are easy to recognize if all of the different features value for similar action has a standard representation for generalized unique features of human action, which consequently makes the recognition process easier. Therefore, illustration of human action features is required to represent the action unique features in a systematic feature representation. The main extracted feature consists of the feature vector that will be used for silhouette classification in its respective classes. As each frame sequence represents a corresponding class of human action, the intra-class variance is small and inter-class variance is larger. Variations between samples are found by calculating the MAE of all samples in the set.

3.3 *Intra-Class and Inter-Class for Human Action Invarianceness*

Invarianceness of human action concepts is proven with the lower variance between features (similarity error) for intra-class (same action image) and higher variance in inter-class (different action image), which is due to the uniqueness of the extracted features. MAE function is one of the statistical functions used for measuring the similarity error by calculating the absolute value of mean variance from the reference data. The equation of MAE is shown as below:

$$MAE = \frac{1}{n}\sum_{i=1}^{f}|(x_i - r_i)|$$

Where n is the number of image frame; x_i is the current image frame; r_i is the reference image frame (first image frame is the reference image); f is the number of features image frame; i is the feature's column of image frame.

The smallest MAE value is considered as the most similar to the original action image, which is the reference image to be compared. On the contrary, the highest MAE value is the most different action. Therefore, the range of MAE between intra-class and inter-class is not a concern as long as it proves the characteristics of human action recognition concepts (the intra-class value must be lower than inter-class values). Mean absolute error (MAE), performed in this work as the similarity measurer in HAI to find the mean of variance between features in a group of data action.

4 Intergration Moment Invariant

New feature extraction of human action using integration moment invariant is presented. The integration moment invariant presented in this paper called Higher United.

Moment Invariant (HUMI). 2D of actions image in video sequence is converted to $f(x, y)$ by translating, rotating and scaling in an action cycle of the video sequence. Higher order Scaled United Moment Invariant (HUMI) has presented by [12]. HUMI can be achieved with the fusion formulation of embedding improvement for scaling factor HGMI from [17] into the UMI from [20]. UMI presented eight formula as shown below.

$$\begin{array}{ll}
\theta_1 = \sqrt{\theta_2}/\theta_1 & \theta_2 = \sqrt{\theta_6}/\theta_1\theta_4 \\
\theta_3 = \sqrt{\theta_5}/\theta_4 & \theta_4 = \sqrt{\theta_5}/\theta_3\theta_4 \\
\theta_5 = \theta_1\theta_6/\theta_2\theta_3 & \theta_6 = \left(\theta_1 + \sqrt{\theta_2}\right)\theta_3/\theta_6 \\
\theta_7 = \theta_1\theta_5/\theta_3\theta_6 & \theta_8 = (\theta_3 + \theta_4)/\sqrt{\theta_5}
\end{array}$$

where θ_1 through θ_6 are unchanged under the image scale, translation and rotation. Meanwhile, θ_7 is for skew invariant that can detect image under reflection. For the θ_8 is change based on the regions, boundaries, and discrete condition. The derivation formulation of the HUMI is given as in Theorem 4.1 below;

Theorem 4.1 To normalise the central moment, the equation is based on GMI using Eq. (1);

$$\eta_{pq} = \frac{\mu_{pq}}{\mu_{pq}^{\frac{p+q+2}{2}}} \tag{1}$$

Where η_{pq} is definition for unequal scaling factor. Then, for the discrete condition, the equation is based on Eq. (2);

$$\eta_{pq} = \left(\frac{\mu_{20}^{(p+1)/2} \mu_{02}^{(q+1)/2}}{\mu_{40}^{(p+1)/2} \mu_{04}^{(q+1)/2}} \right) \mu_{pq} \tag{2}$$

Next for a boundary representation, the equation is based on Eq. (3);

$$\eta''_{pq} = \frac{\mu_{pq}}{(\mu_{00})^{p+q+1}} \tag{3}$$

There are three steps outlines that can be used to explain the integration process of HUMI.

Step 1: As an example, consider θ_1 in UMI, where $\theta_1 = \sqrt{\theta_2}/\theta_1$. From Hu Moment Invariant, θ_2 as given in Eq. (1), $\theta_2 = (\eta_{20} + \eta_{02})^2 + 4\eta_{11}^2$. Substituting θ_1 into θ_2, where $\theta_1 = \eta_{20} + \eta_{02}$ will result in Eq. (4):

$$\theta_2 = \left(\frac{\mu_{20} + \mu_{02}}{\mu_{00}^2} \right)^2 + \frac{4\mu_{11}^2}{\mu_{00}^4} \tag{4}$$

Then substituting Eq. (1) into Eq. (4) will result Eq. (5):

$$\sqrt{\theta_2} = \frac{\sqrt{(\mu_{20} + \mu_{02})^2 + 4\mu_{11}^2}}{\mu_{00}^2} \tag{5}$$

And

$$\theta_1 = \frac{\mu_{20} + \mu_{02}}{\mu_{00}^2} \tag{6}$$

Therefore,

$$\sqrt{\theta_2}/\theta_1 = \frac{\sqrt{(\mu_{20}+\mu_{02})^2+4\mu_{11}^2}}{\mu_{20}+\mu_{02}} = \theta_1 \tag{7}$$

Step 2: Consider θ'_1. From the Hu formulation, $\theta_2 = (\eta_{20}+\eta_{02})^2+4\eta_{11}^2$, as expressed in Eq. (1), then θ'_1 is given by Eq. (8):

$$\theta'_1 = \frac{\mu_{20}}{\mu_{00}^3} + \frac{\mu_{02}}{\mu_{00}^3} \tag{8}$$

Therefore,

$$\sqrt{\theta_2}/\theta_1 = \theta''_1 = \frac{\sqrt{(\mu_{20}+\mu_{02})^2+4\mu_{11}^2}}{\mu_{20}+\mu_{02}} \tag{9}$$

Hence, it can be concluded that

$$\theta_1 = \theta'_1 = \theta''_1 \tag{10}$$

The moment invariant based on the changes of scaling and orientations are recorded, and the results of the image frame variation are compared.

5 Experimental Results

5.1 Action Invarianceness

This section compares and discusses the results obtained from all techniques in order to evaluate the capability of the proposed technique and to determine the best technique for human action recognition. Similarity errors for intra-class (same actions) is lower than inter-class (different actions). The experiments are divided into two categories, which are intra-class and inter-class. Both classes are necessary in order to find the best solution for human action in recognition. In this study, the results of intra-class analysis based on MAE are analyzed. The INRIA Xmas Motion Acquisition Sequence is known as IXMAS is used during experiment. IXMAS dataset contains 13-daily motions perform by 10 actors. This publicly available dataset is frequently updated and accessible online. The experiment are based on 30 frames, 120 frames and 300 frames of action sequence for 13 human actions, which are "check watch" (cw), "cross arms" (ca), "scratch head" (sh), "sit down" (sd), "get up" (gu), "turn around" (ta), "walk" (wk), "wave" (wv), "punch" (pc), "kick" (kc), "point" (pn), "pick up" (pu), and "throw" (th).

Table 1 illustrates the comparison MAE values for GMI and HUMI based on intra-class through 13 human action, respectively. Meanwhile, Table 2 show the result comparison between GMI and HUMI based on inter-class invarianceness. Then, The graph comparison of the result based intra-class invarianceness and inter-class invarianceness is presented in Figs. 2, 3 and 4. To get more significantly about the invarianceness of the proposed integration method, the result based on standard deviation of intra-class invarianceness using 13 human action is presented. The result of the standard deviation is shown at Table 3 respectively. And the graph is illustrated at Fig. 5.

From the experimental results obtained for intra-class analysis, it is noted that most of MAE values produced by HUMI for the 13 human actions are the smallest compared to conventional Geometrical Moment Invariant (GMI) technique. Hence, HUMI is suitable for use in intra-class analysis of frame images. However, HUMI is

Table 1 Intra-class based on MAE for human action invarianceness for 13 human action

Action	Moment invariant	Intra-class 30 frames	Intra-class 120 frames	Intra-class 300 frames
Check watch	GMI	0.71190	0.26952	0.11869
	HUMI	**0.76465**	**0.37515**	**0.14954**
Cross arm	GMI	0.56374	0.3466	0.13622
	HUMI	**0.44890**	**0.24031**	**0.10264**
Scratch head	GMI	**0.67004**	**0.33300**	**0.15363**
	HUMI	0.91191	0.46336	0.19284
Sit down	GMI	0.74628	0.33256	0.13599
	HUMI	**0.37159**	**0.18421**	**0.06949**
Get up	GMI	0.67391	0.29768	0.10690
	HUMI	**0.33691**	**0.21259**	**0.08732**
Turn around	GMI	0.66343	0.35452	0.17316
	HUMI	**0.46589**	**0.21395**	**0.08326**
Walk	GMI	1.33916	0.71876	0.26423
	HUMI	**0.39808**	**0.19687**	**0.06584**
Wave	GMI	0.80022	**0.40578**	0.17034
	HUMI	**0.78470**	0.40793	**0.16852**
Punch	GMI	0.72187	0.45279	0.16460
	HUMI	**0.48509**	**0.22589**	**0.09218**
Kick	GMI	0.63828	0.33489	0.11701
	HUMI	**0.50500**	**0.25015**	**0.09789**
Point	GMI	0.88025	0.38384	0.18573
	HUMI	**0.48268**	**0.2555**	**0.10081**
Pick up	GMI	0.52520	0.40925	0.13275
	HUMI	**0.46549**	**0.23416**	**0.08529**
Throw	GMI	0.31406	0.48527	0.22810
	HUMI	**0.59449**	**0.30951**	**0.12230**

Table 2 Inter-class human action invarianceness

Action	GMI	HUMI
Check watch versus cross arm	0.85908	**0.81130**
Check watch versus scratch head	0.73443	**0.09188**
Check watch versus sit down	0.85779	**0.61469**
Check watch versus get up	1.25152	**0.85655**
Check watch versus turn around	0.66440	**0.63327**
Check watch versus walk	0.27865	**0.36521**
Check watch versus wave	**0.41178**	0.62015
Check watch versus punch	**0.42613**	0.51980
Check watch versus kick	1.34289	**0.12576**
Check watch versus point	0.40706	**0.23099**
Check watch versus pick up	0.80411	**0.34086**
Check watch versus throw	0.72230	**0.49944**

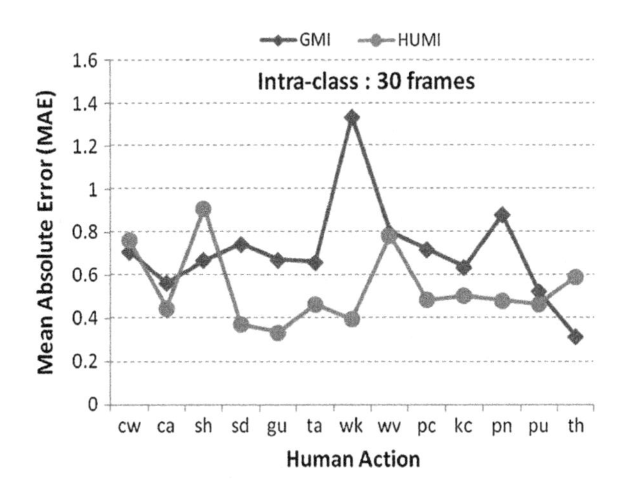

Fig. 2 Result intra-class of 13 human actions for 30 image frames

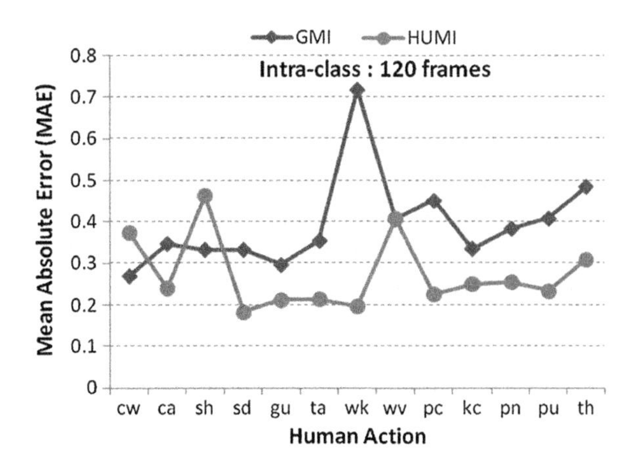

Fig. 3 Result intra-class of 13 human action for 120 image frames

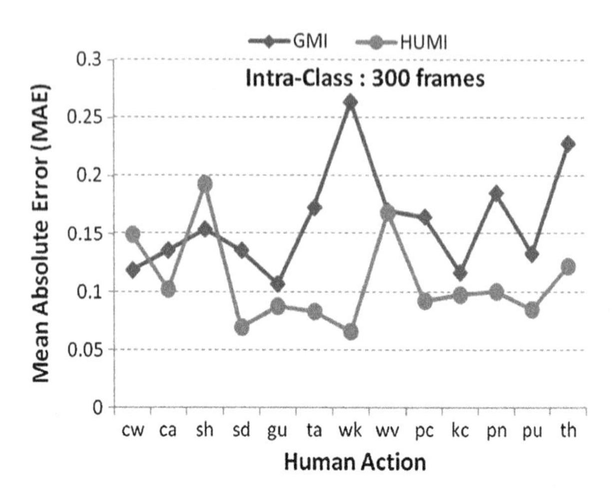

Fig. 4 Result intra-class of 13 human action for 300 image frames

Table 3 Intra-class based on standard deviation for human action invarianceness

Action	Moment invariant	Std 30 frames	Std 120 frames	Std 300 frames
Check watch	GMI	0.86728	0.70973	0.66322
	HUMI	**0.50038**	**0.46493**	**0.46457**
Cross arm	GMI	0.59935	0.66787	0.68083
	HUMI	**0.42066**	**0.47832**	**0.48434**
Scratch head	GMI	0.69179	0.69200	0.74791
	HUMI	**0.47228**	**0.46698**	**0.44943**
Sit down	GMI	0.70661	0.69922	0.71469
	HUMI	**0.45302**	**0.45292**	**0.42183**
Get up	GMI	0.80391	0.71797	0.74050
	HUMI	**0.33812**	**0.39949**	**0.45563**
Turn around	GMI	0.63573	0.66597	0.71471
	HUMI	**0.51916**	**0.50438**	**0.48372**
Walk	GMI	0.71534	0.68511	0.70619
	HUMI	**0.47682**	**0.44480**	**0.44231**
Wave	GMI	1.48413	0.73821	0.73313
	HUMI	**0.56980**	**0.49634**	**0.44804**
Punch	GMI	0.76629	0.67610	0.66119
	HUMI	**0.49726**	**0.45679**	**0.47508**
Kick	GMI	0.71324	0.65820	0.70292
	HUMI	**0.48121**	**0.46900**	**0.46776**
Point	GMI	0.64418	0.68283	0.71128
	HUMI	**0.55886**	**0.49197**	**0.50260**
Pick up	GMI	0.62807	0.75146	0.75111
	HUMI	**0.46650**	**0.44311**	**0.45395**
Throw	GMI	0.82591	0.75641	0.72395
	HUMI	**0.46864**	**0.43034**	**0.44606**

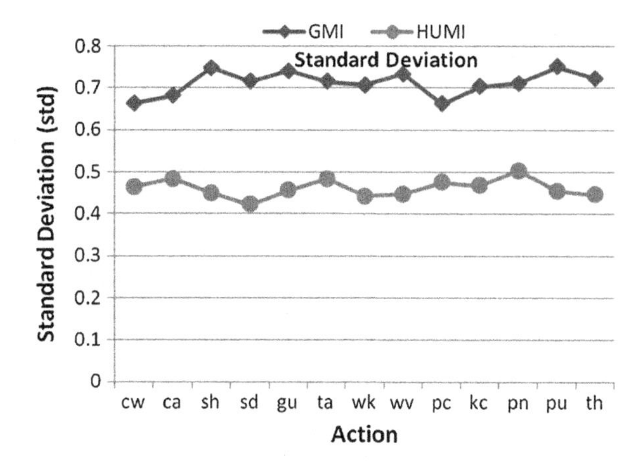

Fig. 5 Result intra-class of 13 human action based on standard deviation

not suitable for in inter-class analysis of the actions as it produces MAE values that are small, which are close to one another, due to the almost similar shape of the silhouettes in terms of boundary and region of the frame images. As for standard deviations analysis of all frames based on the 13 types of actions show that HUMI produces small MAE values, making it suitable for feature representation.

5.2 *Human Action Recognition Accuracy*

The experiment has been conducted to evaluate the recognition performance by implementing several learning methods include Wavelet, Principal Component Analysis (PCA), Expectation Maximization, Normalization, Pre-Discretization (original feature) with the proposed invariant discretization (post- Discretization). The experiments have also been conducted by performing difference classifiers to evaluate the recognition performance. The classifiers used include J-48 (C 4.5) tree, Functional Tree and Best First using WEKA Toolkit. The purpose of this analysis is to identify the most significant features of each class of human action when the data have highly similar features.

Based on HUMI feature vector representation result as shown in the previous analysis, the recognition performances of the six learning algorithms obtained from the experiments for Data Sets A and Data Sets B are shown in Table 4, respectively. Referring to Table 4, even though not producing 100 % recognition rate for Data Set A, but still the proposed Post-Discretization algorithm achieved a recognition rate of over 99 % for J-48, FT and BF classifiers. Meanwhile, the other learning methods achieve lower recognition rate, which are in the range of 69.60 to 72.05 % with J-48 classifier, 33.53–37.65 % with FT classifier, and 43.80–56.69 % with BF classifier. Similarly, though not producing 100 % recognition rate for Data Set B, the proposed Post-Discretization algorithm achieved a recognition rate of over 99 % for J-48, FT and BF classifiers. The other learning methods achieve lower

Table 4 Comparison of learning methods with HUMI

Data Set	Methods	Recognition rate (%)		
		J-48	FT	BF
Data set A: 2730 Training data (70 %) 1170 Testing data (30 %)	Wavelet	72.05	37.65	56.69
	PCA	69.60	33.53	43.80
	EM	70.69	36.33	50.58
	Norm	70.53	36.71	53.34
	Pre-Dis	70.69	36.33	50.58
	Post-Dis	**99.79**	**99.60**	**99.79**
Data set B: 2340 Training data (60 %) 1560 Testing data (40 %)	Wavelet	72.03	39.66	28.22
	PCA	70.67	37.87	54.12
	EM	71.94	36.12	56.88
	Norm	70.39	33.19	49.67
	Pre-Dis	70.84	34.84	56.72
	Post-Dis	**99.87**	**99.79**	**99.80**

recognition rate, which are 70.39–72.03 % with J-48 classifier; 33.19–39.66 % with FT classifier, and 28.22–56.88 % with BF classifier.

The recognition accuracy of post-discretize data are higher compare to pre-discretize data and other learning method based on the HUMI feature vector of human action. This is due to the variance between features that have been improved by implementing discretization technique subsequent to feature extraction with the HUMI. These features are clustered into the same cut that explicitly corresponds to the same action. The lower variation of intra-class and higher inter-class contributed to the better recognition performance.

6 Conclusion

This paper proposed the human action invarianceness method based on the Higher United Moment Invariant (HUMI) in order to validate the feature representation for human action recognition in video surveillance. The experiment of HUMI is performed to validate the human action invarianceness, and extracted features are discretized for better recognition. The results confirm that the invarianceness of human action is still preserved.

Acknowledgements The authors would like to thank Universiti Teknologi Malaysia (UTM) for the support in Research and Development. We also acknowledge the support of Soft Computing Research Group and UTM Big Data Center for the inspiration in making this study a success and our sincere thanks to other researchers who paved the way for this work. This work is supported by The Ministry of Higher Education, Malaysia under Fundamental Grant Research Scheme RJ130000.7809.4F802—Psystem for Breadth First Searching in Shortest Path Algorithm.

References

1. Kaiqi, H., Yeying, Z., Tieniu, T.: A discriminative model of motion and cross ratio for view-invariant action recognition. IEEE Trans. Image Process. **21**, 2187–2197 (2012)
2. Sadek, S., Al-Hamadi, A., Krell, G., Michaelis, B.: Affine-invariant feature extraction for activity recognition. ISRN Mach. Vis. (2013)
3. Wang, S., Yang, Y., Ma, Z., Li, X., Pang, C., Hauptmann, A.G.: Action recognition by exploring data distribution and feature correlation. In: IEEE Conference on Computer Vision and Pattern Recognition (CVPR), pp. 1370–1377. IEEE, (2012)
4. Shabani, A.H., Zelek, J.S., Clausi, D.A.: Multiple scale-specific representations for improved human action recognition. Pattern Recognit. Lett. **34**(15), 1771–1779 (2013)
5. Mokhber, A., Achard, C., Milgram, M.: Recognition of human behavior by space-time silhouette characterization. Pattern Recognit. Lett. **29**, 81–89 (2008)
6. Megavannan, V., Agarwal, B., Venkatesh Babu, R.: Human action recognition using depth maps. In: International Conference on Signal Processing and Communications (SPCOM), pp. 1–5. IEEE (2012)
7. Zhihu, H., Jinsong, L.: Analysis of Hu's moment invariants on image scaling and rotation. In: 2nd International Conference on Computer Engineering and Technology (ICCET), vol. 7, pp. 476–480. IEEE (2010)
8. Lihong, L., Dongli, J., Xiangguo, C., Lixin, S.: A fast discrete moment invariant algorithm and its application on pattern recognition. In: The Sixth World Congress on Intelligent Control and Automation (WCICA), pp. 9773–9777 (2006)
9. Wong, W.-H., Siu, W.-C., Lam, K.-M.: Generation of moment invariants and their uses for character recognition. Pattern Recognit. Lett. **16**, 115–123 (1995)
10. Raveendran, P., Jegannathan, S., Omatu, S.: New regular moment invariants to classify elongated and contracted images. In: Proceedings of 1993 International Joint Conference on Neural Networks. 2083, pp. 2089–2092 (1993)
11. Feng, P., Keane, M.: A new set of moment invariants for handwritten numeral recognition. Proc. IEEE Int. Conf. Image Process. **151**, 154–158 (1994)
12. Muda, A.K., Shamsuddin, S.M., Darus, M.: Invarianceness of higher order united scale invariants. Int. J. Adv. Comput. Sci. Eng. **1**, 105–118 (2007)
13. Muda, A.K., Shamsuddin, S.M., Darus, M.: Invariants discretization for individuality representation in handwritten authorship. In: Srihari, S., Franke, K. (eds.) Computational Forensics. 5158, 218–228. Springer Berlin Heidelberg (2008)
14. Palaniappan, R., Raveendran, P., Omatu, S.: New invariant moments for non-uniformly scaled images. Pattern Anal. Appl. **3**, 78–87 (2000)
15. Pamungkas, R.P., Shamsuddin, S.M.: Weighted central moment for pattern recognition: derivation, analysis of invarianceness, and simulation using letter characters. In: AMS'09. Third Asia International Conference on Modelling & Simulation, pp. 102–106 (2009)
16. Raveendran, P., Omatu, S., Poh Sin, C.: A new technique to derive invariant features for unequally scaled images. In: 1997 IEEE International Conference on Systems, Man, and Cybernetics, 1997. Computational Cybernetics and Simulation, 3154, pp. 3158–3163 (1997)
17. Shamsuddin, M.S., Sulaiman, N.M., Darus, M.: Feature extraction with an improved scale-invariants for deformation digits. Int. J. Comput Math. **76**(1), 13–23 (2000)
18. Shamsuddin, M.S., Sulaiman, N.M., Darus, M.: Invarianceness of higher order centralised scaled-invariants undergo basic transformations. Int. J. Comput. Math. **79**(1), 39–48. Taylor and Francis, Abingdon, ROYAUME-UNI (2002)
19. Balslev, I., Doring, K., Eriksen, R.D.: Weighted central moments in pattern recognition. Pattern Recognit. Lett. **21**(5), 381–384 (2000)
20. Yinan, S., Liu, W., Wang, Y.: United moment invariants for shape discrimination. In: Proceedings of IEEE International Conference on Robotics, Intelligent Systems and Signal Processing. 1, pp. 88–93. IEEE (2003)

21. Theodorakopoulos, I., Kastaniotis, D., Economou, G., Fotopoulos, S.: Pose-based human action recognition via sparse representation in dissimilarity space. J. Vis. Commun. Image Represent. **25**(1), 12–23 (2013)
22. Sadek, S., Al-Hamadi, A., Michaelis, B., Sayed, U.: Human Activity Recognition: A Scheme Using Multiple Cues. In: Bebis, G., Boyle, R., Parvin, B., Koracin, D., Chung, R., Hammound, R., Hussain, M., Kar-Han, T., Crawfis, R., Thalmann, D., Kao, D., Avila, L. (eds.) Advances in Visual Computing, 6454, pp. 574–583. Springer, Berlin Heidelberg (2010)
23. Blank, M., Gorelick, L., Shechtman, E., Irani, M., Basri, R.: Actions as space-time shapes. In: Tenth IEEE International. Conference on Computer Vision (ICCV). 2, pp. 1395–1402 (2005)

Part III
Data Mining Algorithms and Their Applications

Parallel Shallow Water Simulations by Finite Volume Method with CUDA

Kedsararat Khawyuen, Worasait Suwannik and Montri Maleewong

Abstract This paper presents the approach to estimate the water height and its speed in a dam-break problem using the shallow water equations. The finite volume model is implemented using a CUDA program to run on a GPU. CUDA is a parallel programming model developed by NVIDIA. The results have been tested on the large scale problems, i.e., 48×48 and 240×240 cells. The results have been shown that the CUDA program can expedite the computational process by 13 times when comparing with the traditional sequential calculations.

Keywords CUDA · Finite volume method · Shallow water equations · Dam-break

1 Introduction

The shallow water equations [1] have been applied to model and predict flow pat-terns for various real flow, for examples, river flows, or tsunami wave propagation. The equations are derived from the principles of mass conservation and momentum conservation. In the work, we apply the finite volume method [2] FVM for solving the equations in two dimensions. One advantage of using the shallow water equation is that it is a simple calculation when using parallel computational concept due to the repetition of the same equation in every cell. With this reason, the significant computational time can be saved in the case of large domain.

K. Khawyuen (✉) · W. Suwannik
Department of Computer Science, Kasetsart University, Bangkok, Thailand
e-mail: kedsararat.k@gmail.com

W. Suwannik
e-mail: worasait.suwannik@gmail.com

M. Maleewong
Department of Mathematics, Kasetsart University, Bangkok, Thailand
e-mail: montri.m.@ku.ac.th

P. Meesad et al. (eds.), *Recent Advances in Information and Communication Technology 2016*, Advances in Intelligent Systems and Computing 463,
DOI 10.1007/978-3-319-40415-8_10

The parallel computation using CUDA (Computer Unified Device Architecture) developed by NVIDIA is a suitable choice when the finite volume method is applied, [3]. In this work, we have made parallel versions of the finite volume method that is modified from the Sander's calculation, [4]. Some numerical results will be shown with speed up increased. We have increased the efficiency of CUDA by specifying the characteristic and optimal dimension and number of threads per blocks for the execution configuration.

The shallow water equations (SWE) are the set of partial differential equations based on the basic physical conservation laws that describe mass conservation and momentum conservation. Numerical methods can be used as a calculation tool to solve the problem numerically.

The shallow water equations can be expressed as

$$\frac{\partial h}{\partial t} + \frac{\partial (uh)}{\partial x} + \frac{\partial (vh)}{\partial y} = 0, \tag{1}$$

$$\frac{\partial (uh)}{\partial t} + \frac{\partial (u^2 h + \frac{1}{2} g h^2)}{\partial x} + \frac{\partial (uvh)}{\partial y} = 0, \tag{2}$$

$$\frac{\partial (vh)}{\partial t} + \frac{\partial (uvh)}{\partial x} + \frac{\partial (v^2 h + \frac{1}{2} g h^2)}{\partial y} = 0, \tag{3}$$

where h is averaged water depth, u and v are the averaged water velocities in the x and y directions, t is time, and g is the acceleration due to gravity.

The governing equation can be written in a hyperbolic conservation form as

$$\frac{\partial U}{\partial t} + \frac{\partial F(U)}{\partial x} + \frac{\partial G(U)}{\partial y} = 0, \tag{4}$$

where U is the vector of conserved variables while F and G are the vector of conservative fluxes in the x and y directions which can be further written in the vector forms as follows:

$$U = \begin{pmatrix} h \\ uh \\ vh \end{pmatrix}, \tag{5}$$

$$F(U) = \begin{pmatrix} uh \\ u^2 h + \frac{1}{2} g h^2 \\ uvh \end{pmatrix}, \tag{6}$$

$$G(U) = \begin{pmatrix} vh \\ uvh \\ v^2 h + \frac{1}{2} g h^2 \end{pmatrix}. \tag{7}$$

2 Finite Volume Method for Shallow Water Equations

Finite volume method can be used to describe partial differential equations in the form of algebraic equations at any small volume in the domain of interest [2]. First, the problem domain is decomposed into small control volumes as shown in Fig. 1, then dividing the governing equations to solve for the solutions.

In each rectangular area with width Δx and height Δy, $U^n_{i,j}$ represents the vector at co-ordinate (x, y) within the domain of interest, then performing the integral calculation on a set of differential equations for speed and velocity vectors of flow in the direction of the x-axis and the y-axis for each particular cell,

$$U^n_{i,j} = \begin{pmatrix} h^n_{i,j} \\ h^n_{i,j} u^n_{i,j} \\ h^n_{i,j} v^n_{i,j} \end{pmatrix}. \tag{8}$$

In this paper, the Roe Average is used to estimate fluxes along four cell interfaces. Therefore, Eq. (4), can be written in the different form as follows:

$$U^{n+1}_{i,j} = U^n_{i,j} - \left(\frac{\Delta t}{\Delta x}\left(F^n_{i+\frac{1}{2},j} - F^n_{i-\frac{1}{2},j}\right) + \frac{\Delta t}{\Delta y}\left(G^n_{i,j+\frac{1}{2}} - G^n_{i,j-\frac{1}{2}}\right)\right), \tag{9}$$

$$F^n_{i+\frac{1}{2},j} = \frac{1}{2}\left(F^L + F^R - \hat{R}|\Lambda|\Delta\hat{V}\right), \tag{10}$$

where F^L and F^R are actual fluxes on the right and the left of cell face between cells and they can be estimated using

$$F = \begin{pmatrix} h\hat{u} \\ hu\hat{u} + \frac{1}{2}gh^2 \\ hv\hat{u} \end{pmatrix}, \tag{11}$$

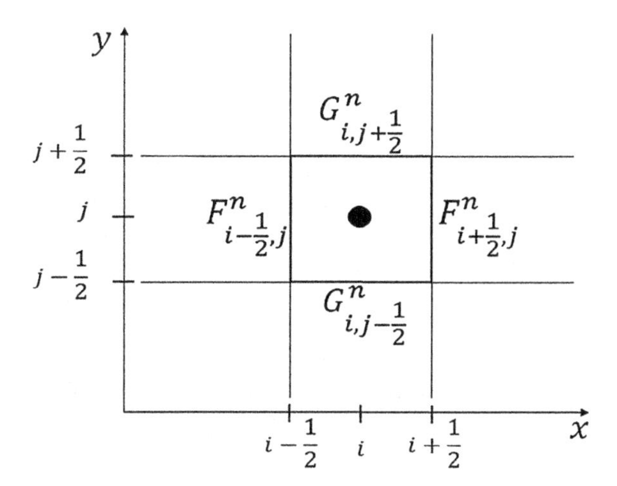

Fig. 1 Discretized domain in the finite volume method

here $\hat{R}$, $|\Lambda|$ and $\ddot{A}\,\hat{V}$ are given by Eqs. (12), (13), and (14). Each column of matrix $\hat{R}$ is associated with eigenvector

$$\hat{R} = \begin{pmatrix} 1 & 0 & 1 \\ \hat{u}-\hat{c} & 0 & \hat{u}+\hat{c} \\ \hat{v} & 1 & \hat{v} \end{pmatrix}, \tag{12}$$

and $|\Lambda|$ is the diagonal matrix is given by

$$|\Lambda| = \begin{pmatrix} \hat{u}+\hat{c} & 0 & 0 \\ 0 & |\hat{u}| & 0 \\ 0 & 0 & |\hat{u}+\hat{c}| \end{pmatrix}, \tag{13}$$

$\ddot{A}\,\hat{V}$ contains the characteristic variables

$$\Delta\hat{V} = \begin{pmatrix} \frac{1}{2}\left(\Delta h - \frac{\hat{h}\Delta u}{\hat{c}}\right) \\ \hat{h}\Delta v \\ \frac{1}{2}\left(\Delta h + \frac{\hat{h}\Delta u}{\hat{c}}\right) \end{pmatrix}. \tag{14}$$

Here,

$$\hat{u} = \frac{\sqrt{h_R}u_R + \sqrt{h_L}u_L}{\sqrt{h_R} + \sqrt{h_L}}, \quad \hat{v} = \frac{\sqrt{h_R}v_R + \sqrt{h_L}v_L}{\sqrt{h_R} + \sqrt{h_L}},$$

$$\hat{h} = \sqrt{h_R} + \sqrt{h_L},\ \hat{c} = \sqrt{\frac{1}{2}g(h_R + h_L)}$$

$$\Delta u = u_R - u_L, \Delta v = v_R - v_L \, and \, \Delta h = h_R - h_L$$

Finally, flux $G^n_{i,j+\frac{1}{2}}$ can be calculated similar to flux $F^n_{i+\frac{1}{2},j}$.

3 CUDA

CUDA (Compute Unified Device Architecture), developed by NVIDIA company, is a software architecture that is suitable for large scale computational workload as it is able to perform parallel computing processes by assigning the workload based on GPU (Graphic Processing Units) capability. CUDA usually divides the workload into blocks, while within each block will be further divided into thread, which each and every thread can be simultaneously processed on GPU with shared memory among memory blocks. With these reasons, the computational time can be significantly reduced. CUDA can be written in C/C++ with APIs for managing elements and memory. It works in two parts: (i) Host for sequential coding for processing on CPU, and (ii) Device for parallel coding for processing works on GPU, which we will refer to as a kernel—a function that allows multi-thread process.

4 Finite Volume Method with CUDA

The computational domain is shown in Fig. 2. We divide the domain into smaller rectangular cells of $N_x \times N_y$ cells with $\Delta x = \Delta y = 1$, and $\Delta t = 0.25$. We simulate the dam-break problem with initial water depth 1.0 contained in the dam and depth 0.1 outside the dam. Initial velocites are zero. Boundary conditions are set as transmissive boundaries, [2].

With the ability of parallel processing, we specify the property of thread block to work in kernel function in two programs. Program (1) is one parallel process that has 11×11 threads per block. It would append additional border cells and copy the value of border cells in these newly added cells. The domain size is $N_{xnew} = N_x + 2$ and $N_{ynew} = N_y + 2$. When dividing into 2-D threads, there will be 11×11 threads per block and the total number of thread blocks is $\frac{N_{xnew}}{11} \times \frac{N_{ynew}}{11}$ as shown in Fig. 3. Program (2) is the parallel process that has 121 threads per block with the number of thread blocks $\frac{N_{xnew} \times N_{ynew}}{121}$ as shown in Fig. 4.

Each thread computes U by copying the vectors U, and fluxes to the memory of GPU using cudaMemcpy command to compute fluxes in each individual cell of Eq. (10). Then the algorithm uses such values as a starting point to calculate the next time step of U. The overview of our CUDA implementation is shown in Fig. 5.

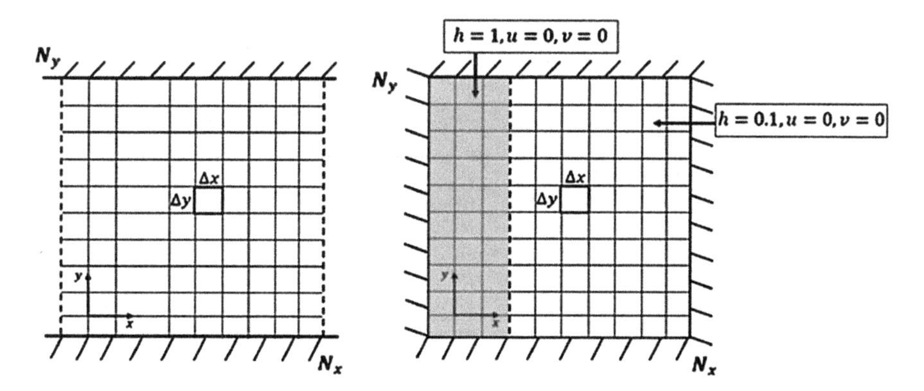

Fig. 2 Computational domain and initial conditions

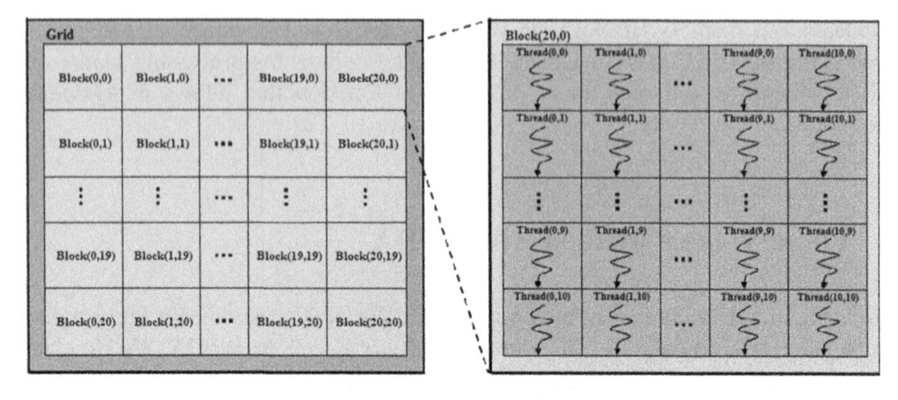

Fig. 3 CUDA thread organization 2 dimensions 11 × 11 threads

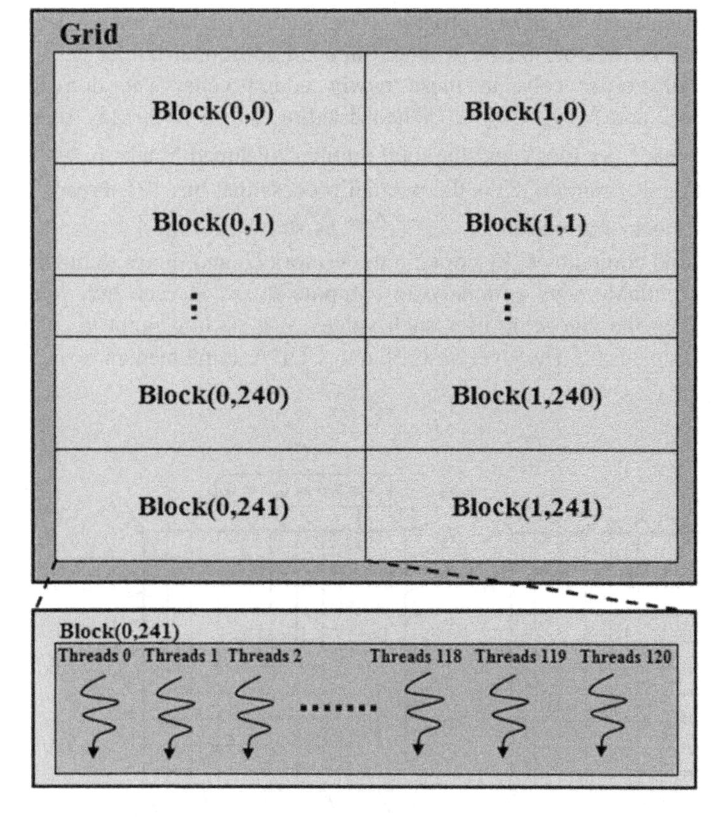

Fig. 4 CUDA thread organization 1 dimension 121 threads

```
Allocate memory on CPU;
    Initialize h (fluid height),u (fluid velocity), v (fluid velocity).
Allocate memory on device
Copy data U,F,G from host to device
    for( tstep=0 ; tstep < NT ; tstep++ )
        Compute function fluxes (U,F,G)
        Compute corrector(U,F,G)
Copy data U from device to host
Free GPU memory
Free CPU memory
Program END
```

Fig. 5 Pseudo code of CUDA FVM

5 Experiment

This work compares the running time of using sequential program with two CUDA programs (called Parallel1 and Parallel2) for simulation of dam-break problem. Parallel1 and Parallel2 use 2D and 1D GPU grids respectively. The source code can be downloaded from https://github.com/ somlim 2529/cudafvm.git. The sequential program runs on a 2.26 GHz Intel Core i5 CPU. The parallel program runs on a 1.53 GHz NVIDIA GeForce 310 M GPU. This GPU has 16 CUDA cores, which is very small compared to 5760 cores in NVIDIA GeForce Titan Z GPU.

6 Result

The surface plots of the water height in various time step are shown in Fig. 6. The accuracy of numerical solution is verified by comparing the numerical result with the exact solution from [5]. The exact solution is shown in Eq. (15). Figure 7 compares the exact result with the FVM result at 100 time steps. These graphs show that our program can provide accurate result.

$$h(x,t) = \begin{cases} 1, & x \leq t \\ \frac{1}{9}\left(2 - \frac{x}{t}\right)^2, & t < x \leq \left(0.226 - \sqrt{0.786}\right) \cdot t \\ 0.786, & \left(0.226 - \sqrt{0.786}\right) \cdot t < x \leq 0.953 \cdot t \\ 0.6, & 0.953 \cdot t < x \end{cases} \tag{15}$$

Running time of all implementations is shown in Table 1. Table 2 shows the speed up for parallel implementations which use 2D and 1D grids. The speed up results are $12\times$, and $13\times$ respectively over the serial implementation in the largest problem. The speed up of 1D to 2D grid is approximately $1.14\times$ for the largest domain.

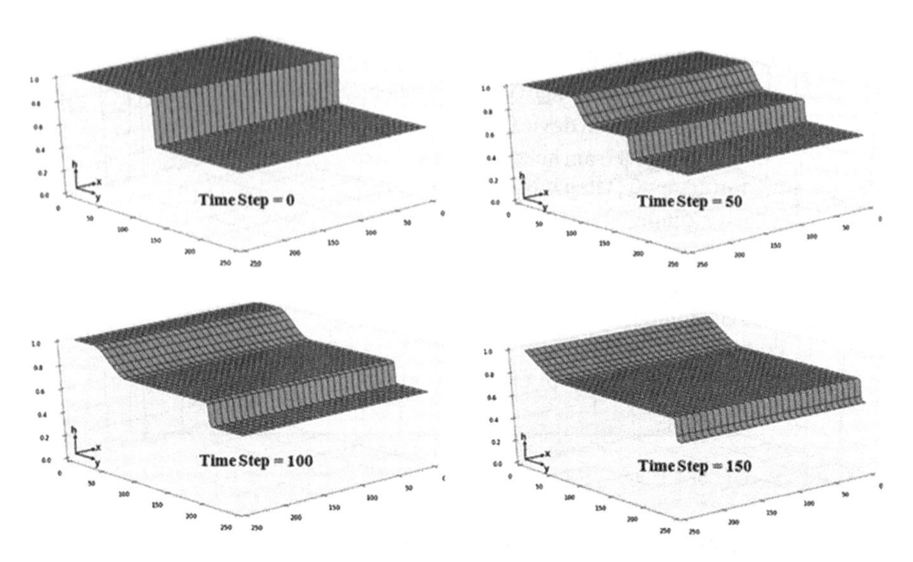

Fig. 6 Surface plot of water height for 0 to 150 time steps

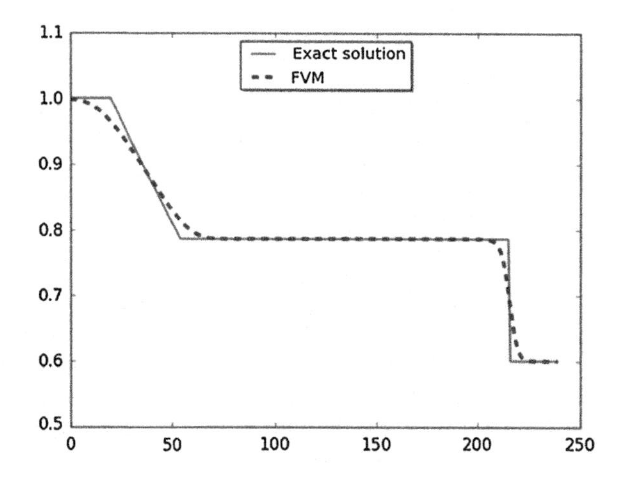

Fig. 7 FVM result and the exact solution along the flow direction at 100 time step

Table 1 The running time of a sequential, a parallel 1, and a parallel 2 for 500 time steps

Domain Size	Sequential (ms)	Parallel 1 (ms)	Parallel 2 (ms)
48 × 48	1312.00	227.98	210.14
126 × 126	8899.00	662.74	636.99
198 × 198	21987.00	1970.60	1738.65
240 × 240	32418.00	2787.48	2451.62

Table 2 Speed up parallel 1 (P1) and parallel 2 (P2) comparing with sequential (S)

Domain size	S/P1	S/P2	P1/P2
48 × 48	5.75	6.24	1.08
96 × 96	13.43	13.97	1.04
144 × 144	11.16	12.65	1.13
240 × 240	11.63	13.22	1.14

7 Conclusions

We have implemented three computer programs to simulate a dam-break problem. Two of them are parallel CUDA program written in C. The simulation result is compared with the analytical result to ensure its correctness. The first parallel program, which uses 2D grid blocks, is approximately 13× faster than our sequential program. We have developed the parallel program further and found that the second parallel version, which uses the 1D grid blocks, is approximately 1.14× faster than our first parallel implementation. The next step of our research is to improve the speed up for other shallow water flow problem, especially very large domain of wet/dry case.

References

1. Shallow Water Equations, https://www.mathworks.com/moler/exm/chapters/water.pdf
2. Maleewong, M.: Modified predictor-corrector WAF method for the shallow water equations with source terms. Math. Probl. Eng. 1–17 (2011)
3. 2D Wet-Bed Shallow-Water Solver Using Local Time Stepping, http://sanders.eng.uci.edu/matlabcodes.html
4. NVIDIA CUDA: CUDA C Programming Guide 7.0 (2015)
5. Stoker, J.J.: Water Waves The Mathematical Theory With Applications (1957)

Intrusion Detection System Based on Cost Based Support Vector Machine

Md. Rafiul Hassan

Abstract In this paper, a novel intrusion detection system (IDS) is developed using a cost based support vector machine (SVM). While developing an IDS, due to the imbalanced characteristics it is very difficult to differentiate the attack events from a non-attack (normal) event in any network environment. The cost based SVM facilitates to put much weight to one pattern over another ones to differentiate attack and non-attack cases with a high accuracy. The same can be applied on a multiclass attack problems by using cost factor to each ratio of different types of attacks. In this study, the cost based SVM has been applied to classify DARPA99 intrusion detection dataset. The experimental results show that the cost based SVM can outperform standard SVM while attempting to differentiate a case as either attack or non-attack (normal). Furthermore, we applied the cost based SVM with an RBF kernel to a multiclass attack problem. Experimental result achieved about 99 % detection accuracy when it was applied to detect the type of attacks as either of Normal, DOS, Probe and R2L from DARPA99 dataset.

Keywords IDS · Cost based SVM · Imbalanced data

1 Introduction

Intrusion Detection is very much essential these days to protect information systems security, especially in the view of worldwide increasing incidents of cyber attacks. Identification of unauthorized use, misuse and attacks on information system is defined as intrusion detection. It is needed because traditional firewalls can't provide full protection against security breaches. An Intrusion Detection System (IDS) doesn't prevent an intrusion, it only detects it and informs the operator.

Md.R. Hassan (✉)
Department of Information and Computer Science, King Fahd University of Petroleum and Minerals, Dhahran 31261, Saudi Arabia
e-mail: mrhassan@kfupm.edu.sa

P. Meesad et al. (eds.), *Recent Advances in Information and Communication Technology 2016*, Advances in Intelligent Systems and Computing 463,
DOI 10.1007/978-3-319-40415-8_11

It detects a hacker breaking into the system or a genuine user exploiting the system resources.

Primary measurement criteria for an IDS are as follows

- False Positives i.e. an event being incorrectly identified as an intrusion when none has occurred.
- False Negatives i.e. an event which IDS fails to identify as an intrusion when it really occurs.
- True positive i.e. an event being correctly classified as an intrusion when one has occurred.
- True Negative i.e. an event being not classified as an intrusion when none has occured.
- Accuracy i.e. how efficient the IDS is in detecting intrusions when it has really occurred.

It is essential to analyze the audit data (generated by the operating systems and networks) in order to estimate the extent of damage occurred, specially in attack trace and listing the attack pattern for future prevention. This makes an IDS a real time detection and prevention tool as well as forensic analysis tool [1].

Artificial intelligence techniques have been widely used by the IDS researchers worldwide due to their generalization capability that help in identifying known intrusions as well as unknown intrusions. Neural Networks have been used to identify both misuse and anomalous patterns [2–4]. Support vector machines (SVMs) have been emerged as a new and powerful technique for learning from data and in particular for solving classification and regression problems. It has been also proved that SVM yields very good result in different domains for the purpose of classification.

SVMs and their variants have been proposed and used by a number of studies for detection of intrusion. Xiao et al. [5] proposed a combined technique of Ad hoc technology and SVM for effective detection of intrusion. Yendrapalli et al. [6] also applied a biased SVM (BSVM) as an IDS tool on DARPA dataset. Recently, Aburomman and Reaz [7] used an ensemble of SVM with other optimization (Particle Swarm Optimization) and clustering method (k-nn) to detect intrusion. A brief discussion about application of SVM as IDS is provided in Sect. 2.

In this paper our aim is to use a cost-based SVM for intrusion detection. Even though the previous studies have evaluated performance of SVM in detecting intrusion, no study has yet explored the efficacy of cost-based SVM to detect intrusion. The available bench mark data in intrusion detection is highly imbalanced in the sense that number of intrusion samples highly outnumber the number of normal samples indicating imbalance in available data. This motivates the research in this paper to use different cost for different class justifying the use of cost-based SVM as an IDS.

The paper is organized as follows. Section 2, a brief literature review is provided. The learning theory of SVM and cost based SVM is written in Sect. 3. Experimental setup, results and analysis is discussed in Sect. 4. Finally, Sect. 5 concludes the paper.

2 Literature Review

A literature review on application of SVM for intrusion detection has been provided by Kausar et al. [8]. Following the literature review it is found that, Xiao et al. [5] proposed a combined technique of Ad hoc technology and SVM for effective detection of intrusion. They identified an enhanced performance of IDS through selecting feature subset and then optimizing the SVM parameters. A Gaussian Kernel SVM could produce a better performance compared with other kernels. They evaluated their technique on DARPA 1998 dataset which has four different attacks namely DOS, R2L, U2R and probe.

Yendrapalli et al. [6] applied a biased SVM (BSVM) as an IDS tool on DARPA dataset. The experimental result using leave-one-out validation technique showed that the performance of IDS for differentiating between normal and u2r attacks is better achieved through using a standard SVM while that of between probe and R2L is better for BSVM. They also concluded that, the performance of SVMs as IDS depends on suitable choice of the SVM parameters.

Yuancheng et al. [9] used SVM followed by selection of features using Kernel Independent Component Analysis. They validated their approach on KDDCUP99 dataset and the experimental results showed a very promising performance of intrusion detection. The resultant SVM also could classify new types of attack which was not included in the training dataset. The ultimate aim of that study was to decrease false alarm with a penalty of overall classification performance.

Gao et al. [10] optimized the SVM parameters and applied the SVM as an IDS. The found SVM very time efficient in detecting intrusion types and also the generalization ability of it. The experimental result proved a better performance by SVM compared with that of RBFNN (Radial Basis Function Neural Network).

Rung-Ching et al. [11] used rough set theory (RST) to deselect less influential features from the dataset and then applied SVM to classify the type of attacks. They evaluated the approach on KDDCUP99 dataset and achieved a higher accuracy in terms of false positive rate and attack detection rate compared with that of an entropy based feature selections.

Yuan et al. [12] proposed the application of hypothesis test theory to SVM classifier (HTSVM) as an IDS. The hypothesis test theory was adopted to decrease the impact of penalty factor in SVM and thereby the overall performance of SVM was improved. The experimental results of using HTSVM on KDDCUP99 dataset showed a better intrusion detection performance compared with that of C-SVM. The results also showed a very good generalization ability of HTSVM classifier.

Guan et al. [13] used the concept of agent along with SVM as an IDS. Each agent involved one SVM to classify two different type of attacks. Finally, a majority voting approach was applied to decide about the type of attack. The experimental results on KDDCUP99 showed a better detection accuracy when compared with the performance of artificial neural networks.

Xiaomei et al. [14] proposed an adaptive genetic algorithm (AGA) to obtain an optimal penalty factor for SVM. Then the SVM was used to analyze audit and detect attack type. Their experimental results on KDDCUP99 dataset revealed the applicability of SVM as IDS.

3 Support Vector Machine

Support Vector Machines introduced by Vapnik [15] have been widely used for applications in many classification and regression problems. The underlying theory of SVM finds a hyperplane which is optimal in separating data in either of two classes. We refer to this hyperplane as optimal separating hyperplane (OSH). Further to this, the kernel trick of transforming data into a higher dimensional space makes SVM an efficient classification tool. Thus, an OSH generation becomes easier in this transformed feature space. The data vectors that lie closed to the OSH are called as support vectors (SV). Since these SVs determine the OSH, they are very useful in classifying data [16].

Let us, consider a training set $D = \{(\mathbf{x}_i, y_i)\}_{i=1}^{L}$, where x_i represents ith input $\mathbf{x}_i \in \Re^n$ and the associated class label is $y_i \in \{-1, +1\}$. In order to search for an OSH in SVM, every input pattern $\mathbf{x}$ is first mapped into a higher dimension feature space $\mathscr{F}$ by $\mathbf{z} = \phi(\mathbf{x})$; where, $\phi(x)$ is a non-linear mapping function as ϕ: $\Re^n \rightarrow \mathscr{F}$. In this case the assumption is that, data are not separable using a linear boundary in real feature space and data in the transformed feature space is linearly separable. Thus, there exists a vector $\mathbf{w} \in \mathscr{F}$ and a scalar value b such that the separating hyperplane is:

$$\begin{aligned} &\mathbf{w} \cdot \mathbf{z} + b = 0 \text{ and} \\ &y_i(\mathbf{w} \cdot \mathbf{z}_i + b) \geq 1 - \xi_i, \quad \forall i \end{aligned} \tag{1}$$

where ξ_i (≥ 0) are referred as *slack variables*. The significance of ξ_i is that, it yields to the misclassified data patterns. Thereby, the term $\sum_{i=1}^{L} \xi_i$ is the measure of misclassification during OSH generation. The ultimate aim of OSH generation is to achieve a maximum classification accuracy and minimum training error. The condition of such optimal hyperplane generation considering data $\mathscr{F}$ is:

$$\begin{aligned} &\textit{minimize}: && \frac{1}{2}\mathbf{w}\cdot\mathbf{w}+C\sum_{i=1}^{L}\xi_i \\ &\textit{subject to}: && y_i(\mathbf{w}\cdot\mathbf{z}_i+b)\geq 1-\xi_i, \text{ and } \quad \xi_i\geq 0, \forall i \end{aligned} \tag{2}$$

here C is a constant parameter known as *regularization parameter*. This parameter represents a trade off measurement between the maximum margin and minimum classification error.

The solution of Eq. (2) is a Quadratic Programming (QP) problem. In the process of solving Eq. (2) first a primal Lagrangian transformation is formed and then the primal Lagrangian is transformed into a dual.

From the primal and dual form of Lagrangian the following optimal hyperplane is obtained:

$$\begin{aligned} &\textit{maximize}: && W(\alpha)=\sum_{i=1}^{L}\alpha_i-\frac{1}{2}\sum_{j=1}^{L}\sum_{j=1}^{L}\alpha_i\alpha_j y_i y_j K(\mathbf{x}_i,\mathbf{x}_j) \\ &\textit{subject to}: && \sum_{i=1}^{L}y_i\alpha_i=0 \quad \text{and} \quad 0\leq\alpha_i\leq C, \quad \forall i. \end{aligned} \tag{3}$$

where $\alpha_1, \alpha_2, \ldots, \alpha_L$ are the non-negative Lagrangian multipliers. The data points $\mathbf{x}_i$ corresponding to $\alpha_i > 0$ lie along the margins of decision boundary and are the SVs. The kernel function $K(.,.)$ is an inner product ($K(\mathbf{x}_i, \mathbf{x}_j) = \phi(\mathbf{x}_i) \bullet \phi(\mathbf{x}_j) = \mathbf{z}_i \bullet \mathbf{z}_j$) among pairwise input patterns. One condition of kernel function is that it must satisfy the Mercer's condition [13]. Through determination of the optimum Lagrangian multipliers, optimum solution for weight vector $\mathbf{w}$ is obtained as

$$\mathbf{w}=\sum_{i\in SVs}\alpha_i y_i\mathbf{z}_i \tag{4}$$

where SVs are the support vectors. For any unknown data vector $\mathbf{x} \in \Re^n$, the classification is done by

$$y=f(\mathbf{x})=sign(\mathbf{w}\cdot\mathbf{z}+b)=sign\left(\sum_{i\in SVs}\alpha_i y_i K(\mathbf{x}_i,\mathbf{x})+b\right) \tag{5}$$

In the process of SVM classifier training, one must tune C and choose a suitable kernel function with its parameters. Since, no theory is available about how to choose the best C and kernels, the performance of SVMs for a problem may vary with this choice. Table 1 lists few different types of kernels used in SVMs.

When the number of samples in two classes (positive and negative classes) are grossly unequal, the classification data is regarded as imbalanced. A technique has been proposed by Morik et al. [17] to deal with imbalanced data for SVM learning using different costs (C_+ and C_- instead of single C in Eq. (2)) for each class.

Lacking data for designing a more refined cost model, the cost-factors are chosen so that the potential total cost of the false positives equals the potential total cost of

Table 1 Types of kernel functions in SVM

Kernel function	Mathematical formula
Linear	$\mathrm{K}(\mathbf{x}_i, \mathbf{x}_j) = \langle \mathbf{x}_i, \mathbf{x}_j \rangle$
Polynomial	$\mathrm{K}(\mathbf{x}_i, \mathbf{x}_j) = \left(\langle \mathbf{x}_i, \mathbf{x}_j \rangle + 1\right)^{\mathrm{d}}$, d: degree of polynomial
Radial Basis Function (RBF)	$K(\mathbf{x}_i, \mathbf{x}_j) = \exp\left(-\left\|\mathbf{x}_i - \mathbf{x}_j\right\|^2 / 2\sigma^2\right)$, σ: width of RBF function
Spline (ANOVA)	$\mathrm{K}(\mathbf{x}_i, \mathbf{x}_j) = 1 + \langle \mathbf{x}_i, \mathbf{x}_j \rangle + \frac{1}{2}\langle \mathbf{x}_i, \mathbf{x}_j \rangle \min\left(\langle \mathbf{x}_i, \mathbf{x}_j \rangle\right) - \frac{1}{6}\min\left(\langle \mathbf{x}_i, \mathbf{x}_j \rangle\right)^3$

the false negatives. This means that the parameters C_+ and C_- of the SVM are chosen to follow the ratio in Eq. (6).

$$\frac{C_+}{C_-} = \frac{\textit{number of positive training examples}}{\textit{number of negative examples}} \tag{6}$$

The quantity $\frac{C_+}{C_-}$ is expressed by a quantity j. Considering the above stated equation, Eq. (2) is reformulated as

$$\begin{aligned} \textit{minimize}: \quad & \frac{1}{2} w \cdot w + C_+ \sum_{i = \textit{positive class}} \xi_i + C_- \sum_{\{j = \textit{negative class}\}} \xi_j \\ \textit{subject to}: \quad & y_i(w \cdot z_i + b) \geq 1 - \xi_i \text{ and } \xi_i \geq 0, \forall i \\ & y_i\left(w \cdot z_j + b\right) \geq 1 - \xi_j \text{ and } \xi_j \geq 0, \forall j \end{aligned} \tag{7}$$

4 Experimental Setup and Results

4.1 Dataset

In this paper we used DARPA99 dataset to evaluate the cost based SVM as an IDS. To generate the dataset an environment was set up to acquire raw TCP/IP dump data for a network by simulating a typical US Air Force LAN. The LAN was operated like a real environment, but being blasted with multiple attacks. A connection is a sequence of TCP packets starting and ending at some well defined times, between which data flows to and from a source IP address to a target IP address under some well defined protocol. Each connection is labeled as either normal, or as an attack, with exactly one specific attack type. Each connection record consists of about 100 bytes. For each TCP/IP connection, 41 various quantitative and qualitative features were extracted. Of this database a subset of 494,021 data were used, of which 20 % represent normal patterns and the rest 80 % are attack data.

4.2 Data Preparation

In this experiment, 15,000 random samples have been chosen from the subset of 494,021 samples. The experiment aims to identify intrusion differentiating it from the normal pattern while the second part of the experiment aims to identify different types of intrusion along with the normal pattern. So the first part is essentially a binary classification problem while the second part is a multiclass problem. For the first part we have defined intrusion as +1 (positive symbolizes attack or intrusion) and normal pattern as −1. For the second part we have defined normal pattern as 1, DoS as 2, Probe as 3, R2L as 4, and U2Su as 5. So it symbolizes a multiclass problem.

The following 10 features (urgent, root_shell, su_attempted, num_root, num_file_creations, num_shells, num_access_files, num_outbound_cmds, is_host _login, is_guest_login) have been deleted from the 15,000 random samples because the value of each feature is zero and hence these features doesn't contribute anything to the variation factor. The rest 31 features were finally used for the intrusion detection in the experiment. Out of 15,000 random samples—3000 were normal pattern, 11,850 were DoS, 120 were probe, 29 were R2L and 1 was U2Su. This was done in accordance to their proportion in the original subset of 494,021 samples. In the last phase of data preparation, the data set was divided into 5 equal sets each consisting of 3000 samples for a fivefold cross validation scheme.

4.3 Performance Measures

The following three performance measures (accuracy, sensitivity and specificity) were used to assess the performance of the SVM as an IDS.

$$Accuracy = \frac{TP + TN}{TP + FP + TN + FN} \times 100\,\% \tag{8}$$

$$Sensitivity = \frac{TP}{TP + FN} \times 100\,\% \tag{9}$$

$$Specificity = \frac{TN}{TN + FP} \times 100\,\% \tag{10}$$

where TP is the number of true positives, i.e. the classifier identifies an intrusion that was labeled as intrusion; TN is the number of true negatives, i.e. classifier identifies a normal pattern that was labeled as normal; FP is the number of false intrusion identification; and FN is the number of false normal identification. Accuracy indicates overall detection accuracy for both normal and intrusion patterns, sensitivity is defined as the ability of the classifier to accurately recognize a intrusion pattern whereas specificity would indicate the classifier's ability not to generate a false detection.

In addition to the above measures, classifier's performance was also evaluated in terms of receiver operating characteristic (ROC). ROC curve plots sensitivity against (1-specificity) as the threshold level of the classifier is varied and depicts the performance of a classifier without regard to class distribution. The area under the ROC curve (AUC) summarizes the quality of classification and is used as a single measure of accuracy.

4.4 Experimental Results and Analysis

4.4.1 Intrusion Detection Between Two Classes: Normal Versus Attack

This section provides the results of two-class (intrusion and normal) detection problem. Two kernel functions—Linear and RBF were used.

For Linear kernel, different values of j (j is the cost factor as defined in Eq. (6)) were used to get different results. For each value of j, fivefold cross validation was performed and then the final value was given by averaging the results from different folds. Table 2 shows the IDS performance for linear kernel cost based SVM with varying values of j.

Table 2 clearly depicts that both accuracy and ROC increase with the increase in value of j till j = 2.0. After that they start to drop, within the region studied. When j = 1.0 i.e. equal emphasis is given both to negative class and positive class (may be referred as standard SVM) the accuracy and ROC are 97.94 % and 0.995 respectively. When the value of j is increased to 2.0 i.e. number of negative training examples (i.e. intrusion pattern) is given double the emphasis as compared to number of positive training examples (i.e. not intrusion), both accuracy and ROC improve.

Table 2 IDS performance using linear kernel cost based SVM

Linear kernel function				
j	Accuracy (%)	Sensitivity (%)	Specificity (%)	ROC
5.0	95.41	99.95	79.41	0.992
3.0	96.72	99.95	87.52	0.993
2.0	**98.19**	**98.93**	**94.91**	**0.996**
1.0	97.94	97.47	98.85	0.995
0.5	97.85	97.32	98.97	0.995
0.25	79.75	77.40	94.36	0.848

The cell with the bold highlights the case where the accuracy as well as the ROC is the highest

We also used an RBF kernel to the cost based SVM as an IDS. In this case, for each value of j, fivefold cross validation was performed for a varying values of σ and then the final value was given by averaging the results from different folds. Table 3 summarizes the performance IDS while the classifier used is an RBF kernel with cost based SVM. the table clearly depict that higher value of j and lower value of σ (i.e. when the width of the kernel decreases and it becomes more non-linear) give a better measure of accuracy, ROC and sensitivity.

It is evident from Table 3 that accuracy, sensitivity and ROC increase with the increase in value of j within the region studied. When j = 1.0 (standard SVM) i.e. equal emphasis is given both to negative class and positive class the accuracy and ROC are 95.80 % and 0.975 respectively. When the value of j is increased to 5.0 i.e. number of negative training examples (i.e. intrusion pattern) is given five times emphasis as compared to number of positive training examples (i.e. not intrusion),

Table 3 IDS performance summary for RBF kernel cost based SVM

RBF kernel function					
J	σ	Accuracy (%)	Sensitivity (%)	Specificity (%)	ROC
5.0	1.0	95.54	94.02	100	0.967
	0.5	95.75	94.28	100	0.973
	0.25	95.98	94.54	100	0.979
	0.1	96.14	94.75	100	0.988
	0.001	**98.21**	**97.40**	**100**	**0.996**
3.0	1.0	95.53	94.00	100	0.967
	0.5	95.75	94.28	100	0.973
	0.25	95.97	94.53	100	0.979
	0.1	96.08	94.71	100	0.987
	0.001	98.17	97.29	100	0.996
2.0	1.0	95.46	93.9	100	0.963
	0.5	95.66	94.15	100	0.97
	0.25	95.79	94.30	100	0.975
	0.1	95.97	94.56	100	0.979
	0.001	98.08	97.14	100	0.995
1.0	1.0	95.33	93.72	100	0.959
	0.5	95.45	93.88	100	0.963
	0.25	95.65	94.14	100	0.97
	0.1	95.80	94.31	100	0.975
	0.001	97.97	97.01	100	0.994
0.5	1.0	95.22	93.54	100	0.951
	0.5	95.27	93.6	100	0.955
	0.25	95.33	93.68	100	0.959
	0.1	95.64	94.11	100	0.97
	0.001	97.91	96.8	100	0.994

The cell with the bold highlights the case where the accuracy as well as the ROC is the highest

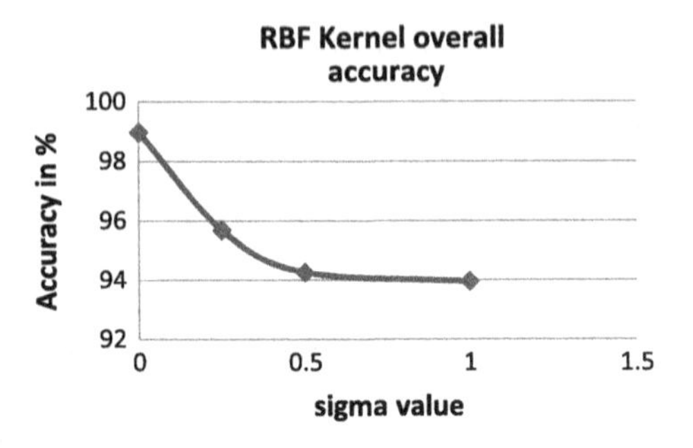

Fig. 1 Overall accuracy RBF kernel with varying σ

both accuracy and ROC improve by a good extent. The improvement in accuracy is almost by 2.5 %, as compared to when equal emphasis is given to both positive and negative classes.

As compared to Linear kernel, there is an improvement in both accuracy and ROC by using RBF kernel. However specificity i.e. the ability of the classifier to identify a normal pattern as normal only increases significantly to 100 % as compared to Linear kernel.

4.4.2 Intrusion Detection Among Multiple Classes

We applied the RBF kernel cost based SVM (as mentioned above RBF kernel provided better performance compared to a linear kernel) to classify the dataset into four attack types: Normal, DOS, Probe, R2L. Figure 1 summarizes the intrusion detection accuracy for $j = 2$ and varying values of σ for the considered data samples.

As shown in the graph, we notice that for a small value of σ and $j = 2$, the performance of RBF kernel with cost based SVM reaches up to 99 %. However, for an increased value of σ, the performance of intrusion detection is not as good as 99 %. This result evidently suggest the importance of choosing SVM parameters for a successful application of SVM as an IDS. Nonetheless, the high detection accuracy and ROC area also encourages the application of cost based SVM for detecting attacks in network environment.

5 Conclusion

In this paper, we proposed the application of a cost based SVM to detect intrusion in a network environment. The experimental results also revealed that, a cost based SVM (i.e. cost factor $j \geq 2$) can outperform the standard SVM (i.e. $j = 1$) while attempting to differentiate whether an event is attack or non-attack (normal). Having proven the efficacy of cost based SVM with RBF kernel, the same method was applied to detect the type of attacks (i.e. Normal, DOS, Probe, R2L) and the experimental results showed an overall accuracy is about 99 %. These high accuracies in attack detection certainly establishes the applicability of cost based SVM as an IDS.

References

1. Mukkamala, S., Sung, A.H., Abraham, A.: Intrusion detection using an ensemble of intelligent paradigms. J. Netw. Comput. Appl. 168–179 (2004)
2. Debar, H., Dorizzi, B.: An application of a recurrent network to an Intrusion detection system. In: Proceedings of the International Joint Conference on Neural Networks, pp. 78–83 (1992)
3. Ryan, J., Lin, M.-J., Miikkulainen, R.: Intrusion detection with neural networks. Advances in Neural Information Processing Systems, pp. 78–83. MIT Press (1997)
4. Mukkamala, S., Janoski, G., Sung, A.H.: Intrusion detection using neural networks and support vector machines. In: Proceedings of IJCNN, pp. 1702–1707 (2002)
5. Xiao, H., Peng, F., Wang, L., Li, H.: Ad hoc-based feature selection and support vector machine classifier for intrusion detection. In: IEEE International Conference on Grey Systems and Intelligent Services (GSIS 2007), pp. 1117–1121 (2007)
6. Yendrapalli, K., Mukkamala, S., Sung, A.H., Ribeiro, B.: Biased support vector machines and kernel methods for intrusion detection. In: Proceedings of the World Congress on Engineering (WCE) 2007, London, U.K (2007)
7. Aborornman, A.A., Reaz, M.B.I.: A novel SVM-kNN-PSO ensemble method for intrusion detection system. Appl. Soft Comput. **38**, 360–372 (2016)
8. Kausar, N., Samir, B.B., Abdullah, A., Ahmad, I., Hussain, M.: A review of classification approaches using support vector machine in intrusion detection. Commun. Comput. Inf. Sci. 1–11 (2016)
9. Yuancheng, L., Zhongqiang, W., Yinglong, M.: An intrusion detection method based on KICA and SVM. In: 7th World Congress on Intelligent Control and Automation (WCICA 2008), pp. 2141–2144 (2008)
10. Gao, M., Tian, J., Xia, M.: Intrusion detection method based on classify support vector machine. In: Proceedings of the 2009 Second International Conference on Intelligent Computation Technology and Automation, pp. 391–394 (2009)
11. Rung-Ching, C., Kai-Fan, C., Ying-Hao, C., Chia-Fen, H.: Using rough set and support vector machine for network intrusion detection system. In: First Asian Conference on Intelligent Information and Database Systems (ACIIDS 2009), pp. 465–470 (2009)
12. Yuan, J., Li, H., Ding, S., Cao, L.: Intrusion detection model based on improved support vector machine. In: Proceedings of the 2010 Third International Symposium on Intelligent Information Technology and Security Informatics, pp. 465–469 (2010)
13. Guan, X., Guo, H., Chen, L.: Network intrusion detection method based on Agent and SVM. In: The 2nd IEEE International Conference on Information Management and Engineering (ICIME), pp. 399–402 (2010)
14. Xiaomei, Y., Peng, W.: Security audit system using adaptive genetic algorithm and support vector machine. In: 3rd International Conference on Advanced Computer Theory and Engineering (ICACTE), pp. 265–268 (2010)
15. Vapnik, V.N.: The Nature of Statistical Learning Theory. Springer, NY (1995)
16. Haykin, S.: Neural Networks—A Comprehensive Foundation. Upper Saddle River (2004)
17. Morik, K., Brockhausen, P., Joachims, T.: Combining statistical learning with a knowledge-based approach-a case study in intensive care monitoring. In: Proceedings of ICML, pp. 268–277 (1999)

Predicting Quality-Assured Consensual Answers in Community-Based Question Answering Systems

Krissada Maleewong

Abstract Although several community-based question answering or CQA systems have been successful at encouraging vast numbers of users to ask and answer questions, and leading to unanticipated explosion of community knowledge, the user-generated contents are confronted with poor quality and untrustworthy problems, while the CQA community deals with conflicts occurred during the question answering process. To tackle such problems, this paper presents a novel approach to predict the best answer as quality-assured consensual answer by simultaneously concerning the content quality and group preference. A set of important features of the answer and its interrelated components as well as social interaction are identified and used to model the predicting function by applying binary logistic regression method. In contrast to the voting-based CQA systems, the proposed model evaluates group preference based on the content quality and community agreement. By training the proposed model using the defined features, the results show that the proposed approach is efficient and outperforms the voting method.

Keywords Community-based Question Answering systems · Community deliberation · User-generated content · Binary logistic regression

1 Introduction

Community-based Question Answering (*CQA*) systems have attracted a large number of people who participate in knowledge creation and sharing. Many CQA websites such as Yahoo! Answers[1] and Stack Overflow[2] allow their users to ask

[1]http://answers.yahoo.com.
[2]http://stackoverflow.com.

K. Maleewong (✉)
School of Information Technology, Shinawatra University, Sam Khok, Pathum Thani, Thailand
e-mail: krissada@siu.ac.th

P. Meesad et al. (eds.), *Recent Advances in Information and Communication Technology 2016*, Advances in Intelligent Systems and Computing 463,
DOI 10.1007/978-3-319-40415-8_12

questions and propose answers to solve the questions. In addition, several CQA websites such as Apple discussion forum,[3] WordReference forum,[4] and Java forum[5] provide a mechanism to support the community deliberation by motivating members to submit an idea or argument to support or oppose a particular answer based on his/her opinion. This deliberation results in a massive volume of the *user-generated content* or *UGC* [1]. However, the quality and reliability of the user-generated contents in the CQA systems vary drastically due to the different skills and expertise of the users [2]. Therefore, the contents in CQA websites result in poor quality and untrustworthy information.

Moreover, many CQA websites have no explicit mechanism to discover the best answer among various candidate answers, while some websites (e.g., Yahoo! Answer and Stack Overflow) applies voting system without community deliberation to identify the best answer. Unfortunately, not all votes are reliable due to the dramatic increase of vote spam phenomenon [3, 4], and the voting system obstructs a user who wants to further develop a proposed answer or search for the reason of accepting or objecting a specific answer.

This paper, therefore, focuses on predicting the best answers in CQA systems that provide the community deliberation. With an emphasis on simultaneously achieve high quality and community accepted answers as the best answer, important features of the best answer concerning the content quality and group preference are identified and applied for classifying a quality-assured consensual answer as the best answer using binary logistic regression method. To evaluate the performance of the proposed approach, an experiment is performed using the deliberation collected from the real-world CQA website.

The organization of this paper is as follows: Sect. 2 discusses the related work. Section 3 presents general features of the deliberation used for modeling the proposed approach. Section 4 conducts an experiment. Section 5 reports the experimental results. Section 6 draws conclusions and future research direction.

2 Related Work

In order to find the best answer, many researches [5–7] have applied a number of content quality metrics (e.g., punctuation and typos) and user interactions (e.g., number of questions asked/resolved by asker/answerer), while several approaches [8–12] have assumed that similar questions provide similar answers, and tried to retrieve similar questions in CQA, and used the answer of the most similar question as the final answer. In addition, many researches [13–16] rank answers based on

[3]http://discussions.apple.com.

[4]http://forum.wordreference.com.

[5]http://forums.sun.com/.

vote numbers. Later, Bian et al. [5] found that not all votes are reliable because users vote for/against a particular answer without much clarification. Hence, a *vote spam attack model* [3] has been proposed to deal with fraudulent votes. However, the model can solve only some common forms of vote spam attacks.

Toba et al. [17] propose a hybrid hierarchy-of-classifiers framework to model the QA pairs. The framework analyzes the question type to guide the selection of the right answer quality model, and use the information from question analysis to predict answer quality score. Tian et al. [18] present a classification model for finding the best answer based on learning from labeled data. The key tasks include designing features measuring important aspects of an answer and identifying the most importance features for the classification. However, the frameworks classify the best answer without concerning group preference data.

To facilitate community deliberation and resolve conflict, many researches have adopted argumentation theories as follows. *World Wide Argument Web* [19] presents a Semantic Web-based system for argument annotation, navigation and manipulation. *Compendium* [20] develops a knowledge management environment to support group deliberation. *Collaboratorium* [21] proposes a collaborative framework to capture discussions as well-structured networks of questions, answers, and arguments. However, they provide no explicit mechanism for selecting the best answer. In the previous works, *Semantic argumentation-based model* [22, 23] has been developed for analyzing community deliberation and discovering the best answer. In contrast, this paper identifies general features of the community deliberation, which then be utilized in the predicting model.

3 Predicting Quality-Assured Consensual Answers

This section generalizes the features of the deliberation in CQA systems and presents the binary logistic regression-based approach to simultaneously learning the content quality and group preference for predicting the best answers.

Let U be the set of users, Q be the set of questions, A be the set of answers, and Arg be the set of arguments. The deliberation can be represented in a graph $G(N, L)$, in which the set of nodes $N = \{U \cup Q \cup A \cup Arg\}$, and the set of links L associated two nodes belonging to different subsets. A user can ask a question or propose a candidate answer. Other members can submit an argument to support or oppose to an answer (or another argument). In general, an answer supported by many arguments is considered as a potential answer to solve the question. On the other hand, an answer made against by many arguments is concerned as a rejected answer. Noted that the deliberation of CQA websites that adopt voting system can be presented into a graph $G(N, L)$, in which a vote score is depicted as an argument.

3.1 Features of Community Deliberation in CQA Systems

With an emphasis on simultaneously achieving high content quality and community accepted answers as the best answer, important features of the deliberation are defined including: (i) the content quality for either answer or argument, and (ii) the group preference, and described as follows:

3.1.1 Content Quality

For a user-generated content (either answer or argument), denoted by $c \in A \cup Arg$, the quality of the content can be measured by considering its: (i) *data quality* and (ii) *user expertise* as follows:

(i) ***Data quality***. To linguistically assess the quality of the content described in textual format, a number of *intrinsic quality metrics* [24] is considered including *misspelling*, *grammaticality*, and *readability*, as follows:

- $misspelling\,(c) \in [0, 1]$ is the average number of spelling mistakes, typos, and out-of-vocabulary words appeared in the data, which can be evaluated by adopting a lexical database such as *WordNet* [25],
- $grammaticality\,(c) \in [0, 1]$ is the average number of sentences that contains no grammar mistake, which can be calculated by applying *link grammar* [26], a widely accepted English syntactic theory, and
- $readability\,(c) \in [0, 100]$ is comprehension difficulty when reading the data content, which can be determined using *Flesch Readability Score* [27], the most popular readability metric applied by many word processing applications such as Microsoft Word and Open Office. In brief, it is calculated by combining the word length and the sentence length; the content with a high score eases user readability.

Thus, an answer with a high content quality implies that it has less number of misspelling and grammar mistakes, and is easy to read.

(ii) ***User expertise***. Intuitionally, a user who has high expertise in a specific area usually proposes a correct answer with higher quality and more precision than a beginner. As a result, the quality of the content should be partially determined based on its user's expertise concerning the *accuracy rate* and *contribution rate* that each user participates in sharing answer and argumentation.

Accuracy rate determines the ability of a user to propose the best answer and to evaluate answers and to assess answers shared by other members, and consequently submit arguments for supporting the best answer and opposing a poor quality one. For a user $u \in U$,

- $accuracy(u_a) \in [0, 1]$ is the ratio of the number of best answers submitted by the user to the number of all answers proposed by the user,

- $accuracy(u_{+arg}) \in [0, 1]$ is the ratio of the number of arguments submitted by the user for supporting any best answer to the number of all arguments submitted by the user,
- $accuracy(u_{-arg}) \in [0, 1]$ is the ratio of the number of arguments submitted by the user for opposing any answer that are later rejected by the community to the number of all arguments submitted by the user.

Moreover, *contribution rate* identifies the contributing experiences of a user. It is computed by means of the number of all answers and arguments submitted by the user. For a user $u \in U$,

- $contribution(u_a) \in [0, 1]$ is the number of all answers submitted by the user, and
- $contribution(u_{arg}) \in [0, 1]$ is the number of all arguments submitted by the user.

3.1.2 Group Preference

In social network analysis, nodal *indegree* quantifies the "*popularity*" of a node received from the nodes that are adjacent to it. During the deliberation, a user node can present his/her individual preference by submitting either supporting or opposing arguments to a particular answer node and/or argument node. Therefore, the *popularity* or *group preference* on the user-generated content node (either answer or argument) can be evaluated by aggregating its indegrees or the degrees of argument, where the degree of a supporting and opposing arguments are +1 and –1, respectively. However, different users have different skills and expertise, and one man one vote yields several problem as previously discussed, the degree of an argument is determined based on its content quality and group preference.

In order to aggregate the degrees of argument as group preference, *Ordered Weighted Averaging* (*OWA*) operator $\varnothing_Q$ [28] is applied. In brief, OWA operator provides a parameterized class of mean type aggregation operators (e.g., max, min, most). For instance, when applying the aggregation operator "*most*" and if the group preference is nearly to 1, the answer is said to be acceptable by "most" members in the community. On the other hand, if the group preference has a negative value close to –1, the answer is said to be rejected by "most" members in the community.

For a user-generated content $c \in A \cup Arg$, let $\ddot{a}_{arg}(c)$ denote its degree of a supporting or opposing argument (the calculation of the degree of argument is described in Sect. 3.2), and $group\_preference\,(c) \in [+1, -1]$ denote its group preference calculated as follows:

$$group\_preference\,(c) = \varnothing_Q(\delta_{arg}(arg_1), \ldots, \delta_{arg}(arg_n)), \quad (1)$$

where $\{arg_1, \ldots, arg_n\}$ is the set of decreasing ordered arguments submitted to the content c.

3.2 Modeling the Prediction of the Best Answer

Based on the features formally defined in the previous section, the prediction model was developed by adopting *binary logistic regression method*. The method estimates the probability of an occurrence of an event (e.g., dead/alive, like/dislike, or good/bad) denoted by Y based on a set of features denoted by X, by fitting data to logistic function as follows:

$$P(Y|X) = \frac{1}{1+e^{-z}}, \tag{2}$$

where

- $e = 2.71828$ is the base of the system of natural logarithms,
- z is a measure of the total contribution of all the features used in the function.

The variable z, also called *logit*, is defined as

$$z = \beta_0 + \beta_1 x_1 + \beta_2 x_2 + \cdots + \beta_i x_i,$$

where

- β_0 is the intercept term,
- $\beta_1, \beta_2, \ldots, \beta_i$ are the regression coefficient of features $x_1, x_2, \ldots, x_i \in X$, respectively.

In order to determine the probability of an answer to be the *best answer*, the features of the content quality and group preference are utilized. The prediction model for answer a is as follows:

$$P(Y = best\,answer|X(a)) = \frac{1}{1+e^{-(\beta_0 + \sum_1^n \beta_i X_i)}}. \tag{3}$$

Consequently, an answer that yields the probability value higher than the cut-off value (0.5) is classified as the best answer.

In Eq. (3), the group preference of answer a ($group\_preference\,(a)$) can be evaluated by aggregating its degrees of argument (δ_{arg}). Similar to the *best answer*, a *strong argument* (an argument submitted to support the best answer or oppose an alternative one) is more likely to maintain high quality and group preference. The degree of argument can be estimated based on the probability of argument *arg* to be a strong argument. By fitting the logistic function using the features of the argument and its group preference, the degree of argument *arg* is computed as follows:

$$P(Y = strong\,argument|X_{Arg}(arg) = \frac{1}{1+e^{-(\beta_0 + \sum_1^n \beta_j X_j)}}) = \delta_{Arg}(arg). \tag{4}$$

Therefore, an argument that produces the probability value higher than the cut-off value (0.5) is specified as a strong argument. The computed value could then be used as the degree of argument *arg*, and applied in Eq. (1) for finding the group preference of an answer. Notice that the group preference feature of a leaf node argument (an argument with no supporting/opposing arguments) is 0.

4 Experimental Setup

This section presents an experimental setup including the collection and statistics of the dataset and evaluation metrics as well as the best answer discovering methods used to compare the experimental results.

4.1 Data Collection and Statistics

This experiment is conducted using the data from the extracted from Apple discussion forum. The dataset (e.g., title, data content, and user information) was collected from answered questions. The extracted information was semi-automatically labeled as questions, answers, and supporting/opposing arguments. For each question, an answer marked as "solved" were labeled as "the best answer", while others were labeled as "alternative answer". A question that contains only one answer was discarded. There were, in total, 273 questions, 901 answers, 2011 supporting and 987 opposing arguments. This whole dataset was randomly separated into two smaller sets with a 7:3 ratio, the first portion consisted of 192 questions, 657 answers, 1355 supporting and 724 opposing arguments for training the models, and the other one had 81 questions, 244 answers, 656 supporting and 263 opposing arguments for testing the models.

4.2 Evaluation Metrics and Methods Compared

Recall that the experiment aims to discover the best answer among various alternative answers. To evaluate the performance of the prediction model, the experiment compares the accuracy rate, F_1 score, and area under the ROC curve (AUC) of baseline method and the three binary logistic regression models using for discovering the best answer as described follows:

- **Baseline model: Vote**. Voting system is adopted as a baseline by calculating the average numbers of supporting and opposing arguments of each answer (positive and negative votes, respectively). If there is no argument for some answers, the vote score is 0.

- **Method I: Quality of answer**. By employing the features of the content quality of an answer, this method evaluates the best answer based on its quality.
- **Method II: Group preference of answer**. This method identifies the best answer by assessing the group preference of each answer. To aggregate all individual preference (the degree of argument), the linguistic quantifier "*most*" is applied and defined by the parameter (0.3, 0.8) [28] for representing the preference of "*most*" users on each answer.
- **Method III: The proposed model (Quality and Group preference)**. This method determines the best answer by using both the features of the content quality and group preference.

5 Experimental Results

In this section, the experimental results are reported and discussed. The binary logistic regression analysis was performed based on the Forward:Likelihood Ratio procedure to obtain the coefficients of the CQA features. The final model for Method III (using Eq. 3) reports that the regression coefficients (β) for most of the features (except the misspelling) of the content quality and group preference are significant and positive, indicating that increasing grammaticality, readability, user expertise, accuracy rate, contribution rate, and group preference are associated with increased odds of achieving the best answer. On other hand, the coefficient for misspelling is significant and negative, indicating that an answer with less number of misspelling is more likely to be the best answer. Furthermore, the *Hosmer and Lemeshow* test showed that the goodness of fit of the equation can be accepted, because the significance of chi-square is larger than 0.05 (0.681). The value of *Nagelkerke* R^2 (0.724) showed that the defined features can explain the occurrence of the best/alternative answer about 72.4 %.

Table 1 summarizes the accuracy of the four methods for predicting the best answer. Obviously, the accuracy of Method III (84.9 %) is significantly higher than that of Baseline and the other methods, which implies that the content quality and group preference are effective features for predicting the best answer. On the other hand, Method I and method II yield about 72.3 and 75.3 % accuracy while Baseline Method gains the lowest accuracy. In addition, the proposed method in Eq. (4)

Table 1 Accuracy, F_1, and AUC of the four methods for predicting the best answer

Method	Accuracy	F_1	AUC
Baseline Method: Vote	0.571	0.557	0.585
Method I (Content quality)	0.723	0.710	0.734
Method II (Group preference)	0.753	0.759	0.783
Method III (Proposed model)	**0.849**	**0.817**	**0.856**

described in Sect. 3.2 for determining the degree of argument yields about 73.5 % accuracy. This implies that in order to strongly support the best answer or strongly oppose alternative answers, a strong argument contains both high content quality and group preference.

The results of the ROC analyses show that Method III produced an outstanding performance, with area under the ROC curve (AUC) of 0.856 and error rate 0.02 with 95 % confidence interval (0.684, 0.790). Also, the AUC is significantly different from 0.5 since *p*-value is 0.000 meaning that the proposed model (Method III) classifies the best answer significantly better than other methods. At the same time, Baseline method, Method I, and II performed with lower AUCs (0.585, 0.734, and 0.783, respectively), and higher error rates (0.033, 0.030, and 0.027, respectively).

6 Conclusions

The study presents that the community deliberation is an important mechanism for enhancing collaboration and collective knowledge creation in CQA websites. To achieve the quality-assured consensual answers, this paper considers the *content quality* and the *group preference* features including *misspelling*, *grammar mistake*, *readability* of the content, and *user expertise*, as well as the *group preference* for predicting the best answer. In addition, the *degree of argument* is developed in order to avoid the vote spam phenomenon. It is estimated based on the probability of an argument to be a strong argument, which could then be aggregated as the group preference using OWA operator. In order to simultaneously learning the content quality and the group preference for discovering the best answer, the binary logistic regression approach is applied using the defined features. The experimental results demonstrate that the proposed prediction model is highly effective when comparing to the voting technique and other methods. Therefore, the deliberation system can be applied to a voting-based CQA website, while the content quality and group preference should be determined in order to select the best answer.

References

1. Chua, A., Banerjee, S.: So fast so good: an analysis of answer quality and answer speed in community question-answering sites. J. Am. Soc. Inf. Sci. Technol. **64**(10), 2058–2068 (2013)
2. Allahbakhsh, M., Benatallah, B., Ignjatovic, A., Motahari-Nezhad, H.R., Bertino, E., Dustdar, S.: Quality control in crowdsourcing systems: issues and directions. IEEE Internet Comput. **17** (2), 76–81 (2013)
3. Bain, J., Liu, Y., Agichtein, E., Zha, H.: A few bad vote too many? Towards robust ranking in social media. In: 4th International Workshop on Adversarial Information Retrieval on the Web (AIRWeb). ACM Press, Beijing, China (2008)

4. Zhu, Y.: Measurement and analysis of an online content voting network: a case study of Digg. In: 19th International Conference on World Wide Web (WWW). ACM Press, Raleigh, NG (2010)
5. Bain, J., Liu, Y., Agichtein, E., Zha, H.: Finding the right facts in the crowd: factoid question answering over social media. In: 17th International Conference on World Wide Web (WWW). ACM Press, Beijing, China (2008)
6. Lou, J., Fang, Y., Lim, K., Peng, J.: Contributing high quantity and quality knowledge to online Q&A communities. J. Am. Soc. Inf. Sci. Technol. **64**(2), 356–371 (2013)
7. Shah, C., Pomerantz, J.: Evaluating and predicting answer quality in community QA. In: 33rd International ACM SIGIR Conference on Research and Development in Information Retrieval, pp. 411–418. ACM, New York (2010)
8. Bernhard, D., Gurevych, I.: Combining lexical semantic resources with question and answer archives for translation-based answer finding. In: 47th Annual Meeting of the ACL and 4th International Joint Conference on Natural Language Processing of the AFNLP, vol. 2, pp. 728–736. Stroudsburg, PA (2009)
9. Cong, G., Wang, L., Lin, C.-Y., Song, Y.-I., Sun, Y.: Finding question–answer pairs from online forums. In: 31st Annual International ACM SIGIR Conference on Research and Development in Information Retrieval, pp. 467–474. ACM, New York (2008)
10. Moschitti, A., Quarteroni, S.: Linguistic kernels for answer re-ranking in question answering systems. Inf. Process. Manage. **47**(6), 825–842 (2011)
11. Wang, X.-J., Tu, X., Feng, D., Zhang, L.: Ranking community answers by modeling question–answer relationships via analogical reasoning. In: 32nd international ACM SIGIR Conference on Research and Development in Information Retrieval. ACM, New York (2009)
12. Zhou, T.C., Lyu, M.R., King, I.: A classification-based approach to question routing in community question answering. In: 21st International Conference on World Wide Web, pp. 783–790. ACM Press (2012)
13. Blooma, M.J., Chua, A.Y., Goh, D.H.-L.: A predictive framework for retrieving the best answer. In: The ACM Symposium on Applied Computing. New York (2008)
14. Liu, Q., Agichtein, E.: Modeling answerer behavior in collaborative question answering systems. In: 33rd European Conference on IR Research (2011)
15. Sakai, T., Ishikawa, D., Kando, N., Seki, Y., Kuriyama, K., Lin, C.-Y.: Using graded-relevance metrics for evaluating community QA answer selection. In: 4th ACM International Conference on Web Search and Data Mining, pp. 187–196. ACM Press (2011)
16. Surdeanu, M., Ciaramita, M., Zaragoza, H.: Learning to rank answers to non-factoid questions from web collections. Comput. Linguist. **37**(2), 351–382 (2011)
17. Toba, H., Ming, Z.-Y., Adriani, M., Chua, T.-S.: Discovering high quality answers in community question answering archives using a hierarchy of classifiers. Inf. Sci. **261**, 101–115 (2014)
18. Tian, Q., Zhang, P., Baoxin, L.: Towards predicting the best answers in community-based question-answering services. In: 7th International AAAI Conference on Weblogs and Social Media, pp. 725–728 (2013)
19. Rahwan, I., Zablith, F., Reed, C.: Towards large scale argumentation support on the semantic web. In: 22nd national Conference on Artificial Intelligence, vol. 2, pp. 1446–1451. AAAI Press, Vancouver, British Columbia, Canada (2007)
20. Buckingham Shum, S., Selvin, A., Sierhuis, M., Conklin, J.: Hypermedia support for argumentation-based rationale: 15 years on from gIBIS and QOC. Rationale Manage. Softw. Eng. 111–132 (2006)
21. Iandoli, L.K., Klein, M., Zollo, G.: Can We Exploit Collective Intelligence for Collaborative Deliberation? The Case of the Climate Change Collaboratorium. MIT Sloan School of Management (2008)
22. Maleewong, K., Anutariya, C., Wuwongse, V.: Analyzing community deliberation and achieving consensual knowledge in SAM. Int. J. Organ. Collect. Intell. **1**(3), 34–35 (2010)

23. Maleewong, K., Anutariya, C., Vuwong, V.: Enabling intelligence in web-based collaborative knowledge management system. Int. J. Syst. Serv.-Oriented Eng. **2**(1), 40–58 (2011)
24. Maleewong, K., Anutariya, C., & Wuwongse, V.: SAM: Semantic argumentation based model for collaborative knowledge creation and sharing system. In: International Conference on Computational Collective Intelligence (ICCCI), vol. 5796, pp. 75–86. Springer, Wroclaw, Poland (2009)
25. WordNet. http://wordnet.princeton.edu
26. Lafferty, J., Sleator, S., Temperley, D.: Grammatical trigrams: a probabilistic model of link grammar. In: AAAI Fall Symposium on Probabilistic Approaches to Natural Language (1992)
27. Flesch, R.F.: The Art of Readable Writing. Harper & Row, New York (1949)
28. Yager, R.R.: Centered OWA operators. Softw. Comput. **11**, 631–639 (2007)

Influence of ERP Employment on Work Skills

Suphan Petyim, Bunthit Watanapa, Nipon Charoenkitkarn and Jidapa Archanainant

Abstract The objective of this research is to investigate the potential of enterprise resource planning (ERP) in enhancing work skills of employees after being used for a while. The study is based on the hypothetical model that ERP vision and/or ERP quality could lever the work skills in the dimensions of conceptual skills, technical skills or interpersonal skills. The multivariate analysis of covariance (MANCOVA) was utilized to control the state covariation of before and after use of ERP. The questionnaire-based survey was conducted in two well developed and successful organizations in Thailand; One is a leading firm in the petrochemical industry and the other is in the construction material industry. The results show that quality of ERP system can significantly affect the enhancement of work skills in many dimensions. Using t-test and regression analysis, further validation and insights are also provided to envision enterprises in planning and managing ERP for enhancing not only the traditional functional success but also individual work skill and intercourse.

Keywords Enterprise resource planning · Work skills · System quality · Information quality · Vision

S. Petyim (✉)
Business Information System, School of Information Technology,
Bangkok, Thailand
e-mail: wow0226@gmail.com

B. Watanapa · N. Charoenkitkarn · J. Archanainant
King Mongkut's University of Technology Thonburi, Bangkok, Thailand
e-mail: bunthit@sit.kmutt.ac.th

N. Charoenkitkarn
e-mail: nipon@sit.kmutt.ac.th

J. Archanainant
e-mail: jidapaarcha@yahoo.com

P. Meesad et al. (eds.), *Recent Advances in Information and Communication Technology 2016*, Advances in Intelligent Systems and Computing 463,
DOI 10.1007/978-3-319-40415-8_13

1 Introduction

Nowadays, the Enterprise Resource Planning (or ERP, in abbreviated form) system has become a strategic information system for organizations to manage enterprise-wide operations to enhance and maintain competitiveness against competitors.

Successfully implementing the ERP system can cause a mild or radical change in business processes and their intercourses so as to achieve planned effectiveness in functional operations and leveraging the level of cooperation and collaboration amongst employees (ERP's strategic vision). However, the eventual success of ERP is also subject to another integral part, namely the quality of the information system (IS) which determines the usability and effectiveness of the logical design in the eyes of the users. A valid vision and practical IS quality shall lead to ERP adoption by employees [1].

Hypothetically, once ERP is introduced for a certain period of time, the successful adoption of best practices shall reinforce positive changes in the employees' attitude and work skills. For example, they may find that they can work faster, have less errors and gain supportive information from others more easily. They should be happier at work and more cooperative and willing to share information. As it has been fully realized that human capital is the most important asset of a company, this would create sustainable success of ERP and reinforce the increase in productivity and profits of the organization.

This research aims at studying the hypothetical effect of ERP vision and quality to change in employee work skills after using ERP for a controlled period of time. For this purpose, the multivariate analysis of covariance (MANCOVA) will be applied to measure the post-state impact of ERP system in-use.

In the next sections, we deliver research background, research methodology, experimental results, business insights and discussion, and the conclusion that includes the direction of future work.

2 Research Background

ERP system is an enterprise-wide application that integrates all data and streamlines business processes in an organization. ERP system is usually based on best practices and standard business procedures. The functional data are consolidated or integrated into a centralized database which enables all departments to effectively share data and enable enterprise-wide collaboration [2]. ERP system has various functions that support various departments such as inventory control, manufacturing, accounting and finance and human resources [3].

In 2003, Hunton et al. pointed out that most businesses view the ERP system as a strategic tool yielding great business value such as improving productivity, streamlining collaboration, reducing costs, and supporting effective decisions, just

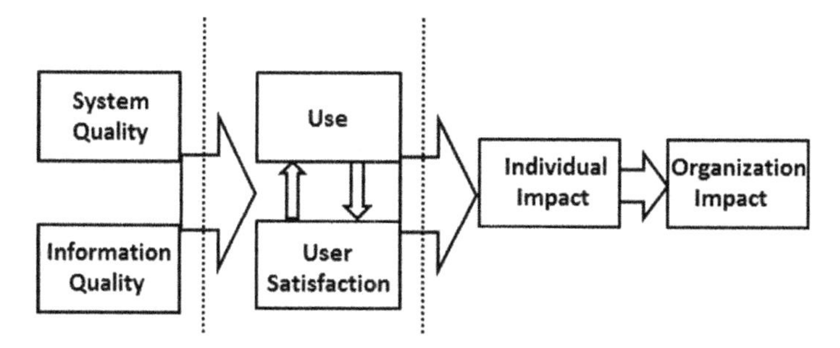

Fig. 1 IS success model source: Delone and McLean (1992)

to name a few [4]. The report also mentioned the possibility that ERP system can induce changes in organizational behavior, e.g. employee relationships, culture, and individual work skill and attitudes.

In fact, a change in organizational behavior could reinforce the sustainable success of ERP system usage, according to the IS success model of Delone and McLean (referred as D&M from this point onward) [5] as shown in Fig. 1. The hypothesis of the model is that the IS quality assures the usability which correlates with user satisfaction. The satisfaction influences the users to accept the system and finally impacts the whole organization.

The D&M model has identified system quality and information quality as key dimensions of IS quality which drive the system usage and user satisfaction. In the higher order of relationships, quality could influence individuals and organizational performance.

System quality is the technical characteristics of a system in producing information output [6]. It could be measured in terms of system flexibility, ease of use, reliability, short response time and usefulness [7]. Information quality is defined as the quality of information deliver to users [8]. Potential measures of this dimension could be accuracy, timeliness, completeness and consistency [9].

Alternatively, IS quality of ERP system is viewed as a degree of meeting its intended purpose or effectiveness. This means that to achieve maximum benefit of ERP system, a well-thought ERP vision aligned with the business vision of a company needs to be established and treated as integral part of fulfilling the organizational goals and mission [10]. In 2008, Ifinedo also mentioned that business vision and top management supports are key determinants of ERP success [10]. This study also mentioned that the vision could impact performance of individual users [11].

Vision is to project organization in the future that is matched with the long-terms target and duty base on its ground truth. Information Technology vision could be viewed as a strategic approach to manage Information Technology to serve business goals [12] (Tables 1 and 2).

Many organizations adopt the ERP system because it could support their business vision, e.g; gain strategic advantages and improve customer service [13, 14].

Table 1 The framework of ERP vision

ERP vision	Definition
Operational	Change processes from manual to automatic
Managerial	Usage of data for planning, monitoring and supporting managers for decision making
Strategic	System ability to support business goals
IT infrastructure and Organizational support	Efficient work patterns and communication, Organizational learning, staff empowerment and moral

Table 2 Description of samples

Attribute	Value	N (59)	Percent (%)	N (63)	Percent (%)
		Petrochemical		Cement	
Gender	Male	14	23.7	21	33.3
	Female	45	76.3	42	66.7
Position	Operation officer	53	89.8	49	77.8
	Manager	3	5.1	12	19.0
	Director	–	–	2	3.2
	Other	3	5.1	–	–
Working Experience (Year)	1–3	13	22	30	47.6
	4–6	20	39.9	5	7.9
	7–9	13	22	5	7.9
	More than 10	13	22	23	36.5
Function (The duty of a person)	Accounting & Finance	34	57.63	29	46.03
	Marketing & Sales	6	10.17	18	28.57
	Production & Material	17	28.81	15	23.81
	Other	2	3.39	1	1.59

In 2002, Shang and Seddon [15] proposed a visionary framework that classifies types of benefits an organization could obtain by using the ERP system. As shown in Table 3, they include the vision of organizational support, e.g. organizational learning and employee empowerment in the framework. This vision is intangible and has not been researched much as yet.

From Table 1, this implies that most of ERP system strategically introduces best practices for performing relevant business processes that require changes of work procedures, collaborative structures, and organizational culture. This links to the D&M model and suggestion of Mahdavian [16] that ERP system deployment can be claimed as a success only when it effectively impacts individuals at a personnel level which is the most important asset of an organization. To measure the impact on personnel by ERP, The classic model of Katz [17], that classifies the skill set of individual into the three basic skill abilities of technical, conceptual, and human, is integrated into this model of study.

Table 3 Results of hypotheses test

Independent variable	Dependent variable	Sig	Hypo. no.	Correlation	Hypo. test
Vision	Technical skill	0.041	H1	0.304	Accept
	Conceptual skill	0.207	H2	0.329	Not accept
	Human skill	0.083	H3	0.268	Not accept
Quality	Technical skill	0.003	H4	0.429	Accept
	Conceptual skill	0.009	H5	0.418	Accept
	Human skill	0.001	H6	0.427	Accept
Vision	Quality	0.000	H7	0.598	Accept

A technical skill is defined as a competence in doing specific tasks or activities. For instance, to proficiently use certain functions in the finance model of the ERP system and resolve some problems related to a work process is an advanced technical skill.

A conceptual skill can refer to knowledge, integrative and systematic thinking skill that enable a person to understand the need of work collaboration, decide and take measures against the root cause for an incurring problem with integrative systematic and proactive way.

A human skill involves the ability to work with people. While a person with a technical skill is adept at using things to do specific task, a person with a human skill is equipped with abilities to get along with people, communicate and comfortably work in teams and collaborate across functions.

Next, we discuss the hypotheses of the study, model, and the methodology.

3 Model and Methodology

Since the study focuses on the impact of ERP system usage to the change in work skills of users. A hypothetical model is established as shown in Fig. 2.

We define the effectiveness of ERP system usage based on the factors of IS quality and vision as discussed in Sect. 2. Work skills are as defined by Katz [17] where skills are divided into 3 basic elements: conceptual, technical and human skill.

Conceptual skill enables one to formulate ideas, understand how parts of an organization interact, systematically think, and resolve problems [17, 18]. Technical skill involves ability to do specific activities and understand methods, procedures and technical aspects of specialized tasks [17]. Lastly, human skill is about people relationship skills. This skill enables people to communicate effectively and work well in a team [17].

Based on the given model, we have derived 7 hypotheses to be investigated.

H1 Vision for the ERP system has positive impact on changes in technical skill
H2 Vision for the ERP system has positive impact on changes in conceptual skill
H3 Vision for the ERP system has positive impact on changes in human skill
H4 Quality for the ERP system has positive impact on changes in technical skill

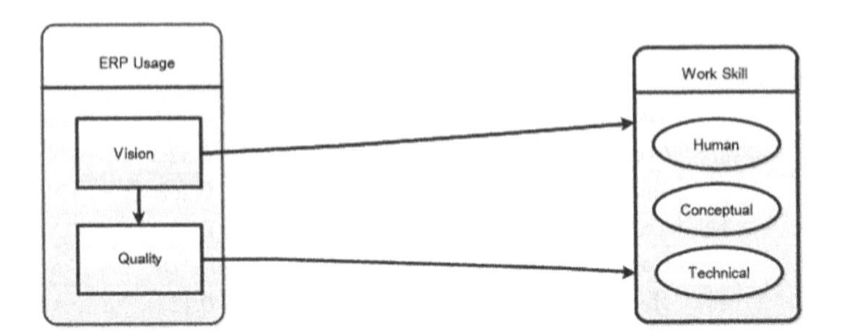

Fig. 2 Hypothetical model for impact of ERP usage to work

H5 Quality for the ERP system has positive impact on changes in conceptual skill
H6 Quality for the ERP system has positive impact on changes in human skill
H7 Vision has positive impact on the quality of the ERP system

3.1 Experiment Design

The model was tested on a targeted population. The sample group was specified as those who have had at least one year experience using the ERP system and have worked in two selected well-known companies in Thailand, one is in a petrochemical business and another is in a cement business. The respondents were asked to give demographic information and answer questions involving ERP vision, ERP quality and work skills. Apart from the demo-graphic part, all questions were measured using 5-point Likert scales.

The data collection method was conducted using a questionnaire. To ensure the reliability of the survey, we had a preliminary test on 10 samples and an expert review. The obtained overall Cronbach alpha was 0.857. For each part of the questionnaire, the Cronbach scores were 0.915 for vision, 0.920 for quality, and 0.741 for work skill.

In this study, the number of effective respondents are 122 out of 150 prospects. The response rate was about 81 %. There were 59 respondents from the company in the petrochemical business and 63 respondents from the company in cement industry. Most of them held operation officer positions and had varied experience (1–6 years) in using the ERP system.

3.2 Data Analyses

The hypothesis test was mainly performed using the Multivariate Analysis of Covariance (or MANCOVA in short) which enables (1) looking up the influence of many independent variables on many dependent variables and (2) treating the experiment with a control of pretest and posttest measurements as required by the study. We also used other descriptive statistics like the t-test and regression analysis for post-analyses. The t-test was for post-comparison between the work skills before and after usage of the ERP system. The regression analysis gave more insight into the vision and quality as determinants of work skills.

4 Results of the Experiment

Next, the results of the hypotheses test are presented with the supportive statistical analyses to give insights into the experiment.

The results of the research model are shown in Table 3. At the significant level (Sig.) of 0.05, hypotheses H1 and H4-7 are accepted. This can be interpreted that the quality of ERP systems significantly impact the improvement on human skills in the three dimensions of technical, conceptual, and human relationships. Denying H2 and H3 means vision does not directly influence the skill of personnel. However, H7 acceptance (vision influence quality) helps to reinforce the importance of vision that can indirectly influence the improvement of human skills through the quality of the ERP system.

See the conceptual diagram in Fig. 3 for a better understanding of the results of the hypotheses test.

To gain insight into industry relevance, Table 4 shows results of hypotheses acceptance in each business type. The main interpretation is that both industries commonly show the high correlation between the quality of ERP system and the improvement of technical skills (H4) and human relationship skills (H6). However, only the petrochemical firm showed a positive influence of quality on conceptual skill. This effect may be described by the longer time of the mode of working experiences (4–6 years versus 1–3 years of the subjects in the petrochemical and cement firms, respectively). Vice versa, the effect of correlation between vision and

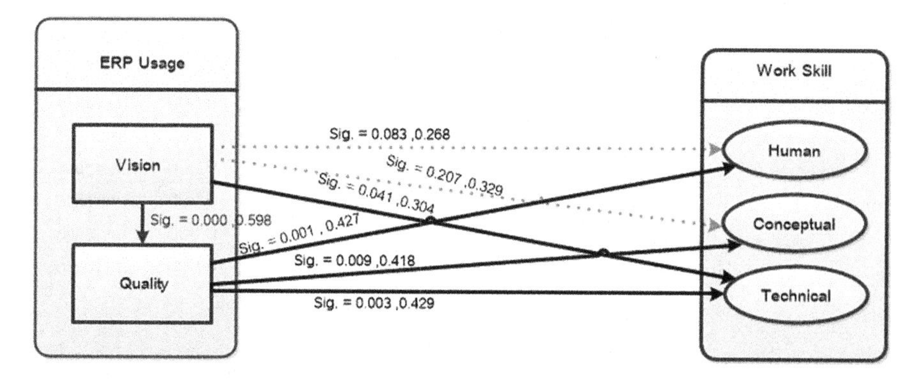

Fig. 3 Conceptual diagram of resulting hypotheses test

Table 4 Hypotheses acceptance in each business type

Business	Independent variable	Dependent variable	Sig	Hypo no.
Petrochemical	Quality	Technical skill	0.004	H4
		Conceptual skill	0.035	H5
		Human skill	0.044	H6
Cement	Vision	Technical skill	0.049	H1
	Quality	Technical skill	0.035	H4
		Human skill	0.001	H6

technical skill is only valid in the cement industry could be explainable. The direction and enforcement according to the vision of ERP could be more actively dominating the work climate when the ERP system is in the initial go-live phase.

To support the results of MANCOVA as above, a direct *t*-test analysis was performed to validate the significance of human skill changes (before and after the ERP system usage comparative measure) see Table 5. The results show that all skill types in both business types are significantly changed in terms of scores after using ERP system for more than 1 year with a *p*-value of 0.000. The largest change (more than 60 %) incurred was in the technical skill and the least change was the human skill. The change pattern was the same for both business types.

In addition, the regression analysis was performed to indicate the casual relationship between the vision or quality (as an independent variable) to each skill type (as a dependent variable). In Table 6, the results show that quality is the best determiner for the change in any skill (R^2 at about 0.18 and *p*-value at about 0.000.) Note that an R^2 value lower than 50 % is expectable when predicting human behavior since human behavior is very subjective [18]. Vision is beaten by quality (for being a determiner of human skill) in every measure (R^2 and coefficient), although the significance is acceptable (about 0.000–0.003).

Table 5 T-test value of before—versus-after usage of ERP system

Business	Skills	Before usage system	After usage system	Difference (%)	Sig	Level changing
Petrochemical	Technical	2.17	3.62	66.82	0.000	Increase
	Conceptual	3.22	3.64	13.04	0.000	Increase
	Human	3.37	3.68	9.20	0.000	Increase
Cement	Technical	2.12	3.58	68.87	0.000	Increase
	Conceptual	3.25	3.75	15.38	0.000	Increase
	Human	3.37	3.76	11.57	0.000	Increase

Table 6 The regression

Independent variable	→	Dependent variable	Sig	R^2	Standardized coefficients
Vision	→	Technical skill	0.001	0.093	0.304
Vision	→	Conceptual skill	0.000	0.108	0.329
Vision	→	Human skill	0.003	0.072	0.268
Quality	→	Technical skill	0.000	0.184	0.429
Quality	→	Conceptual skill	0.000	0.175	0.418
Quality	→	Human skill	0.000	0.182	0.427

5 Conclusion and Future Research

This research studied the impact of ERP system usage on work skills and relied on subjects who worked in specified petrochemical and cement businesses in Thailand. The model hypothesis is vision and quality of the ERP system are determiners of the three work skills: technical, conceptual, and human relationship. MANCOVA was performed to validate the seven hypotheses of the model.

The results show that only the quality of the ERP system has significant impact on the change in work skills in positive way. Vision was found to positively influence the quality and hence it implicitly influenced the change in work skills via the quality. The t-test confirmed the change in work skills after using ERP system and regression analysis helps to confirm that the importance of quality of ERP system is a cause of change in work skills.

In conclusion, the results of this study suggest that (1) having a clear vision of ERP objectives and measures leads to a better quality ERP system and (2) deploying a good quality ERP system can help to enhance work skills of the personnel which is the most important asset of organizations.

For the future research, we plan to extend the measure of the impact of ERP usage to the change of personality traits which can represent the deeper aspects of characteristics of an individual. Additionally, the analysis at the level of business type implies hypotheses that the length of working experience (or time duration of ERP in production state) is potentially the second-ordered factor that controls (1) the relation of the ERP system quality and the change in conceptual skill and (2) the relation of the ERP vision and the change in technical skill. The experiments on these two hypotheses may yield further insights into the realm of ERP systems and human skills enhancement.

References

1. Kalema, B.M.: ERP systems usage: an extended unified theory of acceptance and use of technology. Inf. Sci. Comput. Telecommun. (PACT) 92–98 (2013)
2. Tanis, C., Markus, M.L., van Fenema, P.C.: Multisite implementations. Commun. ACM (2000)
3. Dabenport, T.H.: Putting the enterprise into the enterprise system. Harvard Bus. Rev. **76**, 121–131 (1998)
4. Hunton, J.E., Lippincott, B., Reck, J.L.: Enterprise resource planning systems: comparing firm performance of adoptersand non-adopters. Int. J. Account. Inf. Syst. **4**(3), 165–184 (2003)
5. Delone, W.H., McLean, E.R.: Information systems success: the quest for the dependent variable. Inf. Syst. Res. 60–95 (1992)
6. Guvence, C.: Information systems success and expectations for information technology in-vestment: case study Memory to obtain the Master of Science. In: Information System Memory to obtain the Master of Science. Information System Department, Middle East Technical University (2005)
7. Zhang, Z., Leeb, M.K.O., Huang, P., Zhang, L., Huang, X.: A framework of ERP systems implementation success in China: an empirical study. Int. J. Prod. Econ. **98**, 56–80 (2005)
8. Petter, S., Delone, W., McLean, E.: Measuring information systems success: models, dimensions, measures, and interrelationships. Eur. J. Inf. Syst. **17**(3), 236–263 (2008)
9. Ballou, D., Pazer, H.: Modeling data and process quality in multi-input, multi-output information as vision positively influence the quality of ERP systems. Manag. Sci. **31**(2), 150–162 (1985)
10. Ifinedo.: Impacts of business vision, top management support, and external expertise on ERP success. Bus. Process Manag. J. **14**(4), 551–568 (2008)
11. Kim, E., Lee, J.: An exploratory contingency model of user participation and MIS use. Inf. Manag. **11**(2), 87–97 (1986)
12. Davenport, T.: Putting the enterprise into the enterprise system. Harvard Bus. Rev. **76**(4), 121–131 (1998)
13. Schein, E.H.: The Role of CEO in the management change: the case of information. In: Kochan, T.A., Useem, M. (eds.) Transforming Organization, Oxford University Press, Oxford (1992)
14. Federici, T.: Factors influencing ERP outcomes in SMEs: a post-introduction assessment. J. Enterp. Inf. **22**(1/2), 81–98 (2009)
15. Shang, S., Seddon, P.B.: Enterprise system benefits: how they should be assessed, pp. 1229–1240 (2002)
16. Mahdavian, M., et al.: Developing a model to measure the skills of ERP implementation team. In: International Computer Science and Engineering Conference (ICSEC): ICSEC 2013 English Track Full Papers, pp. 64–67 (2013)
17. Katz, R.L.: Skills of an effective administrator. Harvard Bus. Rev. (January-February 1995) 33–42 (1974)
18. Collins, J.C., Porras, J.I.: Building your company's vision. Harvard Bus. Rev., **74**(5), 65–77 (1996)

Electrocardiogram Identification: Use a Simple Set of Features in QRS Complex to Identify Individuals

Tuerxunwaili, Rizal Mohd Nor, Abdul Wahab Bin Abdul Rahman, Khairul Azami Sidek and Adamu Abubakar Ibrahim

Abstract This paper presents a Multilayer Perception Neural Network developed to identify human subjects using electrocardiogram (ECG) signals. We use the amplitude values of Q, R and S as a features for our experiments. In this study, a total of 87 dataset were collected among 14 subjects from the Physikalisch-Technische Bundesanstalt (PTB) database. Out of the 14 subjects, Q-R-S feature points were taken from different day and time sessions to perform classification with MLP. Out of this data, 66 % is used as training dataset while the remaining 34 % is used for testing. Our method yields 96 % accuracy and demonstrates that the use of three fiducial points is sufficient to identify a subject despite the common practice of taking more feature points.

Keywords Electrocardiogram · QRS peaks · Multi-Layer perceptron · Neural networks · Biometric identification · PTB database

---

Tuerxunwaili (✉) · R.M. Nor · A.W.B.A. Rahman · A.A. Ibrahim
Kulliyyah of Information and Communication Technology,
International Islamic University Malaysia, Kuala Lumpur, Malaysia
e-mail: tuerxunwaili@iium.edu.my

R.M. Nor
e-mail: rizalmohdnor@iium.edu.my

A.W.B.A. Rahman
e-mail: abdulwahab@iium.edu.my

A.A. Ibrahim
e-mail: adamu@iium.edu.my

K.A. Sidek
Kulliyyah of Engineering, International Islamic University Malaysia,
Kuala Lumpur, Malaysia
e-mail: azami@iium.edu.my

P. Meesad et al. (eds.), *Recent Advances in Information and Communication Technology 2016*, Advances in Intelligent Systems and Computing 463,
DOI 10.1007/978-3-319-40415-8_14

1 Introduction

Electrocardiogram (ECG) is a continuous wave formation of electrical signals from the heart. The ECG consists of three main components: P wave, QRS complex, and T wave [1]. The P wave is formed by the actions of atrial depolarization, the QRS complexes is the result of ventricular depolarization and the T wave is due to ventricular repolarization [2]. Since 1903, Electrocardiography is used to check cardiovascular diseases [3]. Subsequently, it was later found that an ECG signal contains more than just physiological and Psychological information of individuals but also includes identity information [4]. Hence, in the past 15 years, a variety of research pertaining to extracting identity information from ECG signal flourished [5].

The ECG signal has the following characteristics: (1) universality, which means that it can be obtained from humans. (2) Collectability, that is an ECG signal can be easily measured compared to other biological signals [6]. (3) Uniqueness, which implies that an ECG signal is unique (4) Permanence, which means it is permanent and finally (5) Performance, which is similar to biometric system, it is secure, efficient and accurate [7, 8].

The difficulty to mimic an ECG signal is one of the major reasons why it offers an attractive alternative to other traditional biometrics [9] like voice based biometric, finger print based biometric and many various forms of biometric identification. For example, voice based biometric can be easily copied with a microphone, fingerprint based biometric can be easily extracted and printed on silicon surfaces, while an iris based biometric can be easily be copied onto a contact lens. In theory, it is possible to compromise or bypass an ECG biometric system, by synthesizing an ECG recording using measured features. However, in practice, it is extremely difficult to replicate an ECG signal at the sensor level [10].

Based on the features that are extracted from an ECG signals, one can classify an ECG biometric system as either fiducial-based, non fiducial-based, or a hybrid method. The fiducial-based methods extract time domain characteristic, amplitude, area, and angle from an ECG signal. An example of feature points that can be extracted are such as the amplitudes of the P, R, and T waves, the temporal distance between wave boundaries (onset and offset of the P, Q, R, S, and T waves), the area of the waves, and the slope information [11]. The non fiducial-based methods do not use the characteristic point as features, instead it dwells on features like wavelet coefficients and power spectral density. This research focus on fiducial dependent features. In one cardiac cycle, as many as 28 fiducial features can be extracted. Figure 1 illustrates the one cardiac cycle with corresponding 28 fiducial points. However, in a single cardiac cycle there could be more than 28 fiducial points.

The published approaches studied in this paper, using different or similar pattern extraction and matching techniques, perform their tests based on 7 or more features [12] to reach an accuracy is 90 % and above.

Inspired by Khalil Ibrahim's work [13] in using 23 points from an ECG signal, this paper attempts to improve Khalil's work in reducing the number of points. The researchers studied, 23 points taken from a section that starts from the offset of a P

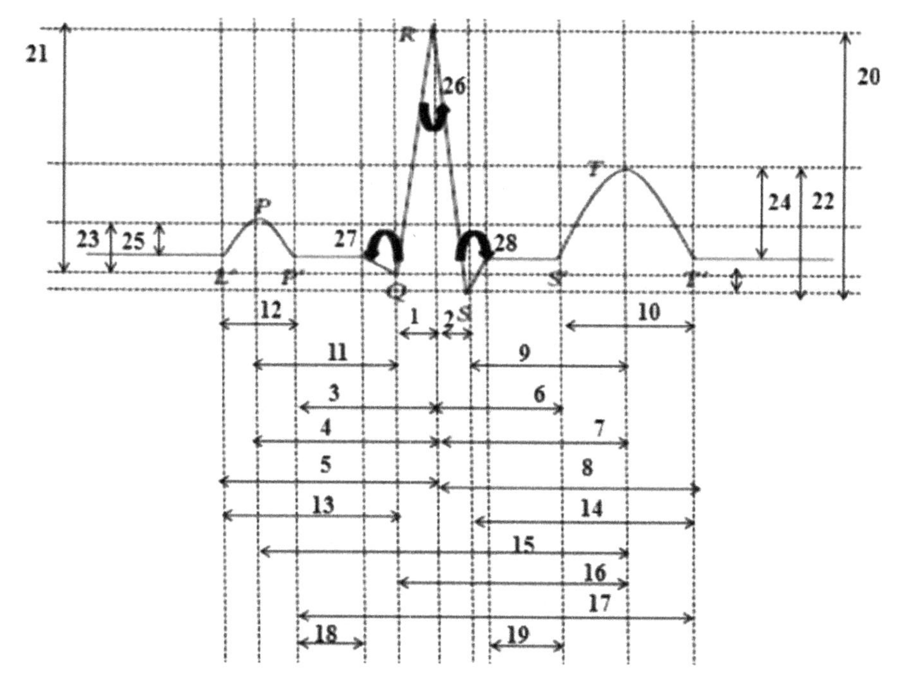

Fig. 1 One cardiac cycle with 28 fiducial points [11]

wave till onset of the T peaks. In Khalil's work, 18 ECG signals (NSRDB) from 18 subjects was obtained in a day and used to extract feature sets. The researchers employ Multilayer perceptron classifier to achieved 96 % accuracy of classification. In this paper, we reduce the amount of points taken to use only 3 points (amplitudes of Q, R and S peaks) in QRS complex.

The rest of this paper is organized as follows: The next Sect. 2 reviews past works related to multilayer perceptron neural network which employs fiducial points in QRS complex. In Sect. 3, we describe the methodology that is being used in this research and present our results for our experiment in Sect. 4. Finally, section presents our summary and conclusion of our research.

2 Related Work

In the study of ECG biometrics, feature selection based on QRS complex and MLP is often used. Most researchers are attracted to studying QRS complexes due to its stability against heart rate variability and as for MLP, Multiplayer perceptron neural network is most used because it is capable of finding meaningful matches in complex structures and irregular data. Furthermore, MLP has better generalization ability on noisy data, [14].

Research on ECG Biometric started with the pioneering work of Biel et al. [15] who purposed the use of amplitude, distance of waves as features to identify a subject. He also proves that a single channel ECG contains enough discriminant information for recognition. Fatema et al. [16] used QRS amplitudes and their differentiated values to form Cardioid graph, which is then used as a feature to Multi-layer perceptron to achieve above 94 % accuracy. In 2012, Vu Mai and Ibrahim Khalil extracts 23 features from QRS complexes, and executed it in Radial Basis function for classification. The authors' achieved an accuracy of 98 %. The authors managed to capture 324 QRS complexes gathered from 18 subjects [17]. In [18], the researchers extracted parts of ECG signal which starts from QRS complex to the end of T wave. This is later used to attain 20 coefficients after Discrete Cosine Transformation (DCT). Similarly, a MLP with one hidden layer is used for classification. An experiment on 22 subjects by the authors yields 99 % accuracy. In the work of Ching, Chun and Yen et al. [9], features of chaotic characteristics, features of ECG were extracted by applying Lyapunov exponents and correlation dimension. A MLP which has two hidden layers with 20 and 5 neurons respectively are utilized, accuracy is 97 %. Takoa Hamdi employees 6 slopes and 3 angles in PQRST wavelet to achieve 96 % accuracy. A total 100 subjects participated to provide the data, and MLP was used for classification [1].

In 2011, Tawfiq use MLP with DCT coefficients to achieve more than 97.7 % accuracy. The DCT coefficients are taken from a section starting from Q peak up to the end of the T peak. A test set of 550 lead I ECG traces recorded from 22 healthy individuals measured at different times are used to validate the system. "The proposed system extracts special parts of the ECG signal starting from the QRS complex to the end of the T wave". Two different approaches are used to compensate for the change in the signal duration with the change in Heart Rate, by the first approach time domain normalization according to Framingham correction formula is used. By the second approach the QT signal to a fixed length is applied. DCT coefficients obtained from the normalized signals were introduced to a MLP classifier [19].

Shen et al. [10] extracted seven temporal and amplitude features from the QRST wave. These are then introduced to decision based neural network (DBNN) as an input and experiment results shows a of gain 80 % accuracy within the population of 20 people. Ghaforani et al. [20] uses Auto regressive (AR) coefficients along with mean of power spectral density (PSD) as an features. In order to assess the effectiveness of the proposed combination, other features including autoregressive (AR) coefficients, Higuchi dimension, Lyapunov exponent, and approximation entropy (A pEn) were extracted from ECG Multi-Iayer-perceptron (MLP), probabilistic neural networks, and k-nearest neighbor (KNN) classifiers were used to classify the extracted features. In addition, simple combination of the features was considered for further improvement in verification rate. The achieved results (100 % accuracy).

Multilayer Perceptron (MLP) neural network is the most frequently used feedforward network, because it can work with small training sets, and it is easier to implement [21]. The MLP and many other neural networks uses a learning algorithm called back propagation. By means of back propagation, the input data is continually presented to the neural network. Through the help of each presentation, an error is calculated by comparing the obtained output of the neural network to the preferred outcome. This error is then fed back (back propagated) to the neural network and is used to modify the weights in such a way that the error decreases repetitively with each iteration and the neural model output approaches the desired outcome [21].

3 Methodology

This paper puts forward a very simple biometric identification system that can achieve high accuracy with minimum numbers of features. The proposed system model is depicted in Fig. 2, and elaborated further in the later sections.

3.1 Experimental Data

The dataset for this experiment is taken from an ECG of 14 individuals obtained from the PTB database with several recordings from different days. Patient are anonymously identified by an identifier numbered as 17, 11, 180, 90, 85, 65, 77, 233, 101, 69, 80, 87, 5, and 293. The ECG of the first eleven subjects is a recording from two different days, while the other two have recordings from 3 different days. One of the samples has recordings from a single day. All of the samples take mde up to 87 samples.

3.2 Preprocessing of ECG Signal

After an ECG acquisition, we normalize each ECG signal by determining the mean of the signal, and subtract that value from all the entries:

$$\mathrm{ECG} = \mathrm{ECG} - \mathrm{mean}(\mathrm{ECG})$$

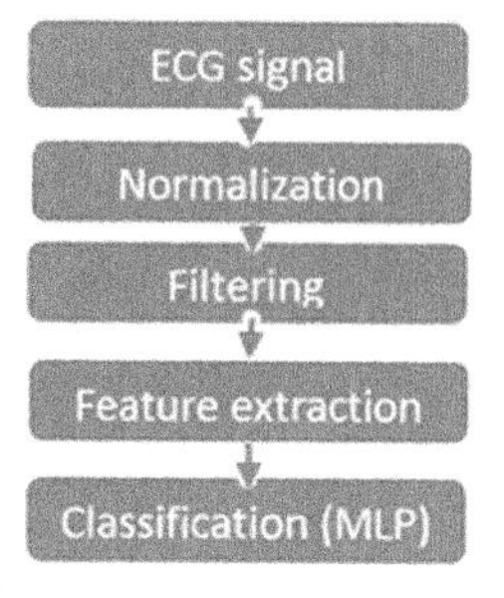

Fig. 2 The methodology Flowchart

Next is the noise removal stage. An ECG signal contains three sources of noise

1. Baseline wondering due to perspiration, respiration, body movements, and poor electrode contact with spectral content usually below 1 Hz.
2. Power line interference.
3. High frequency noise: appears within individual Heart Beats.

To eliminate these noises, cascaded zero phase FIR Low Pass and High Pass filters used: 70 dB 0.05–40 Hz 1 dB ripple.

3.3 Feature Set Selection from QRS Complex

As with many pattern recognition problem, pre-processing was followed by feature extraction and classification [22]. The features we used are simple statistical measures of amplitude.

After pre-processing, we identify Q, R and S peaks by using Pan Tompkins algorithm. From each session of 29, we took 3 samples randomly. The samples do not need to be adjacent, and can be within distances with each other. Each sample consists of 12 set of QRS points. From each sample, we take 12 QRS complex. So, altogether, there are 36 points to be extracted. Figure 3 shows samples taken from one session.

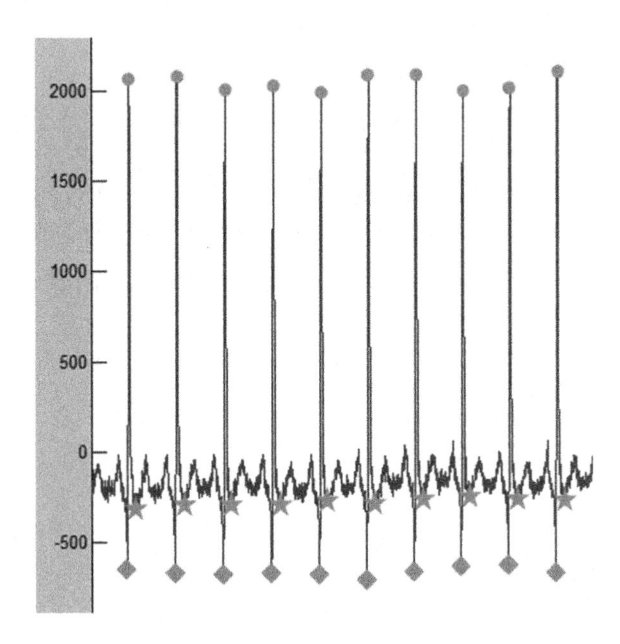

Fig. 3 Feature points taken from one session of ECG signal

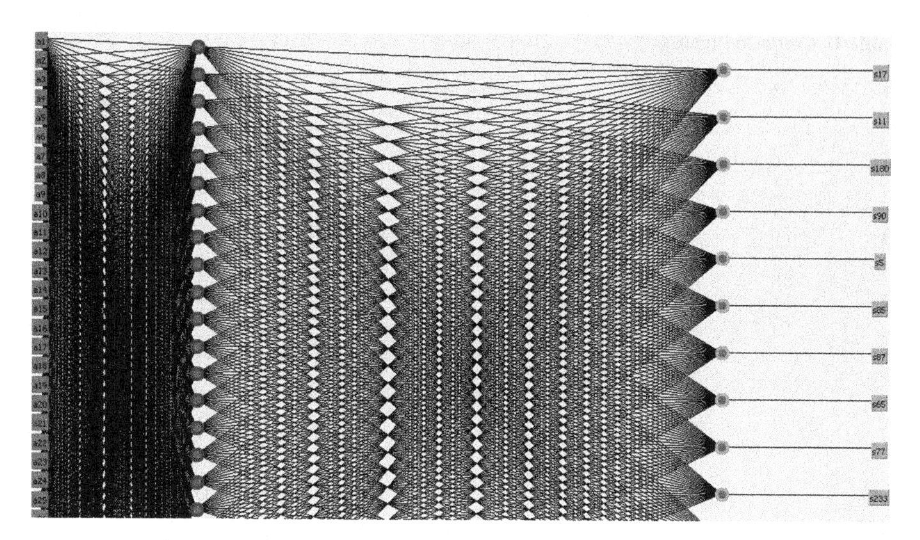

Fig. 4 MLP network topology

3.4 Classification

Multilayer Perceptron is a feed forward artificial neural network model with one or more layers between input and output layer in a directed graph [23]. The main advantage is that MLP is its stability and ease of use. It can approximate any input/output mapping [24].

For this research, we employ MLP with one hidden layer of 37 neurons. They are responsible to find co-relations between input data and output data. The input, is 37 attributes, 36 of them are Y values of QRS, 1 is a Class, it is non-numeric value. Output data is 14 subjects from PTB. This net is trained 500 times to map input to output. Figure 4 is an illustration of the Neural Network.

4 Experiments Analysis and Presentation of the Results

The experiment is done in WEKA, an open source machine learning tool from Waikato University. The machine is a laptop of TOSHIBA satellite L510, 3 GB RAM, running Windows 7 (32) operating system.

An input data given to WEKA as a table of 87 * 37 matrix. 66 % of it used as training data set. 34 % used as testing data set. Table 1 shows the confusion matrix of the experiment. Basically, the diagonal values from top left to downright indicate accuracy, the higher means more accurate.

Table 1 Confusion matrix

a	b	c	d	e	f	g	h	i	j	k	l	m	n	<-- classified as
6	0	0	0	0	0	0	0	0	0	0	0	0	0	a = s17
0	6	0	0	0	0	0	0	0	0	0	0	0	0	b = s11
0	0	6	0	0	0	0	0	0	0	0	0	0	0	c = s180
0	0	0	6	0	0	0	0	0	0	0	0	0	0	d = s90
0	0	0	0	5	0	0	2	0	1	0	1	0	0	e = s5
1	0	0	0	0	5	0	0	0	0	0	0	0	0	f = s85
0	0	0	0	0	0	9	0	0	0	0	0	0	0	g = s87
0	0	0	0	0	0	0	6	0	0	0	0	0	0	h = s65
0	0	0	0	0	0	0	0	6	0	0	0	0	0	i = s77
0	0	0	0	0	0	0	0	1	5	0	0	0	0	j = s233
0	0	0	0	0	0	0	0	0	0	6	0	0	0	k = s101
0	0	0	0	0	0	0	0	0	0	0	6	0	0	l = s69
0	0	0	0	0	0	0	0	0	0	0	0	6	0	m = s80
0	0	0	0	0	0	0	0	0	0	0	0	0	3	n = s293

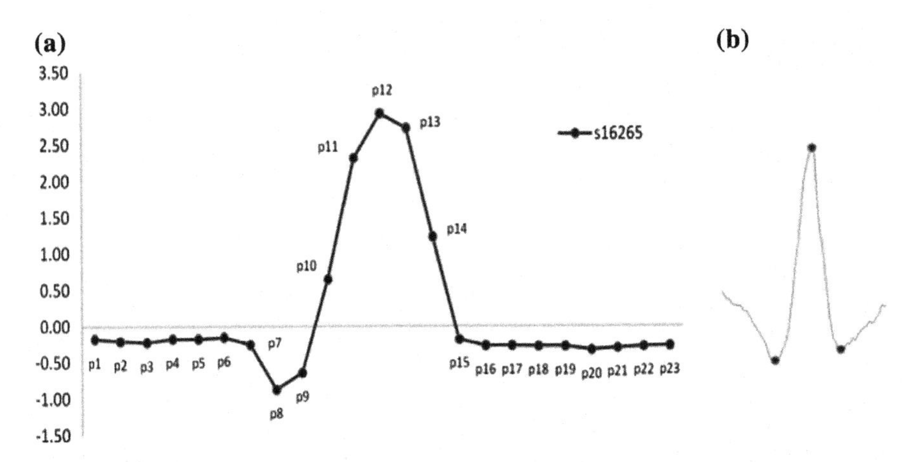

Fig. 5 Feature selection comparison between Khalil's method (**a**) and feature selection technique in our paper (**b**)

The result of the entire experiment shows that our method could achieve 96 % of accuracy. This is similar to the level of accuracy obtained by Khalil [13]. A comparison between Khalil's feature extraction method and our feature extraction method is illustrated in in Fig. 5.

5 Conclusion

In this paper, a new set of features with Multilayer perceptron neural network is tested for identification purpose. We achieved 96 % classification accuracy with 14 subjects of PTB database. The new features are an amplitudes of Q, R and S peaks in QRS complex. We utilized 29 one minute ECG signals recorded in multiple days and multiple sessions. The results from the research demonstrate that even with 3 fiducial points, it is possible to reach high level of identification and authentication accuracy.

References

1. Sidek, K.A., Mai, V., Khalil, I.: Data mining in mobile ECG based biometric identification. J. Netw. Comput. Appl. **44**, 83–91 (2014)
2. Odinaka, I., Lai, P.H., Kaplan, A.D., O'Sullivan, J.A., Sirevaag, E.J., Rohrbaugh, J.W.: ECG biometric recognition: A comparative analysis. IEEE Trans. Inf. Forensics Secur. **7**(6), 1812–1824 (2012)
3. Sao, P., Hegadi, R., Karmakar, S.: A literature review on approaches of ECG pattern recognition. Int. J. Inf. Sci. Intell. Syst. **3**(2), 79–90 (2014)
4. Choudhary, T., Manikandan, M.S.: A novel unified framework for noise-robust ECG-based biometric authentication. In: 2nd International Conference on Signal Processing and Integrated Networks (SPIN), pp. 186–191 (2015)
5. Sidek, K., Khali, I.: Biometric sample extraction using Mahalanobis distance in Cardioid based graph using electrocardiogram signals. In: 34th Annual International Conference of the IEEE EMBS, pp. 3396–3399 (2012)
6. Tashiro, F., Aoyama, T., Shimuta, T., Ishikawa, H., Shimatani, Y., Ishijima, M., Kyoso, M.: Individual identification with high frequency ECG: Preprocessing and classification by neural network. In: Annual International Conference IEEE Engineering Medicine and Biology Society, pp. 2749–2751 (2011)
7. Fratini, A., Sansone, M., Bifulco, P., Cesarelli, M.: Individual identification via electrocardiogram analysis. Biomed. Eng. Online **14**(78), 1–23 (2015)
8. Unar, J.A., Seng, W.C., Abbasi, A.: A review of biometric technology along with trends and prospects. Pattern Recognit. **47**(8), 2673–2688 (2014)
9. Chen, C.K., Lin, C.L., Lin, S.L., Chiu, Y.M., Chiang, C.T.: A chaotic theorectical approach to ecg-based identity recognition [Application Notes]. IEEE Comput. Intell. Mag. **9**(1), 53–63 (2014)
10. Tsao, Y.T., Shen, T.W., Ko, T.F., Lin, T.H.: The morphology of the electrocardiogram for eevaluating ECG biometrics. In: 9th International Conference on e-Health Networking, Application and Services, pp. 233–235 (2007)
11. Tang, X., Shu, L.: Classification of electrocardiogram signals with RS and quantum networks neural. Int. J. Multimedia Ubiquit. Eng. **9**(2), 363–372 (2014)
12. Nasri, B., Guennoun, M., El-Khatib, K.: Using ECG as a measure in biometric identification systems. In: IEEE Toronto International Conference Science and Technology for Humanity (TIC-STH), pp. 28–33 (2009)
13. Sidek, K.A., Mai, V., Khalil, I.: Data mining in mobile ECG based biometric identification. Data Min. Mob. ECG Based Biometric Ident. **44**, 83–91 (2014)
14. Loong, J.L.C., Subari, K.S., Besar, R., Abdullah, M.K.: A new approach to ECG biometric systems: a comparative study between LPC and WPD systems. World Acad. Sci. Eng. Technol. **68**(8), 759–764 (2010)
15. Hamdi, T., Slimane, A.B., Khalifa, A.B.: A novel feature extraction method in ECG biometrics. In: International Image Processing Applications and Sytems Conference (IEEE IPAS'14), pp. 1–5 (2014)
16. Iqbal, F., Sidek, K.A., Noah, N.A., Gunawan, T.S.: A comparative analysis of QRS and Cardioid graph based ECG biometric recognition in different physiological conditions. In: Proceeding of the IEEE International Conference on Smart Instrumentation, Measurement and Applications (ICSIMA), pp. 25–27 (2014)

17. Mai, V., Khalil, I., Meli, C.: ECG biometric using multilayer perceptron and radial basis function neural networks. In: 33rd Annual International Conference of the IEEE EMBS, pp. 2745–2748 (2011)
18. Fatemian, S.Z., Hatzinakos, D.: A new ECG feature extractor for biometric recognition. In: 16th International Conference on Digital Signal Processing (2009)
19. Tawfik, M.M., Selim, H., Kamal, T.: Human identification using time normalized QT signal and the QRS complex of the ECG. In: 7th International Symposium on Communication Systems Networks and Digital Signal Processing (CSNDSP), pp. 755–759 (2010)
20. Ghofrani, N., Bostani, R.: Reliable features for an ECG-based biometric system. In: 17th Iranian Conference of Biomedical Engineering (ICBME), pp. 3–4 (2010)
21. Iqbal, F., Sidek, K.A., Noah, N.A., Gunawan, T.S.: A comparative analysis of QRS and Cardioid graph based ECG biometric recognition in different physiological conditions. In: Proceedinf of the IEEE International Conference on Smart Instrumentation, Measurement and Applications (ICSIMA), pp. 25–27 (2014)
22. Singh, Y., Gupta, P.: ECG to individual identification. In: 2nd IEEE International Conference on Biometrics: Theory, Applications and Systems, pp. 1–8. Washington, DC (2008)
23. Kaur, G., Singh, G., Kumar, V.: A review on biometric recognition. Int. J. Bio-Sci. Bio-Technol. **6**(4), 69–76 (2014)
24. Gawande, P.S., Ladhake, S.A.: Biometric authentication of an individual using multilayer perceptron and support vector machine. Int. J. Comput. Sci. Mob. Comput. **4**(8), 259–263 (2015)

The Research on Improving Algorithms for Hilltop to Improve Search Quality

Peng Lu and Xiao Cong

Abstract The Hilltop algorithm has played a very important role in the search results sort of Google. In this paper, we depth analysis the main ideas of Hilltop, and discussed the problems of the algorithm, such as part of the related documents are excluded from the result set, and when it did not find sufficient the expert documents and do not return any results, etc. To solve these problems, we propose appropriate improvements.

Keywords Web search • Hilltop • Link analysis • Ranking algorithm • Search quality

1 Introduction

Most traditional Web search engines are based on keywords matching, and the result returned is the pages that contain the query keywords. Since the beginning of the search engines have been produced using a variety of sorting algorithms to arrange relevant pages, and strive to make the most valuable and most relevant pages to the user query in the front of the search results. However, when a query is composed by a number of query terms, and search engine always returns a list of documents which made users cannot read the full results due to its length. The research shows that users generally only see the information about 10–20 top results [1]. So, a lot of information would be inadvertently ignoring, but there are some with a lot of valuable contents to be ignored in these information [2].

P. Lu (✉)
Department of Media Technology and Communication,
Northeast Dianli University, Jilin, China
e-mail: Peng.lu2008@gmail.com

X. Cong
College of Science, Northeast Dianli University, Jilin, China
e-mail: lup595@nenu.edu.cn

P. Meesad et al. (eds.), *Recent Advances in Information and Communication Technology 2016*, Advances in Intelligent Systems and Computing 463,
DOI 10.1007/978-3-319-40415-8_15

The nature of the network is a hyperlink, and use it can greatly improve the quality of search results [3, 4]. Based on the idea hyperlink analysis, Sergey Brin and Lawrence Page proposed PageRank algorithm in 1998 [5, 6]. This method to rank the results transferred to the link factors outside page content, and improve the precision and quality, so that the pages more relevant to the query page have a good ranking.

Because regardless of the user's query, the biggest drawback of PageRank is that it cannot distinguish authoritative pages which are general sense or related to the query. However, the page contents on the Internet cover millions or even more topics, users are often looking for some information with a particular topic in the actual retrieval. Chakrabarti [7] studies have shown that the web links on the subject is sensitive. That is, the links of the pages are usually trying to point the other pages which often have the same topics. However PageRank decisions importance of the pages simply based on the number of back links of pages, the quality of pages, and the rank links to the source pages, and ignore the relevance between the links pages and the query topics, the consequence is that the topics of some pages have no relationship with the query conditions, However, they would get a good ranking in the search results due to it sometimes involves a query keywords, this will affect the relevance and accuracy of search results. Meanwhile, some who use the site to optimize the disadvantage of this algorithm, at the same time, some SEO use the disadvantages, such as designing a lot of spam links to a particular site, of this method to enhancing the site's ranking in search results. These all affect the PageRank used.

To overcome the shortcomings of PageRank algorithm, Bharat and other Google software engineer invented the Hilltop algorithm in 2000 [8]. The idea of the algorithm is based on the query and the use of "expert documents", and search the target pages which linked by expert documents, then sort the target pages to enhancing the quality of the search pages and take the pages of high authority have a good sort results. But also to avoid the effects of human reason to improve page PageRank value. Hilltop algorithm plays a key role in the search engine [9], therefore received widespread attention.

2 The Overview of Hilltop Algorithm

The basic idea of Hilltop and PageRank is the same which based on the same assumption, namely, the evaluation criteria of current page of the quality is the quantity and quality of the other pages' link to it. However, different from PageRank is that only the pages linked by expert documents are under consideration as the most important pages for a user's query. For a certain subject point to the other sites which have the same topic, and these sites do not have affiliation with the current site, all such pages are called expert documents. For the user's query, find a series of expert documents related topic, and calculation the authoritative score of these expert documents. And then, select the relevant links in these expert documents to find the target pages. The level value of the target pages was decided by the number and relevance of non-affiliation expert documents which point to them, and the target pages must be linked by non-affiliation expert documents at least two or more. Similarly, the value of the target page also reflects the collective point of view of the expert documents. When there is not enough expert Document, Hilltop algorithm does not return any search results.

2.1 *The Flow of Hilltop Algorithm*

2.1.1 Selected and Indexed Expert Documents

Hilltop algorithm process the pages' database of search engine pages, and pick out the pages which all the forward links exceeds a certain threshold k for a particular subject, and these URLs point to the m standalone hosts which have no affiliation relationship. Such pages are considered to be expert documents and index the key phrases, such as title, heading and anchor.

2.1.2 Calculate the Scores of Expert Documents

Hilltop algorithm calculate the match score between each candidate expert documents and the query that in order to reflect the quantity and quality of key phrases contain query keywords and the matching degree of these key phrases as well as the query. As shown in Eq. 1.

$$S_i = \sum \{ \textit{key} \quad \textit{phrases} \quad p \} \textit{LevelScore}(p) * \textit{FullnessFactor}(p, q) \tag{1}$$

The objective of Hilltop algorithm is that expert documents to be able to match all the query keywords, so the weight of S_0 is largest, S_1 followed and S_2 lightest, the total score of the expert documents is consists by three parts, as shown in Eq. 2:

$$\textit{Expert\_Score} = 2^{32} S_0 + 2^{16} S_1 + S_2 \tag{2}$$

2.1.3 The Sort of Target Pages

First of all, according to the scores of expert documents to rank, and then detected expert documents of top N which they point to the pages called the target pages. A target page to be linked at least two non-affiliated expert documents. The score of target pages is calculated by the following three steps:

Step One for each expert document E, if exist a link from E to the target page T, then draw a line $Edge(E, T)$ from the expert document E to the target page T. For a keyword w of query, set $occ(w, T)$ is the number of key phrases in the expert document E which contain w and restriction it appear in the line $Edge(E, T)$, it is calculated as follows:

$$Edge\_Score(E, T) = Expert\_Score(E) \sum \{w\} occ(w, T) \quad (3)$$

Step Two check the affiliated of expert documents which point to the same target page, and if there are two or more expert documents have then retain it has high value of *Edge_Score*.

Step Three the value of *Target_Score* is calculated with sum of *Edge_Score* the lines linked this target page.

Step Four finally, the rank of target pages in accordance with the *Target_Score*, and return results.

2.2 The Main Problems

Hilltop algorithm provides an objective way to measure the quality of target pages by expert documents and using the links between them, and improve the relevance and precision of the search results and effectively reducing the possibility of artificial manipulation of the rank. That is why this method plays a key role in the search and has been widespread concern. However, through further research and analysis can be found that the algorithm has a number of shortcomings and deficiencies:

- Cannot completely avoid the phenomenon of topic drift. According to the query, Hilltop algorithm use target pages which relevant to the expert documents as the search results returned to the user. Then the expert document may be only one link that matches the query phrase and then find qualified target pages, but these target pages are not necessarily related to the query topic. According to the algorithm, if there are two or more expert documents point to the same target page, then this target page will be returned as the result. In fact, the high possibility of such situation in the network environment. The reason is that the all links of expert documents meet one topic does not necessarily point to the same subject.
- Increased noise. In the part of return target pages does not contain key phrases of queries, and even regardless of the query subject, such pages cannot be returned as the query results to the user. Similarly, even the pages including the query phrase are also relating to irrelevant with the query in a part. Based on the analysis above, it will increase the noise in the list of results returned.
- Part of the related documents are excluded from the result set. The results of A.Z. Broder show that [4]: The Web has the shape of the structure is a bow, the left part of the bow is the pages which have a lot of forward-link without backward links, and there is only a small part of pages have backward links. Then use the Hilltop algorithm for retrieval, these pages were not being retrieved

according to Hilltop algorithm because they are not linked by sufficient expert documents or even most of them do not have an expert document. In addition, part of the expert documents should also be part of the search results, but most of this part is excluded from the search results [9].

- When did not find sufficient expert documents it does not return any results. Such an approach is unreasonable. In fact, when there are not enough expert documents does not mean that there are no results related to the query of the indexed pages. For example, some of the navigation pages are basically at the left side of the bow structure. In the Hilltop algorithm, in order to improve the accuracy of search results only the target pages which two or more expert documents linked will be used as the retrieved results, and part of the expert documents that satisfy the query does not return as result set. But in fact, the target page is necessarily required to meet at least two expert documents does not affect the results of the sort. According to the target score formula to calculate the score, those target pages is only one expert document pointed to be natural behind other well results of the entire set. More reasonable approach is that to the expert documents were not found of query, at this should be combined with the traditional information retrieval document scoring methods and PageRank of other sort methods returns a comprehensive list of results based on the query [10].

3 Improvement Program

Through the above analysis, to ensure the accuracy of the query while avoiding the loss of the query results, reducing noise in the result set and the elimination of the topic drift may occur, proposed the following improvement program for the Hilltop algorithm, called Intelligent Hilltop Algorithm, I-Hilltop. The algorithm is divided into two main parts, the first part is to scan the pages' database which crawl and parse by the web crawler in offline situations, the PageRank value is calculated for each page, and the expert documents also pick out and create an inverted index of its key fields, such as title, heading and anchor etc. The second part is in the process of the query, for a particular query q, first check whether have expert documents and their number, and calculated authoritative score for these expert documents, then found target pages linked by them. The score of target documents are calculated according to the similarity between the content of the target pages and the query q, as well as combined with score of expert documents, finally, according to the *Target_Score* combines the PageRank in descending return to the target document list. When did not find the expert documents, the search results in accordance with the score of content that match the query q combines the PageRank.

3.1 Preprocess

3.1.1 Detecte the Affiliation of Hosts

I-Hilltop is similar to the original Hilltop method, we should first check whether the host has an affiliation, only the authority value of the target pages higher that does not have the affiliation of the host. For two hosts, if one or both of the following conditions are met that the two hosts that has affiliation:

- They have the same IP address of the first three segments;
- The rightmost special markers are same of host name. For example: *.com* and *.edu* are the domain suffix categories appear in the large number of hosts;
- If the two hosts, different of respective categories suffix but other marks are the same, it can also think they have an affiliation.

For example: Compare www.peng.com and *peng.edu*, the rightmost resulting marks are *peng* when they were ignored category suffix *.com* and *.edu*, therefore, they are considered to have an affiliation.

Transitive dependencies: If A and B have the affiliation and B and C also has affiliation, so even if A and C do not have obvious direct contact, but also that A and C have affiliation.

3.1.2 Select Expert Documents

I-Hilltop algorithm processed pages database which through crawled and parsed by search engine, then pick out a collection of pages of high-quality for a particular topic, examine all the pages with out-degree exceeds a certain threshold $k(k=5)$, detect the URL of these pages whether points to independent host $m(m \leq k)$ without affiliation, and each page meets the above criteria are considered to be an expert document.

3.1.3 Expert Document Index

In order to find the expert documents meet the user's query, I-Hilltop algorithm to create an inverted index to storage key phrases appear in the expert document. The key phrase means a piece of text that defining one or more URL of the expert documents. Each key phrase has a range in the document, and the URL of target page found by key phrases. Such as title, heading and anchor text are all recognized of the key phrases. The title is defined all URLs contained in this expert documents, the heading is defined all URLs until appear equally or more important titles, and anchor text is defined only correspond URL.

The match position in expert documents is organized by inverted index list, and each match position corresponding to the particular keywords in key phrases of an

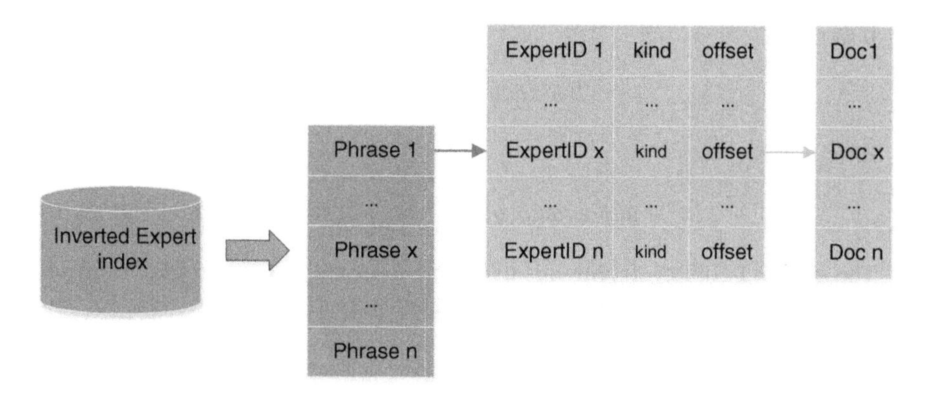

Fig. 1 Expert documentation index

expert document. The following information is stored in each of the matched position, the identifier uniqueness of identification of the phrase in the document, the code of phrase category, such as title, heading or anchor, and offset of words within phrases. In addition, the algorithm provides an internal URL list to each expert document, and provided identifier of keyword for each URL, as shown in Fig. 1.

3.2 *The Query Process*

3.2.1 Calculate the Score of Expert Documents

For query *q*, I-Hilltop algorithm calculated the match score between each candidate expert documents and query *q*, in order to reflect the quantity and quality of the key phrases in the expert documents which contains query keywords, as well as matching degree for these phrases and queries.

Set *k* is the number of words in the query *q*, S_i is the score of accurately calculate for $k-i$ query words contain in the key phrases, as shown in the following equation.

$$S_i = \sum TypeScore(p) * FullFactor(p, q) \tag{4}$$

where, $TypeScore(p)$ is the score of type of the key phrase p, and set title is 16, heading is 6, and anchor text is 1. This score determines set based on what the subject about expert documents, and the assumptions is that title more useful than heading, and heading more useful than anchor text. (2) $FullFactor(p, q)$ is the score of terms appear in the *q* of *p*, set *len* is the length of *p*, m is the number of terms appear in the *p* but not in the *q*, $FullFactor(p, q)$ is calculated as follows:

If $m \leq 2$,

$$FullFactor(p,q) = 1; \tag{5}$$

Else, $m > 2$,

$$FullFactor(p,q) = 1 - (m-2)/len. \tag{6}$$

The goal of I-Hilltop algorithm is to be able to match all the query keywords of expert documents, so the score of expert document is as follows:

$$Expert\_Score = 2^n S_i \tag{7}$$

where, according to the match degree, n is taken of 32, 15 and 0, respectively.

3.2.2 Calculate the Score of Target Page

According to the scores to sort all expert documents, then detected expert documents of top N which they linked are called target pages. Calculate the scores of target pages which reflects both linked the expert documents of number or relevant and the key phrases of restrictions. The score of target page is calculated by the following three steps:

Step 1 For each expert document E. If there is a link from the expert document E to the target page T, then draw a line $Edge(E,T)$ between them. For a query keyword w, set $occ(w,T)$ is the number of key word w which in the key phrases and appear to the $Edge(E,T)$ of expert document E, and $Edge\_Score(E,T)$ is calculated as follows:
For any query keywords, if $occ(w,T)$ is 0,

$$Edge\_Score(E,T) = 0; \tag{8}$$

Else,

$$Edge\_Score(E,T) = Expert\_Score(E) \sum \{w\} occ(w,T) \tag{9}$$

Step 2 check affiliated of expert documents which point to the same target page. If there are two or more expert documents have affiliation and only retain the high value of line which from expert document to target document.

Step 3 By sum the $Edge\_Score$ to calculate the $Target\_Score$ of target pages.

There are two expert documents E_1 and E_2 point to target page T, the line between E_1 and target page T is $Edege(E_1,T)$,there are two links to the target page T in E_1, so the value of $occ(w,T)$ is 2. And the line between E_2 and target page T is $Edege(E_2,T)$, there only one link to the target page T in E_2, and the value of $occ(w,T)$ is 1. As shown in Fig. 2. The final score of target page as shown follows:

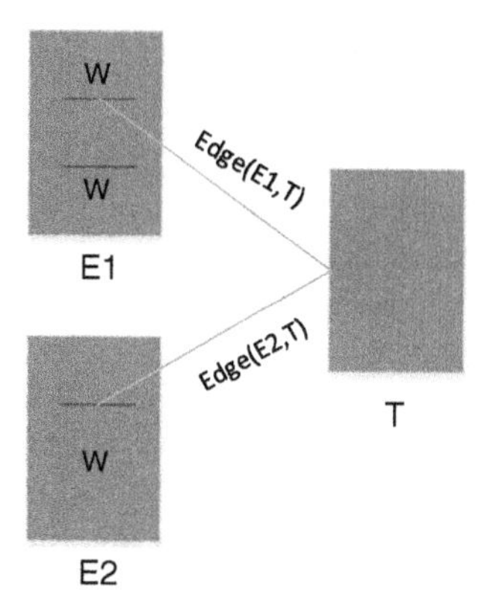

Fig. 2 The calculation of target score

$$Target\_Score = Expert\_Score(E_2) * Edege(E_2, T) + 2Expert\_Score(E_1) * Edege(E_1, T) \quad (10)$$

3.2.3 Sort Results

- The final score of results. If the expert documents are not in the target pages, then take these expert documents as part of the result set, and take the *Expert_Score* of expert documents as the *Target_Score*. This would solve the problems of loss of query results or query results insufficiency.
- Using traditional information retrieval methods of content analysis, such as TFIDF, to calculated similarity score *Match_Score* between query q and target documents and combines PageRank score, obtain the final score for each target page, as shown in the following equation:

  $$Final\_Score = \alpha Match\_Score + \beta Target\_Score + \gamma PageRank\_Score \quad (11)$$

  Where, the experience value of coefficients α, β, and γ are 0.2, 0.5, and 0.3, respectively.
  This will not only avoid the topic drift phenomenon, but also improve the precision of search results. According to the *Final_Score* output the search results in descending order at end, and complete the retrieval process.
- When did not find the expert documents, then in accordance with the conventional information retrieval method and calculate the *Match_Score* of match degree of the query q and each page and combined the *PageRank_Score*, and return to the list of results. It is calculated as shown in the following equation:

  $$Final\_Score = \alpha Match\_Score + \beta PageRank\_Score \quad (12)$$

  Where, the value of coefficients of α and β is 0.5 respectively.

4 Conclusion

The search engine technology is developing rapidly, although various types of ranking algorithms are based on web content, the link between pages and the behavior of users click at this stage, but the core is still focused on the calculation of the pages level, the level of pages generate a significant impact for the search results.

In future work, we will test the proposed improvement algorithm and comparison with the original Hilltop. Hilltop improve the accuracy of search results, but also have some problems. Based on the analysis of the advantages and disadvantages of the Hilltop algorithm in this paper, and then proposed correspond recommendations for improvement. By some improvements to make up for the shortcomings of it to further improve the retrieval precision and recall. But so far, no one ranking algorithm is perfect, almost all search engines have stopped using only one sorting algorithm, but absorb the advantages of various sorting algorithms and improved. In order to the results more objective and reasonable, it need to continue to do in-depth research in many aspects.

References

1. Henzinger, M.: Hyperlink analysis for the web. J. IEEE Int. Comput. **5**(1), 45–50 (2001)
2. Pierre, B., Paolo, F., Padhraic, S.: Modeling the Internet and the Web: Probabilistic Methods and Algorithms. Wiley Press, Hoboken (2003)
3. Peng, L., Xiao, C.: A topic-expert based ranking algorithms for web search. Adv. Intell. Syst. Comput. **361**, 195–204 (2015)
4. Broder, A.Z., Kumar, S.R., Maghoul, F., Raghavan, P., Rajagopalan, S., Stata, R., Tomkins, A., Wiener, J.L.: Graph structure in the web. J. Int. J. Comput. Telecommun. Netw. Arch. **33** (1–6), 309–320 (2000)
5. Page, L., Brin, S., Motwani, R., Winograd, T.: The pageRank Citation Ranking: Bringing Order to the WEB. http://ilpubs.stanford.edu:8090/422/1/1999-66.pdf (1998). Accessed Jan 1998
6. Brin, S., Page, L.: The anatomy of a large scale hypertextual Web search engine. In: Proceedings of the seventh international conference on World Wide Web, Brisbane, Australia, pp. 107–117 (1998)
7. Soumen, C., Mukul, M.J., Kunal, P., David, M.P.: The structure of broad topics on the web. In: Proceedings of the 11th International World Wide Web Conference, pp. 251–262. ACM Press, Honolulu (2002)
8. Krishna, B., George, A.M.: When experts agree: using non-affiliated experts to rank popular topics. ACM Trans. Inf. Syst. (TOIS) **20**(1), 47–58 (2002)
9. Serge, T.: PageRank: meet Hilltop. http://isedb.com/20040127-658/pagerank-meet-hilltop
10. Yates, R.B., Neto, B.R.: Moderm Information Retrieval. Addison Wesley, New York, NY, USA (1999)

A Fast Outlier Detection Algorithm for Big Datasets

Duong van Hieu and Phayung Meesad

Abstract This paper proposes a fast outlier detection algorithm for big datasets, which is a combination of a Cell-based method and a rank-difference outlier detection method associated with a new weighted distance definition. Firstly, a Cell-based method is used to transform a dataset having a very large number of objects into a significant small set of weighted cells based on predefined lower bound and upper bound sizes. A weighted distance function is defined to measure distances between two cells based on their coordinates and weights. Then, a rank-based outlier detection method with different depths is used to calculate outlier scores of cells. Finally, cells are ranked based on scores, outlier objects are identified from ranked cells and eliminated from the provided dataset. Based on experiment results, this proposed method is appropriate for datasets that have a very large number of objects.

Keywords Cell-based · Rank-based outlier detection · Rank-based with different depths · Outlier detection in big datasets

1 Introduction

Outlier analysis plays an important role in data mining and analysis because of its applicability to various applications such as credit card fraud detection, insurance, health care, security intrusion detection, fault detection in safety systems, and military surveillance [1]. Hawkins [2] defined "*An outlier is an observation which deviates so much from the other observations as to arouse suspicious that it was generated by a different mechanism*". In data mining and statistics, outliers are also

D. van Hieu (✉) · P. Meesad
Faculty of Information Technology, King Mongkut's University of Technology North Bangkok, Bangkok 10800, Thailand
e-mail: duongvan.hieu@email.kmutnb.ac.th; dvhieu@gmail.com

P. Meesad
e-mail: pym@kmutnb.ac.th; phayung.m@it.kmutnb.ac.th

P. Meesad et al. (eds.), *Recent Advances in Information and Communication Technology 2016*, Advances in Intelligent Systems and Computing 463,
DOI 10.1007/978-3-319-40415-8_16

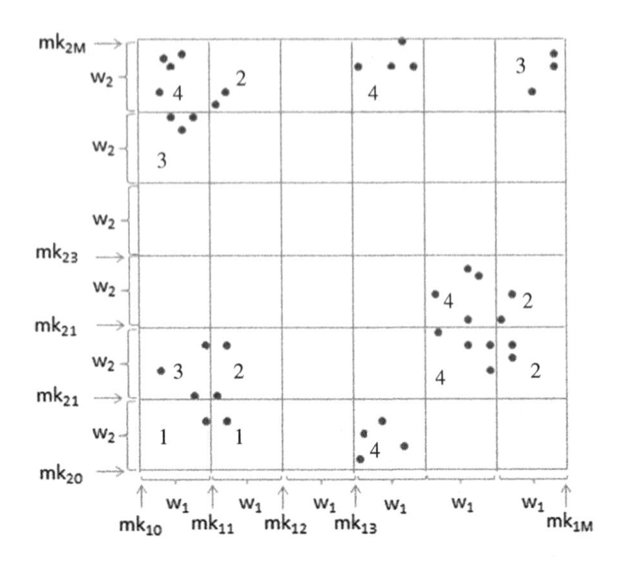

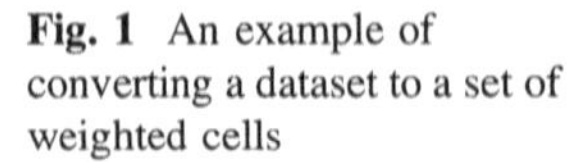
Fig. 1 An example of converting a dataset to a set of weighted cells

called abnormal, deviant, or discordant individuals. Outliers are objects but they are different from the remaining objects [3] (Fig. 1).

Highly accurate outlier detection algorithms are extensively computational. When datasets become very large in size, finding outliers can be a problem [4]. This paper proposes a fast outlier detection algorithm for big datasets, which is considerably better than previous neighborhood-based outlier detection methods in terms of executing time.

Contributions of this paper are very significant to outlier detection for big datasets. Firstly, accuracy level of the proposed algorithm is almost the same as the rank-difference outlier detection method (RDOS) [5]. However, the proposed algorithm is extremely faster than the RDOS method when working with very large datasets. Secondly, proposal of a new weighted distance function to measure distances between two cells based on their coordinates and weights, which is also suitable to work with cell-based methods.

The rest of this paper is organized as follows. Section 2 is the related literature. Section 3 explains the proposed algorithm. Section 4 is the evaluation. And, Sect. 5 covers conclusions and discussions.

2 Related Literature

Outlier detection methods normally focus on the discovery of objects from provided datasets, which match a predefined model. Unfortunately, outliers can be defined as objects if they naturally fit these models. Most outlier detection algorithms produce results in one of two types which are binary label and real-valued outlier scores [3]. In the first type, objects are marked as outliers or non-outliers. On the other hand, the second type assigns a score as a real value to each object and sorts objects in a descending order based on scores, the top k objects are considered as k outliers [6].

This paper mainly focuses on the second type of outlier detection. The first algorithm belonging to this group uses local outlier factor (LOF) [7]. This density-based outlier detection algorithm firstly calculates abnormal degree of each object based on local reachability density and $k-distance$ neighbors. Then, all objects are ranked based on outlier degree. Objects with the highest outlier scores are considered as outliers. The LOF algorithm is extensively computational when working with datasets which have a large number of objects.

To improve the results of the LOF algorithm when working with low density patterns, the connectivity-based outlier factor (COF) has been proposed [8]. The COF of an object **p** is defined as the ratio of its average chaining distance with average chaining distances of its k nearest neighbors. Similar to using LOF, after obtaining the connectivity-based outlier factor values of all objects, objects are ranked in a descending order based on COF values. The objects with highest COF values are considered as outliers. Using COF may be better than using LOF in the case of low density pattern. However, using COF is more computational than using LOF.

To detect outliers in more complex situations, influential measure of outliers (INFLO) was proposed [9]. This method uses reverse nearest relationship to measure outlier degree of an object. The influential measure of an object **p** is defined as ratio of density in its neighborhood with average density of objects in its reverse nearest neighbors. Similar to COF, INFLO is also an extensively computational time consuming algorithm and inadequate to work with large datasets.

Similar to INFLO, rank-based outlier algorithms (RBDA) use the concept of the reverse neighborhood to rank objects [10]. Rank of an object **p** among all its neighbors is firstly calculated. Next, an outlier degree value of an object **p** is defined as the ratio of sum of ranks of **p** among its neighbors. Then, outlier degree values of objects are normalized based on these values and sizes detected. Objects associated with normalized outlier degree, which is greater than L, are considered outliers. Other outlier algorithms based on ranking and clustering including RADA, ODMR, and ODMRD have also been proposed [11]. However, based on executing time, the aforementioned algorithms are inappropriate to work with datasets that have a very large number of objects.

To provide a highly accurate detection algorithm, a precise ranking method was proposed [12]. It is actually a sampling method. Firstly, S subsets having the same size of a provided dataset are arbitrarily extracted. Then, outlier scores of objects belonging to each subset are calculated. And, the final outlier score of each object is calculated as the sum of its outlier scores from S subsets. Finally, objects are ranked based on outlier factor scores and objects with the highest scores are considered as outliers. This method can provide highly accurate results but its executing time is tremendously long when working with big datasets due to the size of subsets also being a large number. Moreover, the number of subsets will be inversed proportionally to the size of subsets.

To overcome the weakness of using a fixed value k of LOF and utilize the reversed nearest relationship to provide highly accurate results, an outlier detection algorithm with rank difference (RDOS) was proposed [5]. The main advantage of this algorithm is to determine an optimal k' which may be less than the k defined by a user. By using this optimal k', the RDOS algorithm produces more accurate results compared to LOF, COF, INFLO, RBDA, and ORMRD. However, the main weakness of this RDOS algorithm is its long executing time compared to LOF, COF, INFLO, RBDA, and ORMRD.

Outlier detection algorithms not only need to handle high dimensional data but also a very large number of objects [13]. Detecting outliers in big datasets is challenging because detected outliers need to be reliably discovered within an acceptable period of time.

3 A New Outlier Detection Algorithm for Big Datasets

3.1 Problem Setting

To successfully improve detection of n outliers in datasets that have a very large number of objects, the following must be achieved:

- Lower executing time compared to the rank-difference outlier detection method.
- Maintain the same accuracy level as the rank-difference outlier detection method.
- And, no difficulties to implement the proposed method on a normal personal computer.

3.2 A Proposed Solution

It is difficult to design an algorithm which is much faster than the rank-difference outlier detection method [5] while maintaining the same high accuracy level. There might only be one solution, the number of objects in a provided dataset should be transformed to a smaller number to reduce calculations and executing time. This paper proposes a new fast outlier detection method which is a combination of a revised version of the cell-based method [14] and the rank-difference outlier detection method [5] to achieve the above goals. The proposed method follows three steps explained below.

Step 1: To transform a big dataset to a small set of weighted cells. The purpose of this step is to convert a dataset with a great number of objects to a significantly small set of weighted cells to reduce the number of calculations. Let $X = (\mathbf{x}_1, \mathbf{x}_2, \ldots, \mathbf{x}_N)$ be a provided dataset having N samples and D attributes, $\mathbf{v}_{min} = (v_{\mathrm{min1}}, v_{\mathrm{min2}}, \ldots v_{\mathrm{minD}})$ and $\mathbf{v}_{max} = (v_{\mathrm{max1}}, v_{\mathrm{max2}}, \ldots v_{\mathrm{maxD}})$ be the minimum and maximum values of attributes. Each segment $[v_{\mathrm{minj}}, v_{\mathrm{maxj}}]$ is equally divided into M intervals to establish new vectors $\mathbf{w} = (w_1, w_2, \ldots, w_D)$ and $\mathbf{m}_j = (m_{j1}, m_{j2}, \ldots, m_{jM})$ with $j = 1, \ldots, D$. Values of $v_{\min j}$, $v_{\max j}$, w_j and m_{jk} are calculated by (2), (3), (4) and (5), respectively.

$$X = \begin{pmatrix} x_{11} & x_{12} & \ldots & x_{1D} \\ x_{21} & x_{22} & \ldots & x_{2D} \\ \ldots & \ldots & \ldots & \ldots \\ x_{N1} & x_{N2} & \ldots & x_{ND} \end{pmatrix} \tag{1}$$

$$v_{\min j} = \min(x_{ij}) \text{ with } i = 1, \ldots, N \text{ and } j = 1, \ldots, D \tag{2}$$

$$v_{\max j} = \max(x_{ij}) \text{ with } i = 1, \ldots, N \text{ and } j = 1, \ldots, D \tag{3}$$

$$w_j = \frac{v_{\max j} - v_{\min j}}{M} \text{ with } j = 1, \ldots, D \tag{4}$$

$$m_{jk} = v_{\min j} + j \times w_j \text{ with } j = 1, \ldots, D \text{ and } k = 0, \ldots, M \tag{5}$$

Let C_L and C_U be predefined lower bound and upper bound numbers of unempty cells, an initial value of M is calculated by (6).

$$M = 1 + (C_L + C_U)^{1/D} \tag{6}$$

Let C be a number of cells created by dividing the provided dataset into a grid with the current value of M obtained from (6), $C_{unempty}$ be a number of unempty cells among C cells, Δ_C be difference between predefined lower bound of unempty cells and the current number of unempty cells calculated by (7), Δ_M be difference between a desired value of M and the current value of M calculated by (8). Values of M will be adjusted by (9) until satisfying (10) or (11).

$$\Delta_C = \left| C_{unempty} - C_L \right| \tag{7}$$

$$\Delta_M = 1 + \Delta_C^{1/D} \tag{8}$$

$$M = \begin{cases} M + \Delta_M, & \text{if } C_{unempty} < C_L \\ M - \Delta_M, & \text{if } C_{unempty} > C_U \end{cases} \tag{9}$$

Let M^-, M^+ be values of M at the previous and the next iterations, the process of finding an expected value of M will stop when $C_{unempty}$, M^- and M^+ satisfy (10).

$$(C_L \le C_{unempty} \le C_U) \text{ or } (M^- = M^+) \tag{10}$$

In a specific case, the value of $C_{unempty}$ is unchanged when values of M are changed a number of times. The process of finding an expected value of M should stop even though $C_{unempty} < C_L$ or $C_{unempty} > C_U$. Let n_{max} be a predefined number such as 10 and n_{loop} be a number of times values of M are successively changed but the value $C_{unempty}$ is unchanged. A new exceptional stopping condition is (11).

$$(n_{loop} \geq n_{max}) \tag{11}$$

Let $\mathbf{Y} = (\mathbf{y}_1, \mathbf{y}_2, \ldots, \mathbf{y}_C)$ be a set of unempty weighted cells created from the provided dataset. $\mathbf{y}_i = (y_{i1}, y_{i2}, \ldots, y_{iD}, w_i)$ is an unempty cell where y_{ij} with $j = 1, \ldots, D$ is coordinates of $\mathbf{y}_i$, and w_i is weight of $\mathbf{y}_i$ which is a number of objects belonging to that cell. The proposed transformation step is depicted by Algorithm 1.

Algorithm 1. Dataset2WeightedCellSet ($\boldsymbol{X}, C_L, C_U$)

Input: A dataset **X**, predefined bound sizes called C_L, C_U
Output: A set of unempty cells called **Y**
Begin

1. Calculate vectors $\mathbf{v}_{min}$ and $\mathbf{v}_{max}$ using (2) and (3), respectively.
2. Calculate an initial value of M using (6).
3. Calculate vectors **w** using (4) and $\mathbf{m}_j$ with $j = 1, \ldots, D$ using (5).
4. Map data objects $\mathbf{x}_i = (x_{i1}, x_{12}, \ldots, x_{iD})$ to cells based on the value of $\mathbf{x}_i$ and the vectors $\mathbf{m}_j$ with $j = 1, \ldots, D$ as in Fig. 1.
5. Calculate a number of unempty cells called $C_{unempty}$.
6. Check stopping conditions
   - If (10) or (11) is satisfied, return a set of unempty cells **Y** and stop.
   - Otherwise, adjust M using (9) and repeat from step 3.

End

Step 2: To sort cells based on outlier scores. The purposes of this step are to calculate outlier scores of cells and rank them in a descending order of scores. Firstly, to measure distances between two weighted cells, this paper proposes a new function to calculate weighted distance by (12). In this formula, $E - distance$ is Euclidean distance, $w(\mathbf{x})$ is weight of **x** or a number of objects belonging to a cell **x**, and w_{max} is the maximum weight of cells.

$$distance(\mathbf{x}, \mathbf{y}) = E - distance(\mathbf{x}, \mathbf{y}) - \frac{w(\mathbf{x}) + w(\mathbf{y})}{2 \times w_{max}} \tag{12}$$

Secondly, the rank-based outlier detection algorithm with various depths [5] is applied on the sets of weighted cells. In this step, $d(\mathbf{p}, \mathbf{q})$ is weighted distance from **p** to **q**, $k - distance(\mathbf{p})$ is weighted distance from **p** to its *kth* nearest neighbor, $N_k(\mathbf{p})$ is a set of k nearest neighbors of **p**, $d_k(\mathbf{p}, \mathbf{q})$ is a set of weighted distances from **p** to $\mathbf{q} \in N_k(\mathbf{p})$, $\Omega_k(\mathbf{p})$ is forward density around **p** at $k - distance(\mathbf{p})$, $R_k(\mathbf{q})$ is reverse ranking of **p** with respect to $\mathbf{q} \in N_k(\mathbf{p})$, $\Omega_R(\mathbf{q})$ is reverse density around **q** at $k - distance(\mathbf{p})$. σ, $h_{optimal}$, and Ω_{Smooth} are calculated according to [5]. This ranking algorithm is depicted as Algorithm 2.

Algorithm 2. CellRanking $(\boldsymbol{Y}, k)$

Input: Unempty cells $\mathbf{Y}$, a number of nearest neighbors k
Output: Ids of cells associated with the ranked outlier scores, and an optimal k'
Begin
1. *//Find sorted distances and cells from each cell to its k nearest cells*
For each cell $\mathbf{p} \in \mathbf{Y}$
    Find $k - distance(\mathbf{p})$
    $d_k(\mathbf{p}, \mathbf{q}) \leftarrow$ ascending sort$(d(\mathbf{p}, \mathbf{q}))$ with $\mathbf{q} \in N_k(\mathbf{p})$
    $N_k(\mathbf{p}) \leftarrow$ascending sort$(N_k(\mathbf{p}))$ by $d(\mathbf{p}, \mathbf{q})$ with $\mathbf{q} \in N_k(\mathbf{p})$
End for
2. *//Find reverse cell ranks and reverse density of each cell with various depths*
For each cell $\mathbf{p} \in \mathbf{Y}$
    For depth k_i from 2 to k
        For each $\mathbf{q} \in N_k(\mathbf{p})$
            $R_{k_i}(\mathbf{q}) \leftarrow$reverse rank $\mathbf{p}$ by $\mathbf{q}$
        End for
    End for
    $\bar{\Omega}_R(\mathbf{p}) \leftarrow median\left(\frac{R_{k_i}(\mathbf{q})}{d_{k_i}(\mathbf{p}, \mathbf{q})}\right)$
End for
3. *// Find median of reverse density*
$\bar{\Omega}_R \leftarrow median(\bar{\Omega}_R(\mathbf{q}))$
4. *//Find optimal k'*
$\sigma \leftarrow \frac{median(|\bar{\Omega}_R - median(\bar{\Omega}_R(\mathbf{q}))|)}{0.6745}$
$h_{optimal} \leftarrow \frac{0.9 \times \sigma}{N^5}$
$\Omega_{max} \leftarrow 0$
$k_{optimal} \leftarrow 0$
For each k_i from 2 to k
    $\Omega_{smooth} \leftarrow \frac{1}{N \times h_{optimal}\sqrt{2 \times \pi}} \sum_{i=1}^{N} e^{-\frac{1}{2}\left(\frac{(\bar{\Omega}_R - \bar{\Omega}_R(\mathbf{q}))^2}{h_{optimal}}\right)}$
    If $\Omega_{smooth} > \Omega_{max}$ then
        $\Omega_{max} \leftarrow \Omega_{smooth}$
        $k_{optimal} \leftarrow k_i$
    End if
End for
5. *//Assign outlier score of each cell*
For each cell $\mathbf{p} \in \mathbf{Y}$
    For each depth k_i from 2 to k
        $\Omega_{k_i}(\mathbf{p}) \leftarrow \frac{k_i}{d_{k_i}(\mathbf{p}, \mathbf{q})}$
        $Score_{k_i}(\mathbf{p}) = \frac{(R_{k_i} - k_i)}{\Omega_{k_i}(\mathbf{p}) \times w(\mathbf{p})}$
    End for
    Score$(\mathbf{p}) = median($Score$_{k_i}(\mathbf{p}))$
End for
6. *//Sort cells in a descending order of scores*
$\mathbf{Y} \leftarrow$descending sort$(\mathbf{Y})$ by $Score(\mathbf{p})$ with $\mathbf{q} \in \boldsymbol{Y}$
7. Return $\mathbf{Y}$ and $k_{optimal}$.
End

Step 3: To obtain outliers from cells. The purposes of this step are to identify outlier objects in the ranked cells and eliminate outlier objects from the provided dataset. Let $\mathbf{Z} = (\mathbf{z}_1, \mathbf{z}_2, \ldots, \mathbf{z}_n)$ be a list of n outliers, $\mathbf{Y}$ is an output of the previous step. The process of retrieving outlier objects from these cells is depicted as Algorithm 3.

Algorithm 3. OutlierRetrieving ($\boldsymbol{X}, n, \boldsymbol{Y}$)
Input: A dataset $\mathbf{X}$, a number of outliers n, set of ranked cells $\mathbf{Y}$.
Output: A set of data object as outliers $\mathbf{Z}$, dataset $\mathbf{X}$ after removing outliers

```
Begin
    found ← 0
    j←1
    Z←∅
    while ((found < n) and (j ≤ n))
        for each object x∈X and x ∈ y_j
            Z ←Z + {x}
            X ← X − {x}
            found← found + 1
        end for
            j←j + 1
    end while
    Return Z and X.
End
```

4 Evaluation

4.1 Experiment Setting

To compare results of the proposed method and the RDOS method, algorithms were implemented using the C programming language, compiled by the TDM-GCC 4.8.1 64 bit release associated with the Dev C++ 5.9.2, running on a normal personal computer. The machine was configured with an Intel processor core i5-24000 CPU 3.10 GHz, 8 GB of RAM, and Windows 7. A predefined $k = 10$ was used as the number of nearest neighbors, 29 pairs of $C_L, C_U = C_L + 1000$ with $C_L = i \times 1000$ where $i = 1, 2, \ldots, 29$ are tested.

4.2 Datasets for Experiments

Datasets used in the experiments are collected from previous researches. They are *Dataset2D* obtained from [14] with 100 outliers added, a part of *TDriveTrajectory* dataset obtained from Microsoft website [15], a part of *Activity Recognition* dataset and a part of *Person Activity* dataset obtained from UCI website [16]. Information of the selected datasets are shown in Table 1.

Table 1 Information of datasets

Datasets	No. of objects	No. of attributes	No. outliers need to be detected
Activity Recognition	162,501	3	10
Person Activity	164,860	3	10
TDrive Trajectory	176,424	2	7
Dataset2D	800,059	2	100

4.3 *Experiment Results*

This part compares the results when using the proposed method (abbreviated to Cell-RDOS) to results from the rank-difference outlier detection method (abbreviated to RDOS). Criteria used for comparison are the number of matching detected objects between the two mentioned algorithms, and the average of executing time.

When working with datasets having fewer than 200,000 objects, results from Table 2 show that unmatching detected outliers between the proposed algorithm and the previous algorithm are only 1 object apart among the number of needed outliers. However, results in Table 3 show that the proposed algorithm can reduce executing time from 34.86 to 99.17 % compared to the RDOS algorithm. The executing time of the dataset *TDrive Trajectory* is reduced only 34.86 % because this dataset has low-density patterns.

When working with the bigger dataset, which has 800,059 objects, the proposed method produces exactly the same outliers compared to the results of the RDOS method. However, the proposed Cell-RDOS algorithm can reduce 99.88 % of executing time compared to the RDOS method. This reduced time is approximately equal to 163.27 h or 6.8 days. In the case of the dataset *Dataset2D*, the previous RDOS algorithm consumed 9,808 min. The proposed Cell-RDOS, meanwhile, consumed only 12 min.

Moreover, the proposed Cell-RDOS algorithm produces the same results with $C_L \geq 27,000\,cells$. The numbers of matching detected outliers are smaller when values of C_L are reduced.

Table 2 Comparison of matching results between two algorithms with $C_L = 29,000\,cells$

Datasets	No. of objects	No. of detected outliers	No. of matching detected objects	Percentage of matching (%)
Activity Recognition	162,501	10	9	90
Person Activity	164,860	10	9	90
TDrive Trajectory	176,424	7	6	85
Dataset2D	800,059	100	100	100

Table 3 Comparison of executing time between two algorithms with $C_L = 29,000\, cells$

Datasets	No. of objects	Executing time (in minutes)		Reduced time (%)
		RDOS	*Cell-RDOS*	
Activity Recognition	162,501	510	4	98.37
Person Activity	164,860	534	8	99.17
TDrive Trajectory	176,424	612	399	34.86
Dataset2D	800,059	9,808	12	99.88

5 Conclusions and Discussions

The experiment results show that the proposed Cell-RDOS algorithm can be considered as a great tool for ranking-based outlier detection for big datasets. Accuracy level of the proposed algorithm matches the accuracy level of the RDOS algorithm. But, the Cell-RDOS method can reduce up to 99 % of executing time compared to the RDOS method when working with very large datasets such as *Dataset2D*. Based on the experiment results, to ensure correct results, the predefined lower bound should be at least 27,000 cells.

References

1. Chandola, V., Banerjee, A., Kumar, V.: Anomaly detection: a survey. ACM Comput. Surveys **41**, 15:1–15:58 (2009)
2. Hawkins, D. M.: Introduction. In: Hawkins, D.M. (ed.) Identification of Outliers, pp. 1–9. Chapman & Hall (1980)
3. Aggarwal, C.C.: Outlier Analysis. In: Aggarwal, C.C. (ed.) Data Mining, pp. 237–263. Springer International Publishing Switzerland (2015)
4. Hodge, V.J.: Outlier detection in Big Data. In: Wang, J. (ed.) Encyclopedia of Business Analytics and Optimization, vol. 5, pp. 1762–1771. IGI Global (2014)
5. Bhattacharya, G., Ghosh, K., Chowdhury, A.S.: Outlier detection using neighborhood rank difference. Pattern Recogn. Lett. **60–61**, 24–31 (2015)
6. Shaikh, S., Kitagawa, H.: Top-k outlier detection from uncertain data. Int. J. Autom. Comput. **11**, 128–142 (2014)
7. Breunig, M.M., Kriegel, H.P., Raymond, T.: Ng, and Sander, J.: LOF: identifying density-based local outliers. ACM. SIGMOD Record **29**, 93–104 (2000)
8. Tang, J., Chen, Z., Fu, A.W.C., Cheung, D.W.L.: Enhancing effectiveness of outlier detections for low density patterns. In: Proceedings of the 6th Pacific-Asia Conference on Advances in Knowledge Discovery and Data Mining (2002)
9. Jin, W., Tung, A.H., Han, J., Wang, W.: Ranking outliers using symmetric neighborhood relationship. In: Ng, W.K., Kitsuregawa, M., Li, J., Chang, K. (eds.) Advances in Knowledge Discovery and Data Mining, vol. 3918, pp. 577–593. Springer, Berlin (2006)
10. Huang, H., Mehrotraa, K., Mohana, C.K.: Rank-based outlier detection. J. Stat. Comput. Simul. **83**, 518–531 (2013)

11. Huang, H., Mehrotra, K., Mohan, C.: Algorithms for detecting outliers via clustering and ranks. In: Jiang, H., Ding, W., Ali, M., Wu, X. (eds.) Advanced Research in Applied Artificial Intelligence, vol. 7345, pp. 20–29. Springer, Berlin (2012)
12. Ha, J., Seok, S., Lee, J.S.: A precise ranking method for outlier detection. Inf. Sci. **324**, 88–107 (2015)
13. Hodge, V.J.: Outlier Detection in Big Data. In: Wang, J. (ed.) Encyclopedia of Business Analytics and Optimization, vol. 5, pp. 1762–1771. Business Science Reference, Hershey (2014)
14. Hieu, D.V., Meesad, P.: A Cell-MST-Based method for big dataset clustering on limited memory computers. In: 7th International Conference on Information Technology and Electrical Engineering, pp. 632–637. Chiang Mai, Thailand (2015)
15. Yuan, J., Zheng, Y., Xie, X. Sun, G.: Driving with knowledge from the physical world. In: Proceedings of the 17th ACM SIGKDD international conference on Knowledge discovery and data mining. San Diego, California, USA (2011)
16. Lichman, M.: Machine Learning Repository. http://archive.ics.uci.edu/ml/datasets.html

Fast Computing of Microarray Data Using Resilient Distributed Dataset of Apache Spark

Ransingh Biswajit Ray, Mukesh Kumar and Santanu Kumar Rath

Abstract Microarray technology is one of the emerging technologies in the field of genetic research that many biologists use to monitor expression levels of genes in a given organism. Microarray experiments are used to investigate genome-wide expression changes in health care aspects. Analysis of large Microarray datasets has become a challenging task, as the number of genes available in commercial probe sets and the number of test samples for an experimental set increases. The colossal amount of raw gene expression data often leads to computational and analytical challenges including feature selection and classification of the dataset into correct group or class. In this paper, statistical method (test), i.e., ANOVA based on Spark framework is proposed to select the pertinent features. After feature selection, various classifiers i.e., Naive Bayes (sf-NB) and Logistic Regression (sf-LoR) based on Spark framework are applied to classify the Microarray dataset. A detail comparative analysis in terms of execution time and accuracy is done on these feature selection and classifier methodologies that are based on Spark framework and conventional system respectively.

Keywords Big Data Analysis · Spark · Resilient Distributed Dataset · Microarray · sf-ANOVA · sf-NB · sf-LoR

1 Introduction

Microarray technology has helped the biologists to measure expression level of thousands of genes in a single experiment. The following three characteristics defines the properties and dimensions of microarray.

R. Biswajit Ray (✉) · M. Kumar · S.K. Rath
Department of Computer Science and Engineering, National Institute of Technology, Rourkela 769008, India
e-mail: ransingh.b.ray@gmail.com

M. Kumar
e-mail: mkyadav262@gmail.com

S.K. Rath
e-mail: skrath@nitrkl.ac.in

P. Meesad et al. (eds.), *Recent Advances in Information and Communication Technology 2016*, Advances in Intelligent Systems and Computing 463,
DOI 10.1007/978-3-319-40415-8_17

- **Volume**: Human genome contains millions of gene, and microarray is used to store and measure the expression level of these huge amount of information.
- **Velocity**: The genomic behaviour changes very frequently with time, as a result of which the repository of microarray data also changes periodically.
- **Variety**: The microarray data is stored and maintained in the various types and formats, which is usually semi structured type.

The multi-step, data-intensive nature of this technology has created an unprecedented informatics and analytical challenge. The curse of dimensionality problem is the major obstacle in microarray data analysis that hinders the useful information of dataset and leads to computational instability [6]. Therefore, the selection/extraction of relevant features (genes) remains as an imperative in the analysis of microarray data of cancer, which is an important step towards effective classification.

Big data applications are gradually becoming the main focus of attention because of the vast increment of data generation and storage that has taken place in the last decade or so. Reducing the execution time is extremely important for these applications, which has a high computational resource demand on memory and CPU time. To counter this, the concept of distributed computing has been adopted, where the data are distributed on various systems in a cluster and processed using various parallel processing paradigm [2, 3, 5, 9]. Hadoop, which is based on MapReduce framework has been highly instrumental in executing large-scale data-intensive applications on commodity clusters. While this data flow programming model is valuable for a vast class of applications, there are few applications which cannot be expressed productively as non-cyclic data flows. This includes numerous iterative machine learning algorithms, and also interactive data analysis tools. A new cluster computing frame-work called Spark, which bolsters applications with working sets while giving similar scalability and fault tolerance properties to MapReduce can outperform Hadoop by running up to 100 times faster in memory, and up to 10 times faster even when running on disk.

In this paper ANOVA statistical test based on Spark framework has been proposed to select relevant features from microarray dataset, and then classification of microarray dataset is performed using classifiers i.e., Naive Bayes and Logistic Regression based on same Spark framework. The performance of the algorithms are tested on Spark cluster with three slave (worker) nodes and one master (driver) system.

The rest of the paper is organized in the following manner: Sect. 2 explains the proposed work for feature selection methodology i.e., sf-ANOVA based on the Resilient Distributed Dataset (RDD) of Spark framework, which is used to select relevant features of microarray dataset. Section 3 highlights the basic concepts of Spark architecture and RDD. The implementation details are provided in Sect. 4. Results and interpretation details are discussed in Sect. 5, which presents the comparative analysis in terms of execution time for the feature selection and classifier methodologies that are based on Spark framework and conventional system respectively. Conclusion and scope for future work are mentioned in Sect. 6.

2 Proposed Work

The presence of a huge number of insignificant features reduces the analysis aspect of diseases like 'cancer'. This issue could be resolved by analyzing the dataset from proper perspective. This section presents an approach for classification of microarray data, which consists of three phases:

i. Missing data imputation and normalization methods, which are used for pre-processing of input data.
ii. Statistical test sf-ANOVA based on Spark framework, which is used for selecting relevant features.
iii. After selecting the relevant features, classifiers, i.e., Naive Bayes (sf-NB), Logistic Regression (sf-LoR) based on Spark framework has been applied to classify microarray dataset into their respective classes (binary/multi-class).

Figure 1 shows the graphical representation of proposed approach.

3 Spark Architecture and Resilient Distributed Dataset

3.1 Spark Architecture

SparkContext object defined in main program (called the driver program) is responsible for coordinating the independent sets of processes running on a Spark cluster [11]. The SparkContext can connect to any of the several cluster managers (Yarn, Mesos or Sparks own standalone cluster) that are responsible for allocating resources across applications. Once the connection is established Spark tries to acquire executors on nodes in cluster that computes, processes and store data for Spark applications. The application code (defined by Python or JAR files passed to SparkContext) is then sent to executors. Finally, SparkContext sends tasks to the executors to run. Figure 2 shows Spark architecture in cluster mode [11].

3.2 Resilient Distributed Dataset (RDD)

RDD is a read-only distributed collection of objects partitioned across multiple machines that can be rebuilt in case a partition is lost. Elements of an RDD do not reside in reliable or physical storage; instead, Spark computes them only in a lazy fashion that is, the first time they are used in an action. This means that RDDs are by default reconstructed each time when an action operation is performed on them. It is the core concept in Spark framework, which is a fault-tolerant collection of elements that can be operated on in parallel. RDD can be viewed as a table in a database. It

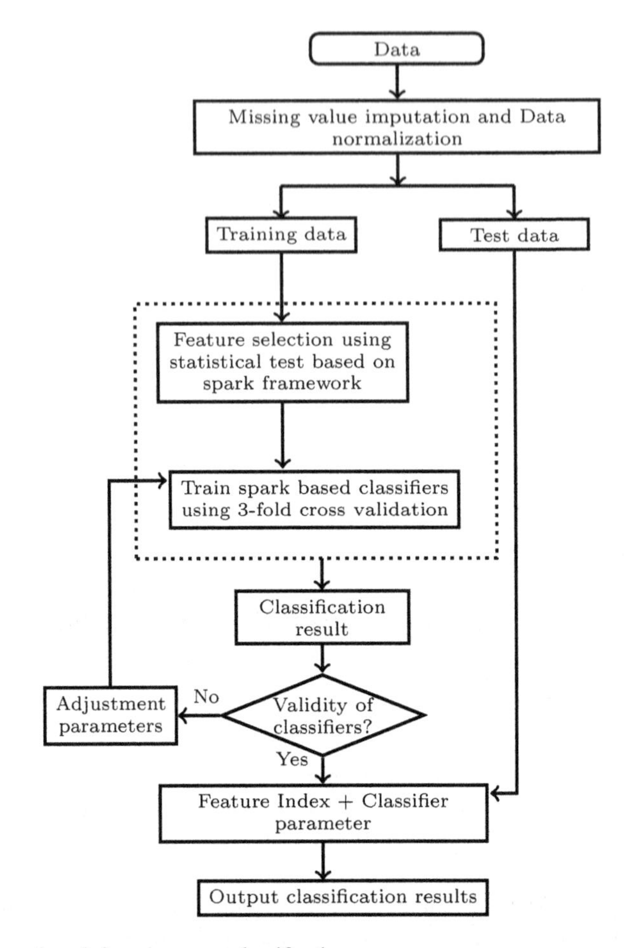

Fig. 1 Proposed work for microarray classification

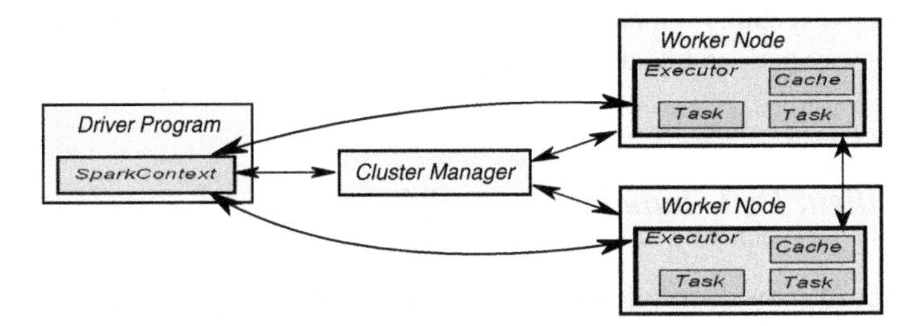

Fig. 2 Spark architecture in cluster mode

can hold any type of data. They help with rearranging the computations and optimizing the data processing. RDDs are also immutable i.e., it can be modified with a transformation, which will return a new RDD whereas the original RDD remains the same. RDD supports two types of operations i.e., Transformation and Action [10].

4 Implementation

4.1 Dataset Used

The dataset used for classification analysis is obtained from National center of Biotechnology Information (NCBI GEO, http://www.ncbi.nlm.nih.gov/gds/). The summary of the datasets is tabulated in Table 1.

4.2 Feature Selection Approach

In this section, the implementation details of the proposed statistical test based on Spark framework are discussed. The input file, which is stored in HDFS is a matrix of the form $N \times M$, where N is the total number of features and M is the number of samples in the dataset. The algorithm is divided into two sections, one is the driver program and another is the method used for RDD transformation. The driver program is responsible for creating a SparkContext, RDDs, and performing transformations and actions. In RDD transformation method, the distributed dataset (RDD) is transformed by reading a line ($featureVector_i$), calculating the required test statistic (s_i) and p-value for each $featureId_i$, and returning the result in form of key-value pairs ($featureId_i$, p-value) satisfying the required condition. Parallel transformation operation on RDDs are performed by worker nodes and the key-value pair results are returned to the driver program. The driver performs RDD action operation by collecting the results from all worker nodes.

ANOVA test has been applied to compare the 'multiple mean' values of the dataset, and visualize whether there exists any significant difference between multiple sample means. The statistic for ANOVA is called the F-statistic [8], which can be calculated using following steps:

1. The variation between the group is calculated as:

$$\textit{Between sum of squares } (BSS) = m_1 \left(\bar{X}_1 - \bar{X}\right)^2 + m_2 \left(\bar{X}_2 - \bar{X}\right)^2 + \cdots \tag{1}$$

$$\textit{Between mean squares } (BMS) = BSS/df \tag{2}$$

Table 1 Microarray dataset used

Dataset	# Samples	# Features	# Classes	Data size (GB)	# Training samples	# Testing samples
GSE13159 [1]	2096	54675	18	9.3	1397	699
GSE13204 [4]	3428	1480	18	9.5	2165	1083
GSE15061 [7]	870	54675	3	3.8	580	290

2. The variation within the groups is calculated as:

$$Within\ sum\ of\ squares\ (WSS) = \left(m_1 - 1\right)\sigma_1^2 + \left(m_2 - 1\right)\sigma_2^2 + \cdots \quad (3)$$

$$Within\ mean\ squares(WMS) = WSS/df_w \quad (4)$$

where df = degree of freedom, $df_w = (M - k)$, σ = standard deviation M = Number of samples, k = Number of groups, and m_k = no. of samples in group k.
3. F-test statistic is calculated as:

$$F = BMS/WMS \quad (5)$$

Algorithm 1 shows the implementation of ANOVA based on Spark framework.

Algorithm 1 sf-ANOVA

Input: $N \times M$ Matrix, where N is number of features and M is number of samples.
Output: Top P features.

```
1: Transform()
2: for each feature f_i do
3:     Calculate the value of BMS using Eq. 2.                ▷ i = 1, 2, ..., N
4:     Calculate the value of WMS using Eq. 4.
5:     Calculate the F-value (F_i = BMS/WMS)
6:     Calculate the p-value (p_i) corresponding to each F-value using F-distribution curve.
7:     if p_i < 0.001 then
8:         Select the feature, called fs_i.
9:         Return (fs_i, p_i).
10:    else
11:        Discard the feature.
12:    end if
13: end for
14: Driver Main():
15: Read N × M Matrix file from HDFS
16: Create RDD of Input N × M Matrix
17: SelectedFeatures ← RDD.map(Transform())
18: for each fs_i in SelectedFeatures.collect() do
19:     Print fs_i
20: end for
```

4.3 *Classification Approach*

Classification is a common form of supervised learning, where algorithms attempt to predict a variable from features of objects using labeled training data. 'Spark' is observed to be suitable for running parallel classification algorithms on a large dataset. The basic classifiers supports only binary classification, but can be extended to handle multiclass classification by using either One-versus-all (OVA) or All-versus-all (AVA). Two classifiers one discriminative i.e., Logistic Regression and another generative i.e., Naive Bayes are considered for analysis purpose.

- **Naive Bayes (sf-NB) based on Spark Framework**
  Naive Bayes method is a supervised learning algorithm applying Bayesian theorem with the assumption of independence between every pair of features. The Bayes theorem sates that

$$P(A|B) = P(B|A) * P(A)/P(B) \tag{6}$$

Thus, for a given sample X with attribute value $X = (a_1, a_2, ..., a_N)$ and class V_j. The probability value that X belongs to class V_j can be calculated as:

$$P(V_j|X = (a_1, a_2, ..., a_N)) = \quad P(X|V_j)P(V_j)/P(X) \tag{7}$$

$$= P(V_j)\prod_{i=1}^{N}(P(a_i|V_j))/P(X) \tag{8}$$

Where $P(V_j)$ is the prior probability for each class and $P(a_i|V_j)$ is the likelihood probability for each attribute, where $i = 1, ..., N$ and $j = 1, ..., C$.
In the testing phase, the likelihood for each attribute for a given class is calculated and a new instance x^t having attributes $(a_1^t, a_2^t, ..., a_N^t)$ is classified as:

$$\hat{P}(a_i^t|V_j) = (phi_{ij}^{a_i^t}) * (1 - phi_{ij})^{(1-a_i^t)}) \tag{9}$$

where, *phi* is the weight of each attribute for a given class; which is calculated as follow:

$$phi = X' * Y/M \tag{10}$$

where X is the matrix of training samples $(M \times N)$, Y is a matrix of dimension $M \times C$, where M is the number of training samples, N is the number of attributes, and C is the number of classes. Each row of Y consists of zeros except for the position that indicates the class label.

After calculating the value of $P(x_i^t|V_j)$, the new instance x^t is labeled as the following equation

$$V_{nb} = \underset{V_j \in V}{argmax}\, P(V_j)\prod_{i=1}^{N}\hat{P}(a_i^t|V_j) \tag{11}$$

- **Logistic Regression (sf-LoR) based on Spark Framework**

Logistic regression is a parametric form for the distribution $P(Y|X)$ where Y is a discrete value and $X = \{x_1, ..., x_n\}$ is a vector containing discrete or continuous values. The parametric model of logistic regression can be written as:

$$P(Y = 1|X) = \frac{1}{1 + exp(w_0 + \sum_{i=1}^{n} w_i X_i)} \tag{12}$$

and

$$P(Y = 0|X) = \frac{exp(w_0 + \sum_{i=1}^{n} w_i X_i)}{1 + exp(w_0 + \sum_{i=1}^{n} w_i X_i)} \tag{13}$$

The parameter W of the logistic regression is chosen by maximizing the conditional data likelihood. It is the probability of the observed Y values in the training data. The constraint can be written as:

$$W \leftarrow \underset{W}{argmax} \sum_{l} lnP(Y^l|X^l, W) \tag{14}$$

5 Results and Interpretation

The results obtained from the proposed algorithms (Sect. 2) based on various datasets are discussed below.

5.1 Analysis of sf-ANOVA Feature Selection Method

Statistical test viz., ANOVA test is applied on training dataset in order to remove irrelevant information and thus avoiding 'Curse of dimensionality problem'. The statistical test consider two hypothesis i.e., Null and Alternate hypothesis based on properties like mean, median, variance etc. Null hypothesis assumes that there exists no significant difference between the groups or classes, while alternate hypothesis assumes the contrary of null hypothesis. By considering 99.9 % confidence interval, if the p-value is less than 0.001, the null hypothesis is rejected and alternate hypothesis is accepted. Sorting these features according to their p-values helps to identify the features with strong representation.

Table 2 Execution details of sf-ANOVA

Dataset	Spark timing (S)	Conv. timing (S)	# Total features	# Features selected	Spark. processing efficiency (S^{-1})	Conv. processing efficiency (S^{-1})
GSE13159	19.76	38.95	54675	37016	2766.71	1403.91
GSE13204	10.15	19.18	1480	1423	145.76	77.16
GSE15601	9.37	16.76	54675	6786	5837.32	3262.80

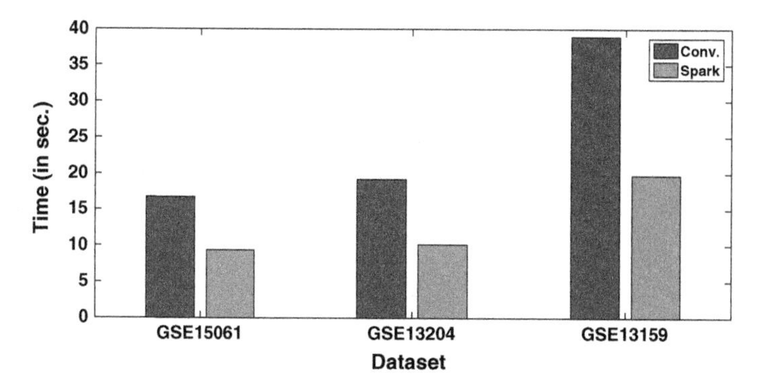

Fig. 3 Feature selection using ANOVA

The proposed feature selection technique is executed on both Spark cluster and Conventional system respectively. The overall execution details of the feature selection technique on all three datasets is tabulated in Table 2. Processing efficiency in Table 2 is defined as the number of features processed per second. Comparison in terms of execution time in both Spark cluster and conventional system is shown in Fig. 3. From Fig. 3, it is evident that the execution time for the proposed feature selection technique is much less in Spark cluster as compared to the same in conventional system.

5.2 *Analysis of Classifiers*

After feature selection, classifiers based on Spark framework i.e., Naive Bayes (sf-NB) and Logistic Regression (sf-LoR) are applied to classify the microarray datasets with reduced number of features. '3-fold cross validation (CV)' technique is applied to assess the performance of classifiers, as it provides a more realistic assessment by generalizing significantly to unseen data. The overall execution details of Naive Bayes and Logistic Regression classifiers based on Spark framework are tabulated in Table 3. Processing efficiency in Table 3 is defined as the number of samples processed per second. The amount of time consumed by Naive Bayes and Logistic Regression classifiers on Spark cluster and conventional system is shown in Fig. 4. From Fig. 4 it is evident that the time taken by the classifiers based on Spark framework, to analyze the datasets is much less as compared to the same in conventional system.

Table 3 Classification results

Classification technique	Dataset	Conv. timing (S)	Spark timing (S)	Train accuracy (%)	Test accuracy (%)	Regularization parameter	Conv. processing efficiency (S^{-1})	Spark processing efficiency (S^{-1})
Naive Bayes	GSE13159	92.99	25.66	80.14	82.19	NA	7.51	27.24
	GSE13204	10.42	5.77	82.73	84.76	NA	103.93	187.69
	GSE15061	7.72	4.84	60.52	60.34	NA	37.56	59.91
Logistic Regression	GSE13159	2851.30	766.75	53.69	56.94	2	0.24	0.91
	GSE13204	80.99	46.84	87.90	90.03	0.0001	13.37	23.12
	GSE15061	17.48	11.97	85.52	87.24	0.5	16.59	24.22

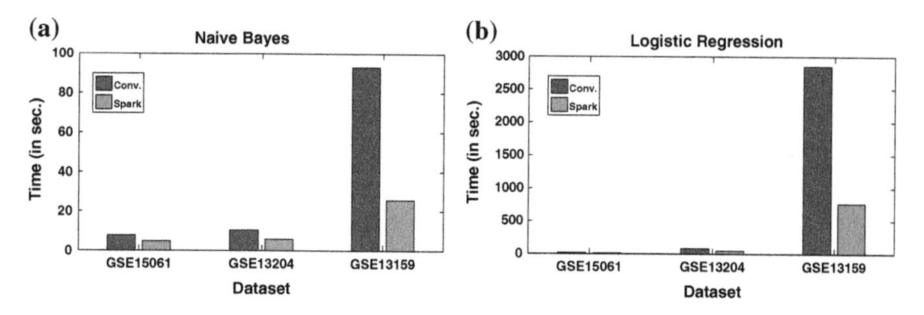

Fig. 4 Comparison of execution on Spark cluster (Spark) and Conventional system (Conv.). **a** Naive Bayes classifier. **b** Logistic Regression classifier

6 Conclusion

In this paper, statistical test (sf-ANOVA) based on Spark framework is proposed to select relevant features from large microarray datasets. After feature selection classifiers, i.e., Naive Bayes and Logistic Regression based on Spark framework are proposed to classify microarray datasets. The proposed approach works in a distributed manner on scalable clusters, and its performance increases with increase in data size, which is observed from Figs. 3 and 4. From the obtained result it is inferred that ANOVA test provides better accuracy with Naive Bayes as compared to Logistic Regression.

The proposed approach is tested on three standard large microarray data, which are publicly available. This approach can be tested on other large microarrays, if and when available in future. Further, this work can be extended by considering the applicability of machine learning techniques, Deep Learning, Decision Tree, etc. using Resilient Distributed Dataset of Spark framework.

References

1. Haferlach, T., Kohlmann, A., Wieczorek, L., Basso, G., Te Kronnie, G., Béné, M.C., De Vos, J., Hernández, J.M., Hofmann, W.K., Mills, K.I., et al.: Clinical utility of microarray-based gene expression profiling in the diagnosis and subclassification of leukemia: report from the international microarray innovations in leukemia study group. J. Clin. Oncol. **28**(15), 2529–2537 (2010)
2. He, Q., Zhuang, F., Li, J., Shi, Z.: Parallel implementation of classification algorithms based on mapreduce. In: Rough Set and Knowledge Technology, pp. 655–662. Springer (2010)
3. Islam, A.T., Jeong, B.S., Bari, A.G., Lim, C.G., Jeon, S.H.: Mapreduce based parallel gene selection method. Appl. Intell. 1–10 (2014)
4. Kohlmann, A., Kipps, T.J., Rassenti, L.Z., Downing, J.R., Shurtleff, S.A., Mills, K.I., Gilkes, A.F., Hofmann, W.K., Basso, G., DellOrto, M.C., et al.: An international standardization programme towards the application of gene expression profiling in routine leukaemia diagnos-

tics: the microarray innovations in leukemia study prephase. Br. J. Haematol. **142**(5), 802–807 (2008)
5. Kumar, M., Rath, S.K.: Classification of microarray using mapreduce based proximal support vector machine classifier. Knowl.-Based Syst. **89**, 584–602 (2015)
6. Lee, G., Rodriguez, C., Madabhushi, A.: Investigating the efficacy of nonlinear dimensionality reduction schemes in classifying gene and protein expression studies. IEEE/ACM Trans. Comput. Biol. Bioinform. **5**(3), 368–384 (2008)
7. Mills, K.I., Kohlmann, A., Williams, P.M., Wieczorek, L., Liu, W.M., Li, R., Wei, W., Bowen, D.T., Loeffler, H., Hernandez, J.M., et al.: Microarray-based classifiers and prognosis models identify subgroups with distinct clinical outcomes and high risk of aml transformation of myelodysplastic syndrome. Blood **114**(5), 1063–1072 (2009)
8. Sheskin, D.J.: Handbook of Parametric and Nonparametric Statistical Procedures. CRC Press (2003)
9. Wang, S., Pandis, I., Johnson, D., Emam, I., Guitton, F., Oehmichen, A., Guo, Y.: Optimising parallel r correlation matrix calculations on gene expression data using mapreduce. BMC Bioinform. **15**(1), 351 (2014)
10. Zaharia, M., Chowdhury, M., Das, T., Dave, A., Ma, J., McCauley, M., Franklin, M.J., Shenker, S., Stoica, I.: Resilient distributed datasets: a fault-tolerant abstraction for in-memory cluster computing. In: Proceedings of the 9th USENIX Conference on Networked Systems Design and Implementation, pp. 2–2. USENIX Association (2012)
11. Zaharia, M., Chowdhury, M., Franklin, M.J., Shenker, S., Stoica, I.: Spark: cluster computing with working sets. In: Proceedings of the 2nd USENIX Conference on Hot Topics in Cloud Computing, vol. 10, p. 10 (2010)

Heuristic Search Algorithms for Constructing Optimal Latin Hypercube Designs

Anamai Na-udom and Jaratsri Rungrattanaubol

Abstract Computer simulated experiments (CSEs) have been extensively used to explore the relationship between input variables and output response in science and engineering applications. Normally, CSE process is time consuming and computationally expensive to run and the output response from computer simulated experiments is deterministic. Consequently the space filling designs, which focus on spreading design points over a design space, are necessary. Latin hypercube designs (LHD) are widely used in the context of CSE. The optimal LHD for a given dimension of problem is constructed by using search algorithms under pre-specified optimality criteria. This paper proposes the methods to enhance the performance of search algorithms which have been widely used in the context of CSE. The results indicate that the proposed enhancement method can improve the performance of the search algorithms for constructing the optimal LHD.

Keywords Optimal designs · Search algorithm · Computer simulated experiments · Latin hypercube design

1 Introduction

In the past three decades, computer simulated experiments (CSE) have replaced classical experiments to investigate physical complex phenomena, especially when classical (physical) experiments are not feasible. For example, the use of reservoir simulator to predict ultimate recovery of oil, the use of finite element codes to predict

A. Na-udom
Faculty of Science, Department of Mathematics, Naresuan University, Phitsanulok, Thailand
e-mail: anamain@nu.ac.th

J. Rungrattanaubol (✉)
Faculty of Science, Department of Computer Science and Information Technology, Naresuan University, Phitsanulok, Thailand
e-mail: jaratsrir@nu.ac.th

P. Meesad et al. (eds.), *Recent Advances in Information and Communication Technology 2016*, Advances in Intelligent Systems and Computing 463,
DOI 10.1007/978-3-319-40415-8_18

behavior of metal structure under stress, and so on [1]. The nature of computer simulated experiments is deterministic [2]; hence identical settings of input variables always produce an identical set of output response. Consequently, space filling designs that aim to spread the design points over a region of interest are required. The most popular class of space filling design in the context of computer simulated experiments is latin hypercube design (LHD). LHD design was originally proposed by Mckay and co-workers [3] in 1979. The optimal LHD can be constructed through combinatorial methods (non-search algorithm) [4] or through search algorithms [5, 6]. The former method generates design with good design properties but it is restricted in terms of a design size. For example methods proposed by Butler [4] are limited to a design size of a prime number. The latter method is based largely on improving design by exchanging between the pairs of design points. Exchange algorithms can be time consuming to implement, however, the generated design are flexible and straightforward. The CSEs are usually complex and consist of many input variables to investigate [7] and hence a large number of design runs are required to estimate the parameters corresponding to the factors of interest in the model. For example, if the problem of interest consists of d input variables and n number of runs, the total number of all possible LHD is $(n!)^d$. Obviously this number explodes exponentially as the values of n and d increase; hence the full space of LHD cannot be explored. In this case we need search algorithms to lead us to a good design with respect to an optimality criterion. The key idea of all existing search algorithms is to use some kinds of exchange procedures to move towards the better designs.

The search based approach for selecting a design is implemented by combining search algorithms and the optimality criterion [8]. For example, Morris and Mitchell [5] adopted a version of Simulated Annealing algorithms (SA) to search for optimal LHDs with respect to ϕ_p criterion. Li and Wu [8] proposed a columnwise-pairwise algorithm (CP) with respect to the D efficiency criterion. It was reported that CP is very simple and easy to implement. Ye and his co-workers [6] adapted CP algorithm to search for symmetric LHD under various optimality criteria such as entropy and ϕ_p criteria. Park [9] proposed a row-wise element exchange algorithm along with IMSE and entropy criteria. Leary et al. [10] adapted CP and SA algorithms to construct the optimal designs within the orthogonal-array based Latin hypercube class by using the ϕ_p criteria. Jin et al. [11] developed an enhanced stochastic evolutionary algorithm (ESE) to search for the best design considering various optimality criteria such as a maximin distance criterion, ϕ_p criterion and entropy criterion. ESE has received wide attention from researchers due to its performance in constructing the optimal LHD. Liefvendahl and Stocki [12] applied a version of Genetic algorithm (GA) to search for the optimal LHD considering ϕ_p and a maximin distance criterion. A similar work can be found in [13] as the authors applied GA for constructing maximin designs. Grosso et al. [14] used the iterated local search algorithm and SA in constructing the optimal LHD under maximin distance and ϕ_p criterion. Vianna et al. [15] proposed the algorithm for fast optimal LHD by using the idea of seed design under maximin distance and ϕ_p criterion. Husslage and Rennen [16] proposed the method to construct the optimal LHD using different starting points. Due to the popularity of SA and ESE along with ϕ_p criteria, this paper presents the enhancement method to improve the capability of SA and ESE under ϕ_p criterion. In the following sections, we describe designs and optimality criteria, followed by search algorithms and its modification, results and conclusion, respectively.

2 Latin Hypercube Design and Optimality Criteria

2.1 Latin Hypercube Design

LHD is a matrix (***X***), which obtains n rows and d columns where n is the number of runs and d is the number of input variables. LHD can be constructed based on the idea of stratified sampling, which can ensure all subregions in the space are sampled with equally probability. A Latin hypercube sampling has

$$X_{ij} = \frac{\pi_{ij} - U_{ij}}{n}, \tag{1}$$

where $\pi_{i1}, \pi_{i2}, \ldots, \pi_{id}$ are independent random permutation of (1, 2, ..., n) and U_{ij} are n x d values of *i.i.d.* uniform $U[0,1]$ random variables independent of the π_{ij}. In practice, LHD can be easily generated by a random permutation of each column which contains the levels (1, 2 ,..., n). Then the d columns are combined together to form the design matrix ***X***. LHD can ensure uniform coverage of each input variable from a different single dimension. This shows a benefit of LHD on deterministic computer experiments.

2.2 The ϕ_P Optimality Criterion

Morris and Mitchell [5] proposed a modification of maximin distance criterion to search for the optimal design. For a given design X, the Euclidean intersite distance between any two design points can be calculated from

$$d(x_i, x_j) = \left[\sum_{k=1}^{d} (x_{ik} - x_{jk})^2\right]^{1/t} \tag{2}$$

By using (2), all intersite distances for every pairs of design points are calculated and can be expressed in a symmetric matrix. Let Euclidean distance list $d_1, d_2, \ldots, d_m$ be the distinct elements list from the smallest to largest, and also define index list $(J_1, J_2, \ldots, J_m)$ which J_j is the number of pairs of sites in the design separated by distance d_j. Thus X is a maximin design if among available designs, it maximizes d_j while J_j is minimized. The scalar criterion can be expressed as

$$\phi_p = [\sum_{j=1}^{m} J_j d_j^{-p}], \tag{3}$$

where p is a positive integer, J_j and d_j specified from X. In this study, the adaptive form of ϕ_p [11] which is simpler than (3) to implement is considered

$$\phi_p = \left[\sum_{i=1}^{n-1} \sum_{j=i+1}^{n} \frac{1}{d_{ij}^p}\right]^{\frac{1}{p}} \tag{4}$$

After ϕ_p value has been calculated, a design that minimizes ϕ_p is considered as an optimal design.

3 Search Algorithms

There are many search algorithms for constructing optimal LHD. This section will firstly discuss two search algorithms Stimulated Annealing (SA) and Enhanced Stochastic Evolutionary (ESE), and then explain how to modify and improve each algorithm for better efficiency.

3.1 Simulated Annealing (SA) Algorithm

SA is based on the analogy between the simulation of annealing solids and the problem solving of large combinatorial optimization problems [9]. Morris and Mitchell [5] adapted SA to construct optimal LHD using ϕ_p optimality criterion. LHD minimizing ϕ_p value is reserved as the best optimal LHD in the search space. The entire process can be described in Algorithm 1.

Algorithm 1: Simulated Annealing (SA)

1. Initialize I_{max}, t_0 and C_t
2. Set X to a random LHD, $X_{best} = X$, $t = t_0$
3. Set $I = 1$, $Label = 0$
4. Construct X_{try} by randomly element exchange in X
5. If $\phi_p(X) - \phi_p(X_{try}) \geq tl$ or accept with probability $e^{-[\phi_p(X_{try})-\phi_p(X)]/t}$ then $X = X_{try}$, $Label = 1$
6. If $\phi_p(X_{try}) < \phi_p(X_{best})$ then $I = 1$, $X_{best} = X_{try}$
   Else $I = I + 1$
7. If $I < I_{max}$ then goto 4
   Else $t = t*C_t$
8. If $Label = 1$ then goto 3
   Else report X_{best}

SA performs searching by obtaining a new LHD (X_{try}) via randomly element-exchange and update X with a better LHD or a worse LHD with satisfying probability, then update X_{best} with a better LHD. The probability of accepting a worse design is controlled by a value of temperature t, a chance of worse LHD acceptance decreases by a cooling system ($t = t * C_t$). SA has several parameters such as t_0, I_{max} and C_t. The discussion of setting SA parameters for LHD construction and how well SA can perform in terms of moving away from a local optimum value of ϕ_p can be seen in [5]. In this study, we use a heuristic method to find the best set of parameters to use in SA as presented in [17].

This paper focuses on improving the efficiency of SA to construct optimal LHD. From Eq. (4), it can be obviously seen that time for ϕ_p calculation mainly depends on the number of run (n), hence it is approximate to the Euclidean distance matrix calculation (n^2). The search process takes a long time mainly because of ϕ_p calculation. Jin et al. [11] has proposed an optimized way to calculate ϕ_p when element-exchange is assigned to LHD. The element-exchange is a swap operation between two selected points. Hence, when exchanging points between rows i_1 and i_2 within column k ($x_{i_1k} \leftrightarrow x_{i_2k}$), only elements in rows i_1 and i_2, and columns i_1 and i_2 in the distance matrix are changed. The ϕ_p of LHD after element-exchange can be calculated with a linear time as follows.

For any $1 \leq j \leq n$ and $j \neq i_1, i_2$ let

$$s(i_1, i_2, k, j) = \left|x_{i_2k} - x_{jk}\right|^t - \left|x_{i_1k} - x_{jk}\right|^t \tag{5}$$

then

$$d'_{i_1j} = d'_{ji_1} = \left[d^t_{i_1j} + s(i_1, i_2, k, j)\right]^{1/t} \tag{6}$$

and

$$d'_{i_2j} = d'_{ji_2} = \left[d^t_{i_2j} + s(i_1, i_2, k, j)\right]^{1/t} \tag{7}$$

Thus new ϕ_p is computed by

$$\phi'_p = \begin{bmatrix} \phi_p^p + \sum\limits_{1 \leq j \leq n, j \neq i_1, i_2} \left[(d'_{i1j})^{-p} - (d_{i1j})^{-p}\right] + \\ \sum\limits_{1 \leq j \leq n, j \neq i_1, i_2} \left[(d'_{i2j})^{-p} - (d_{i2j})^{-p}\right] \end{bmatrix}^{1/p} \tag{8}$$

As shown in (5)–(8), when performing element exchange, only some rows and columns will be updated, hence there is no need to reconstruct the entire distance matrix to calculate ϕ_p, hence the complexity of ϕ_p calculation reduces to a linear time. Therefore, after this modification, the modified SA (MSA) performs effectively by reducing time complexity especially in ϕ_p calculation.

3.2 Enhanced Stochastic Evolutionary (ESE) Algorithm

Jin et al. [11] proposed an algorithm called enhanced stochastic evolutionary (ESE) to construct an optimal design for CSE. The algorithm performs searching in 2 steps, a local search called inner loop and updating a global best and fine tuning probability of accepting a worse design called outer loop. The inner loop performs a local search by constructing a set of LHD and selecting the optimal LHD from the set. The set contains J LHDs formed by element-exchanging at column i mod d of X. If the selected design is better or not better but has good enough probability then the local optimal design (X) will be updated.

Algorithm 2: Enhancement of Stochastic Evolutionary (ESE)

1. Initialize Th, J, M
2. Set X to a random LHD, $X_{best} = X$
3. Set $X_{old\_best} = X_{best}$, $i = 0$, $n_{acpt} = 0$ and $n_{imp} = 0$
4. Randomly create J distinct LHDs by element exchanging X at column(i mod d)
5. Select the best LHDs from J distinct LHDs and assign it to X_{try}
6. If $\phi_p(X)-\phi_p(X_{try}) \geq tl$ or $\phi_p(X_{try})-\phi_p(X) \leq Th*random(0,1)$ then
   $X = X_{try}$, $n_{acpt} = n_{acpt} + 1$
   If $\phi_p(X_{try}) < \phi_p(X_{best})$ then $X_{best} = X_{try}$, $n_{imp} = n_{imp}+1$
7. If $i < M$ then $i=i+1$, goto 4
   Else goto 8
8. If $\phi_p(X_{old\_best})-\phi_p(X_{best}) \geq tl$ then $flag_{imp} = 1$
   Else $flag_{imp}=0$
9. If $flag_{imp}=1$ then
   - If $n_{acpt}/M > \beta_1$ and $n_{imp}/M < n_{acpt}/M$ then
     - $Th = Th*\alpha_1$
   - Else if $n_{acpt}/M > \beta_1$ and $n_{imp}/M = n_{acpt}/M$ then
     - $Th = Th$
   - Else
     - $Th = Th/\alpha_1$

   Else
   - If $n_{acpt}/M \geq \beta_1$ and $n_{acpt}/M \leq \beta_2$ and step=0 then
     - If step=0 or step=1 then $Th = Th/\alpha_3$
     - Else If step=2 then $Th = Th*\alpha_2$
   - Else If $n_{acpt}/M \leq \beta_1$ then
     - $Th = Th/\alpha_3$
     - step =1
   - Else If $n_{acpt}/M \geq \beta_2$ then
     - $Th = Th*\alpha_2$
     - Step =2
10. If $\phi_p(X_{old\_best})-\phi_p(X_{best}) \geq tl$ then $X_{best}=X_{old\_best}$, $i_{out}=1$
    Else $i_{out}= i_{out}+1$
11. If $i_{out} < Max$ then Goto 3
    Else Report X_{best}

From Algorithm 2, the inner loop performs in 4–7, and repeats with the maximum loop (M). The X_{best} and X are updated with the acceptance criteria. The outer loop controls the process by updating the value of temperature Th. Unlike SA, the process of updating the temperature (Th) is not fixed, but is controlled by the performance of searching in terms of the inner loop improvement by the number of improvement (n_{imp}) and number of acceptance (n_{acpt}). There are two processes of updating Th called improving process and exploration process. The process is described in 9, when X_{best} get improved in the inner loop and better than the previous best design (X_{old_best}), the improving process is active otherwise exploration process is active.

In improving process ($flag_{imp} = 1$), Th is adjusted based on the performance in a local best LHD search by considering an acceptance ratio (n_{acpt}/M) and the improvement ratio in (n_{imp}/M), β_1 and β_2 are cutting point values for updating Th, where $0 < \beta_1 < 1$. Jin et al. [4] recommended β_1 to be 0.1. If the improvement ratio is greater than β_1, Th is increased by α_1 ($Th = Th/\alpha_1$), else if the improvement ratio is lower than the acceptance ratio Th is decreased by α_1 ($Th = Th * \alpha_1$), otherwise unchanged, where $0 < \alpha_1 < 1$ and $\alpha_1 = 0.8$ as suggested by Jin et al. [4].

In exploration process ($flag_{imp} = 0$), when a local best LHD performs worse, Th is adjusted to help a search process moving away from a local optimal design by considering only the accept ratio. If the acceptance ratio is between β_1 and β_2 where $0 < \beta_1 < \beta_2 < 1$, Th is increased by α_2. If the acceptance ratio is greater than β_2, Th is decreased by α_3, where $0 < \alpha_2 < \alpha_3 < 1$, $\alpha_2 = 0.9$ and $\alpha_3 = 0.7$, Th is rapidly increased, this means more worse designs could be accepted, then Th is decreased slowly for searching better design after moving away from a local optimal design. Jin et al. [11] recommended 0.1 for β_1 and 0.8 for β_2.

3.3 *Modified Enhancement of Stochastic Evolutionary (MESE)*

This section we present the enhancement method on ESE. The modified version is called MESE. We combine the advantage of SA (i.e. local search process) and the advantage of ESE (i.e. global search process) to improve the search process. MESE still contains 2 nested loops. The outer loop is similar to ESE except a stopping rule and a new variable X_{gbest} to record the best so far LHD, while the inner loop is modified as describe in Algorithm 3. The maximum number of cycles used is replaced by the following condition. If X_{best} is not improved from the global best design (X_{gbest}) for δ consecutive times, the search process is terminated. In this study, we set $\delta = 40$ [6].

We have modified ESE by combining a part of SA into the modified ESE (MESE). Algorithm 3 explains this modification.

Algorithm 3: Modified Enhancement of Stochastic Evolutionary (MESE)

1. Initialize Th, J, M, Max, C_{max}
2. Set X to a random LHD, $X_{best} = X$, X_{gbest}, $i_{out} = 0$
3. Set $X_{old\_best}=X_{best}$, $i=0$, $n_{acpt}=0$, $n_{imp}=0$, $j=0$ and $X_{try}=X$
4. $X_{tmp}=X_{try}$, $c=0$
5. X_{tmp}=Random element exchanging X at column(i mod d)
6. If $\phi_p(X_{tmp}) - \phi_p(X_{try}) \geq tl$ then $X_{try} = X_{tmp}$
   If $c<J$ then $c=c+1$, goto 5
   Else goto 7
   Else goto 5
7. If $\phi_p(X_{try})-\phi_p(X) \geq tl$ or $\phi_p(X_{try})-\phi_p(X) \leq Th*random(0,1)$ then
   $X = X_{try}$, $n_{acpt}= n_{acpt}+1$
   If $\phi_p(X_{try})<\phi_p(X_{best})$ then $X_{best}=X_{try}$, $n_{imp}=n_{imp}+1$, $j=1$
   Else $j = j + 1$
   Else goto 8
8. If $i<M$ OR $j<C_{max}$ then $i = i+1$, goto 4
   Else goto 9
9. Perform step 8 and step 9 in Algorithm 2
10. If $\phi_p(X_{best})-\phi_p(X_{gbest}) \geq tl$ then $\phi_p(X_{gbest})=\phi_p(X_{best})$, $i_{out}=1$
    Else $i_{out}= i_{out}+1$
11. If $i_{out} < Max$ then goto 3
    Else report X_{gbest}

In Algorithm 3, the major enhancement was made in the inner loop; the modification is on 4–6 and 7. Instead of generating J distinct LHDs at one time and select the best one, we modify it by constructing one LHD (X_{tmp}) at a time and keep the best one (X_{try}) for J iterations. This change can decrease the computational time complexity since we keep the optimal LHD while generating it, instead of obtaining the entire distinct J LHDs and deciding the most optimal one. In this study the parameters J is set to be ${}^{n}C_2/5$ but not larger than 50, and the parameter M is in a range of $2 * {}^{n}C_2 * d/J \leq M \leq 100$. The tolerance level ($tl$) is set to 0.0001, from the empirical study, the smaller value does not improve the search process. $C_{max} = d * 10 + 10$ and $Max = 40$. All simulation studies presented in this paper were performed using R program.

4 Result and Discussion

The values of ϕ_p at the termination step of MSA, ESE and MESE for each dimension of problems are presented in Table 1. Each case was repeated for 10 times to consider the effect of different starting points. The descriptive statistics (max, min, mean and sd.) on the ϕ_p values obtained from each search technique are

Table 1 Performance of MSA, ESE and MESE

		ϕ_p				Time (s.)	No. exchange
n × d	method	max	min	mean	sd.	average	average
9 × 2	MSA	4.273538	4.273538	**4.273538**	1.48E-12	16.140	47044.7
	ESE	4.344617	4.273538	4.301970	0.034821	2.854	5760.0
	MESE	4.344617	4.273538	4.294007	0.031286	**1.972**	**5644.8**
19 × 3	MSA	4.936365	4.898022	**4.916390**	0.013472	103.194	140022.5
	ESE	4.935001	4.901203	4.922769	0.010908	53.183	**41040.0**
	MESE	4.958163	4.914072	4.926576	0.011805	**30.942**	42160.0
99 × 7	MSA	5.756299	5.750880	5.753385	0.001473	769.902	193111.8
	ESE	5.745721	5.740989	5.742462	0.001387	954.711	200000.0
	MESE	5.745267	5.740459	**5.742109**	0.001577	**800.993**	**199865.0**
129 × 8	MSA	5.904712	5.901620	5.902779	0.000759	1070.692	201891.2
	ESE	5.891672	5.887732	**5.889600**	0.001383	1217.445	200000.0
	MESE	5.891365	5.888284	5.890195	0.001029	**1063.358**	200000.0
201 × 10	MSA	6.182737	6.179070	6.181468	0.001015	1905.174	218859.0
	ESE	6.164494	6.162494	6.163368	0.000648	1885.965	200000.0
	MESE	6.164341	6.161576	**6.163364**	0.000721	**1717.370**	200000.0
451 × 15	MSA	6.777918	6.775912	6.777281	0.000589	5338.902	231194.8
	ESE	6.754473	6.753710	6.754034	0.000220	4646.026	200000.0
	MESE	6.754603	6.753482	**6.753898**	0.000321	**4404.563**	200000.0
801 × 20	MSA	7.274153	7.272498	7.273293	0.000546	11892.890	251293.7
	ESE	7.248591	7.247829	7.248093	0.000262	9035.538	200000.0
	MESE	7.248234	7.247825	**7.248006**	0.000136	**8880.095**	200000.0

presented. The results indicate that MSA, ESE and MESE perform similarly for small dimension of problem (i.e. d = 2 and 3) in terms of minimization of ϕ_p. Further, the standard deviation values displayed a slightly larger amount of variation over 10 replications in ESE and MESE than that of MSA. This indicates the consistency in the search process for MSA when different starting points are considered. When considering medium dimension of problems (i.e. d = 7, 8 and 10), ϕ_p values from ESE and MESE are slightly lower than MSA. In addition, the standard deviation values obtained from ESE and MESE are smaller than MSA. This indicates that the search process of ESE and MESE is more consistent when the search space is larger. For large dimensions of problem (i.e. d = 10 and 15), both of ESE and MESE perform better than MSA while MESE is slightly better than ESE in terms of minimization of ϕ_p values. Hence if the goal is concerned with a good space filling design property, either ESE or MESE can be used for constructing the optimal LHD.

The results of the performance (efficiency) for MSA, ESE and MESE algorithms are presented in terms of time and number of exchanges for each algorithm to reach the optimal ϕ_p values, as shown in Table 1. As mentioned before, for each dimension of problems the search algorithms are repeated for 10 times, hence all values are presented as the average values. It can be clearly seen that MESE converges much faster than ESE and MSA as the time elapsed is less than that of ESE and MSA. When considering the number of exchange required in the search process, it is observed that number of exchange in MESE is less than the other two algorithms. The similar results are also observed in the case of medium and large dimension of problems as MESE converges much faster than MSA while it performs slightly better than ESE. This indicates that if time constraint is taken into account, MESE could be the better choice for constructing the optimal LHD designs.

5 Conclusion

This paper presents the method to enhance the SA and ESE algorithms in the construction of the optimal LHD. The major enhancement method appears in the calculation of ϕ_p criterion and the tolerance level setting in SA. For MESE, the enhancement is applied by using the combination of SA and ESE especially in the inner loop as shown in Algorithm 3. As presented in the result section, MESE performs better than ESE and MSA in terms of the design property achievement and the efficiency. Hence MESE would be recommended for the construction of optimal LHD for CSE. In order to extend the conclusion, other classes of design can be developed and collaborated with MESE to search for the best design in the class. Further, other types of search algorithm like Particle swarm optimization (PSO) or any type of clever algorithms can be further developed in constructing an optimal LHD. The validation of the approximation model accuracy developed from the obtained optimal LHD could also be further investigated in order to explore the relation of space filling property and prediction accuracy of the model.

References

1. Koehler, J., Owen, A.B.: Computer experiments. Handbook of Statistics, vol. 13, pp. 261–308. Elsevier Science, New York (1996)
2. Sacks, J., Welch, W.J., Mitchell, T.J., Wynn, H.P.: Design and analysis of computer experiments. Stat. Sci. **4**(4), 409–435 (1989)
3. Mackay, M.D., Beckman, R.J., Conover, W.J.: A comparison of three methods for selecting values of input variables in the analysis of output from a computer code. Technometrics **21**, 239–246 (1979)
4. Butler, N.A.: Optimal and orthogonal latin hypercube designs for computer experiments. Biometrika **88**(3), 847–857 (2001)
5. Morris, M.D., Mitchell, T.J.: Exploratory design for computational experiments. J. Stat. Plann. Infer. **43**, 381–402 (1995)
6. Ye, K.Q., Li, W., Sudjianto, A.: Algorithmic construction of optimal symmetric latin hypercube designs. J. Stat. Plann. Infer. **90**, 145–159 (2000)
7. Bates, R.A., Buck, R.J., Riccomagno, E., Wynn, H.P.: Experimental design and observation for large systems. J. Roy. Stat. Soc. B **58**, 77–94 (1996)

8. Li, W., Wu, C.F.J.: Columnwise-pairwise algorithms with applications to the construction of supersaturated designs. Technometrics **39**, 171–179 (1997)
9. Park, J.S.: Optimal latin hypercube designs for computer experiments. J. Stat. Plann. Infer. **39**, 95–111 (1994)
10. Leary, S., Bhaskar, A., Keane, A.: Optimal orthogonal-array-based latin hypercubes. J. Appl. Stat. **30**(5), 585–598 (2003)
11. Jin, R., Chen, W., Sudjianto, A.: An efficient algorithm for constructing optimal design of computer experiments. J. Stat. Plann. Infer. **134**, 268–287 (2005)
12. Liefvandahl, M., Stocki, R.: Study on algorithms for optimization of latin hypercubes. J. Stat. Plann. Infer. **136**, 3231–3247 (2006)
13. Li, Z., Shigeru, N.: Maximin distance-lattice hypercube design for computer experiment based on genetic algorithm. IEEE Explore **2**, 814–819 (2001)
14. Grosso, A., Jamali, A., Locatelli, M.: Finding maximin latin hypercube designs by iterated local search heuristics. Eur. J. Oper. Res. **197**(2), 541–547 (2009)
15. Viana, F.A.C., Venter, G., Balanov, V.: An algorithm for fast optimal latin hypercube design of experiments. Int. J. Numer. Meth. Eng. **82**(2), 135–156 (2010)
16. Husslage, B.G.M., Rennen, G., van Dam, E.R., Hertog, D.D.: Space-filling Latin hypercube designs for computer experiments. Optim. Eng. **12**, 611–630 (2011)
17. Na-udom, A.: Experimental design methodology for modeling response from computer simulated experiments. Ph.D. thesis, Curtin University of Technology (2007)

A Hybrid Multi-objective Genetic Algorithm with a New Local Search Approach for Solving the Post Enrolment Based Course Timetabling Problem

Dome Lohpetch and Sawaphat Jaengchuea

Abstract The post enrolment based course timetabling problem (PECTP) is one type of university course timetabling problem which a set of events has to be scheduled into time slots and suitable rooms according to students' enrolment data. This problem is classified as a NP-complete combinatorial optimization problem and hence it is very hard and highly time-consuming to solve the problem to find an optimal timetable. In this paper we have developed a non-dominated sorting genetic algorithm (NSGA-II) hybridized with two local search technique and a tabu search heuristic for solving the multi-objective PECTP. In addition to the original LS technique will be used in NSGA-II, a new LS technique is also used in NSGA-II in order to further improve the performance of NSGA-II, especially reducing violation of the first soft constraint. The algorithm takes advantage of the exploitation ability of two local search techniques and a tabu search heuristic to improve the results obtained in the exploration phase of the NSGA-II. The proposed hybrid approach is tested on a set of standard benchmark problem in comparison with other methods from the literature, and experimental results show that the proposed hybrid approach is able to find promising solutions for solving the multi-objective PECTP.

Keywords Multi-objective optimization problem · Non-dominated sorting genetic algorithm · Local search · Tabu search · Post enrolment based course timetabling problem

D. Lohpetch (✉) · S. Jaengchuea
Faculty of Applied Science, Department of Mathematics, King Mongkut's University of Technology North Bangkok, Bangkok, Thailand
e-mail: dome.l@sci.kmutnb.ac.th

S. Jaengchuea
e-mail: sawaphat.ja@gmail.com

P. Meesad et al. (eds.), *Recent Advances in Information and Communication Technology 2016*, Advances in Intelligent Systems and Computing 463,
DOI 10.1007/978-3-319-40415-8_19

1 Introduction

The post enrolment based course timetabling problem (PECTP) is a one type of university course timetabling problem (UCTP), and it is the assignment of university courses to suitable room and time slots according to students' enrolment data. The timetable is constructed after student enrolment in a way that all students can attend the events which they want to enroll. The PECTP is based on real world situation where students are given choices of events that they want to attend and a timetable is produced according to these choices. Usually, the UCTP is a *NP-complete* [1] and is classified as a *combinatorial optimization* (CO) problem [2]. As a result, this problem is very difficult to find the optimal solution which often leads to too high computational times. Many works have taken the PECTP as a single objective optimization problem by combining multiple criteria into a single scalar value and then minimizing the weighted sum of constraint violations as the only one objective function [3]. On the other hand, few works have tackled the PECTP as a multi-objective optimization problem. Burke et al. [4] proposed a multi-objective hyper-heuristic approach for solving the multi-objective PECTP. Datta et al. [5] applied a non-dominated sorting genetic algorithm (NSGA-II) as a university class timetable optimizer. Abdullah et al. [6] employed the NSGA-II for solving a quad-objective PECTP. Jat and Yang [3] introduced hybridizing of the NSGA-II with guided search and two local search techniques to solve a tri-objective PECTP.

Recently, a hybrid genetic algorithm (GA) with local search (LS) and tabu search (TS) approach, denoted as *HGALTS* [7], has been proposed for solving the PECTP. HGALTS consists of GA, LS technique, and TS heuristic. HGALTS can bring about the advantage of the exploitation ability of LS techniques and a TS heuristic to improve the results obtained in the exploration phase of the NSGA-II.

In this paper, we extend our investigation from our work in HGALTS [7] with a new LS technique for solving the multi-objective PECTP. Our proposed hybrid algorithm combines NSGA-II with two LS techniques and TS heuristic, denoted as *HNSGA2LTS*. The performance of the proposed hybrid approach was tested on a set of standard benchmark problem and compared results against other methods from the literature. The experimental results indicated that the proposed hybrid approach was a good method for solving the multi-objective PECTP.

2 Multi-objective PECTP

The post enrolment based course timetabling problem (PECTP) model used in this paper was introduced by Socha et al. [8]. In PECTP, a set of events is assigned into time slots and suitable rooms according to students' enrolment data, while each assignment satisfies hard and soft constraints. In this problem model, the following three hard constraints must be satisfied for a feasible solution (no violation of hard constraints):

- **H1**: No student attends more than one event at the same time;
- **H2**: The room is big enough for all the attending students and satisfies all the features required by the event;
- **H3**: Only one event is in each room at any timeslot.

This problem model is also interested in a number of soft constraint violations. The following three soft constraints that should be minimized are as follows:

- **S1**: Students should not attend an event in the last time slot of a day (this is, time slots 9, 18, 27, 36 or 45);
- **S2**: Student should not attend more than two events in consecutive time slots in the same day;
- **S3**: Students should not have to attend only one event in a day.

The three soft constraints mentioned above can be taken as one objective function separately, and then they are minimized the number of violations of each one. As a result, we have three objective functions, $f_1(x)$, $f_2(x)$, and $f_3(x)$ which are associated with the above three kind of soft constraints, respectively, resulting in solving PECTP as the multi-objective PECTP in this paper.

3 Methodology: Proposed Hybrid Approach

We proposed a hybrid NSGA-II [9] to solve the multi-objective PECTP, and it was combined with a LS technique and a TS heuristic as in HGALTS [7]. Moreover, we also add a new LS technique into the hybrid approach to improve the results even further. The pseudo code of the proposed hybrid NSGA-II with two LS techniques and a TS heuristic, denoted as *HNSGA2LTS*, is shown in Fig. 1.

The basic framework of HNSGA2LTS is based on a steady-state NSGA-II which one child solution is generated at each generation. For each solution, each event is assigned a random time slot according to a uniform distribution and is allocated a room by using the matching algorithm. In order to improve the quality of those random solutions, two LS techniques were applied to improve them into feasible or near-feasible solutions. Next, we will describe the details of the main components of the basic framework of HNSGA2LTS including LS1, LS2 and TS, respectively.

3.1 Local Search 1 (LS1)

In HNSGA2LTS, a LS technique used in this research is based on basic framework of LS in [10] and denoted as *LS1* in this paper.

```
Proposed Hybrid Approach - HNSGA2LTS
input:  A problem instance I
g ← 0 set the generation counter
Create archive set to keep non-dominated solutions
for each solution s_i of population do
   s_i ← random initial solution
   s_i ← solution s_i after applying LS1
   s_i ← solution s_i after applying LS2
   s_i ← compute fitness value for each solution s_i
end for
Assign rank and crowding distance for each solution s_i of population
Push non-dominated solutions into archive set
while the termination condition is not reached do
   Select two parents from population by crowded tournament selection
   s ← child solution after applying uniform crossover with a probability p_c
   s ← child solution after applying mutation with a probability p_m
   s ← child solution after applying LS1
   s ← child solution after applying LS2
   s ← child solution after applying TS heuristic
   s ← compute fitness value of child solution
Consider child solution and update archive set
Combine all solutions together and assign ranks and crowding distances for them
Form a new population from the combined population based on rank and crowding
  distance
g ← g + 1
end while
output: Return non-dominated solutions in archive set
```

Fig. 1 The pseudo code for proposed hybrid approach: HNSGA2LTS

3.2 *Local Search 2 (LS2)*

In this paper, we propose a new LS technique, denoted as *LS2* in this paper and the pseudo code for LS2 technique is shown in Fig. 2.

As mentioned before, LS1 works on all events but LS2 works on a set of events. The main aim of LS2 is to reduce the first soft constraint violation or the first objective function that students should not attend an event in the last time slot of a day. The basic idea of LS2 is to move events in the last time slot of a day, i.e., time

```
Local Search Algorithm 2 (LS2)
input: solution s from the population
if the solution s is feasible but soft constraint violations remain then
   Generate a randomly-ordered list of the events in the last time slot
   for each event e_i in the list do
      for each time slot t_j do
         if time slot t_j is not last time slot then
            Calculate a move of event e_i with time slot t_j by N1
           Apply the matching algorithm to assign rooms for event in the time
             slot affected by the move and delta-evaluate the result
           if this move reduces soft constraint violations without introducing a
              hard constraint violation then
              Make the move
              Go to consider the next event in the list
           end if
         end if
      end for
   end for
end if
output: return a possibly improved solution s
```

Fig. 2 The pseudo code for local search algorithm 2 (LS2)

slots 9, 18, 27, 36 or 45, to another time slot. Furthermore, LS2 is applied only if the solution s is feasible solution with remaining soft constraints violation.

LS2 first generate a randomly-ordered list of events in the last time slot of a day. Then, it tries a move each event in the list to another time slot except the last time slot by using neighborhood structure N1 and delta-evaluate the result of the move. If this move reduces soft constraint violation without violating the hard constraints then make the move and the next event in list is considered continuously. Otherwise, we do not apply the moves and try to move this event to the next time slot, then the next time slot, then the next time slot, and so on except the last time slot. If the move in N1 all fail for this event, then the next event in list is also considered in the next sequence. This process is continues until termination condition is reached, e.g., an improvement is reached or the maximum number of steps s_{max} is reached.

3.3 Tabu Search (TS)

The TS technique is a local search technique, and it is usually known to be a powerful tool for solving difficult optimization problems [6, 11]. We applied a tabu search after a local search in order to further improve the quality of the solution in our HGALTS [7]. TS used in this paper is based on the TS heuristic described in [10].

4 Experiments

In this section, we test the performance of the proposed HNSGA2LTS for solving the multi-objective PECTP in comparison with relevant algorithms, and it was coded in GNU C++ version 4.6 under Ubuntu 12.04 LTS running on a virtual machine (KVM hypervisor) with 3.0 GHz processor and 1 GB of memory. All relevant algorithms were tested on 11 benchmark instances proposed by Socha et al. [8] in 2002 using a problem instance generator written by Paechter, developed within the Metaheuristic Network (MN) [12], and these instances, called “MN instances” [13], are divided into three classes: five small instances, five medium instances and one large instance. The characteristics for each class are given in Table 1.

For each size of problem instance, the time limit t_{max} was used as the stopping criteria, by setting their value as in [3]: t_{max} was set to 100 s for small instances with 50 runs of each meta-heuristic on each problem instance, 1,000 s for medium instances with 50 runs of each meta-heuristic on each problem instance and 10,000 s for the large instance with 20 runs of each meta-heuristic on each problem instance. The parameters for all methods were set as follows: the population size N at 50, a crossover probability p_c at 0.6, a mutation probability p_m at 0.5 and size of archive at 20. For two local searches, the maximum number of steps s_{max} was set

Table 1 Characteristics for three classes of MN instance

Class	Small	Medium	Large
Number of events	100	400	400
Number of rooms	5	10	10
Number of features	5	5	10
Number of students	80	200	400
Maximum events per student	20	20	20
Maximum students per event	20	50	100
Approximate features per room	3	3	5
Percentage of the feature used	70	80	90

to 300 for small instances, 1,500 for medium instances, and 2,500 for the large instance, respectively, which is similar to work of Jat and Yang [3]. Last, the termination condition in TS is the maximum number of iterations tss_{max}, setting to 10 for small instances, 20 for medium instances, and 70 for large instance.

4.1 Comparison with Relevant Algorithms

We compare the performance of **HNSGA2LTS** with 2 relevant meta-heuristics on the standard benchmark of MN: **NSGALS** (NSGA-II with LS1) and **NSGA2LS** (NSGA-II with LS1 and LS2). All experiments were run on the same group of computers and used the same set of parameter as mention above, in order to make a fair comparison of the performance between HNSGA2LTS and relevant algorithms.

Table 2 shows the comparison of results for all algorithms in term of the best, average, and standard deviation of three objective values on the 11 problem instances, where **"x%Inf"** means the percentage of runs that failed to find a feasible solution. The best solution for each dataset is highlighted in bold font. From Table 2, it can be seen that the objective function values of HNSGA2LTS on all small problem instances are ties or much smaller than the values of NSGALS and NSGA2LS. On medium instance, it shows that LS2 in NSGA2LS takes the step of reducing violation on the first soft constraint, while integrating two local searches (LS1 and LS2) with Tabu search in the proposed HNSGA2LTS provides better results. Moreover, the proposed HNSGA2LTS was the only one method providing a feasible solution for the large instance. From the results in this table, we can claim plausibly that our HNSGA2LTS can perform better than NSGALS and NSGA2LS in all instances due to assistance of LS2 technique and TS heuristic.

Table 2 Comparison of relevant methods on different problem instances

Dataset	Method	Best			Average			S.D.		
		f1	f2	f3	f1	f2	f3	f1	f2	f3
Small1	NSGALS	**0**	**0**	**0**	0.01	0.02	1.80	0.32	0.31	1.91
	NSGA2LS	**0**	**0**	**0**	0.01	**0.00**	1.76	0.38	0.07	1.84
	HNSGA2LTS	**0**	**0**	**0**	**0.00**	**0.00**	**0.08**	0.00	0.00	0.34
Small2	NSGALS	**0**	**0**	**0**	0.05	0.01	4.08	0.71	0.16	3.00
	NSGA2LS	**0**	**0**	**0**	0.01	0.02	5.11	0.42	0.23	2.82
	HNSGA2LTS	**0**	**0**	**0**	**0.00**	**0.00**	**0.64**	0.00	0.00	0.89
Small3	NSGALS	**0**	**0**	**0**	0.07	0.01	1.97	1.06	0.15	1.96
	NSGA2LS	**0**	**0**	**0**	**0.00**	**0.00**	1.39	0.00	0.00	1.58
	HNSGA2LTS	**0**	**0**	**0**	**0.00**	**0.00**	**0.20**	0.00	0.00	0.53
Small4	NSGALS	**0**	**0**	**0**	0.01	**0.00**	1.22	0.22	0.09	1.56
	NSGA2LS	**0**	**0**	**0**	**0.00**	**0.00**	1.34	0.00	0.00	1.76
	HNSGA2LTS	**0**	**0**	**0**	**0.00**	**0.00**	**0.20**	0.00	0.00	0.45
Small5	NSGALS	0	0	0	0.02	0.00	0.46	0.54	0.09	0.98
	NSGA2LS	0	0	0	0.00	0.00	0.34	0.00	0.00	0.91
	HNSGA2LTS	**0**	**0**	**0**	**0.00**	**0.00**	**0.00**	0.00	0.00	0.00
Medium1	NSGALS	79	96	**0**	138.84	135.86	**5.27**	22.73	15.05	2.20
	NSGA2LS	37	111	**0**	**67.36**	156.41	5.70	14.49	17.52	2.76
	HNSGA2LTS	**30**	**71**	**0**	72.47	**110.90**	5.82	17.89	17.29	2.75
Medium2	NSGALS	82	99	**0**	150.82	132.43	**5.28**	23.42	13.03	2.31
	NSGA2LS	**25**	108	1	**65.83**	156.36	5.90	14.87	14.72	2.42
	HNSGA2LTS	34	**70**	**0**	74.47	**111.49**	5.31	17.54	17.61	2.75
Medium3	NSGALS	126	129	2	199.25	171.19	8.45	25.30	14.85	2.51
	NSGA2LS	**45**	152	1	**96.72**	192.21	8.76	17.83	16.26	2.98
	HNSGA2LTS	48	**90**	**0**	106.62	**135.28**	**6.97**	20.25	16.48	2.71
Medium4	NSGALS	65	96	**0**	122.26	130.86	5.16	22.99	13.86	2.64
	NSGA2LS	20	107	**0**	**53.56**	148.35	**4.94**	13.10	13.55	1.87
	HNSGA2LTS	**16**	**65**	**0**	59.74	**101.28**	5.30	16.24	15.33	2.38
Medium5	NSGALS	112	105	2	188.57	142.72	9.72	29.39	17.08	3.56
	NSGA2LS	63	97	**1**	128.93	155.11	9.36	25.63	23.60	4.15
	HNSGA2LTS	**42**	**59**	**1**	**124.47**	**108.72**	**8.95**	34.96	19.78	3.52
Large1	NSGALS	–	–	–	100 %Inf			–	–	–
	NSGA2LS	–	–	–	100 %Inf			–	–	–
	HNSGA2LTS	425	377	22	587.97	446.32	37.95	69.22	44.87	6.59

4.2 Comparison with Other Algorithm on Multi-objective PECTP

In this section, it provides the comparison between the results of our proposed HNSGA2LTS and the results of Guided search non-dominated sorting genetic algorithm (GSNSGA) by Jat and Yang [3] on the three objective values tested on the 11 MN instances as reported in Table 3. The following items are experimental conditions of HNSGA2LTS and GSNSGA.

1. **HNSGA2LTS** (Hybrid non-dominated sorting genetic algorithm with two local techniques and tabu search): This is our proposed algorithm to solve a three-objective PECTP model for minimizing the soft constraint violation.
2. **GSNSGA** (Guided search non-dominated sorting genetic algorithm): Jat and Yang [3] in 2011 proposed GSNSGA that integrated a guided search technique and two local search techniques into NSGA-II to solve a three-objective PECTP model for minimizing the soft constraint violation. The results were reported out of 20 runs with 100 s per run for small instances, 1,000 s per run for medium instances, and 10,000 s per run for the large instance.

From Table 3, the best solution for each dataset is highlighted in bold font, and it can be seen that the average of the three objective function values of HNSGA2LTS better than the values of GSNSGA on all small instances. Moreover, HNSGA2LTS

Table 3 Comparison of HNSGA2LTS with the relavent algorithm from the literature on the multi-objective PECTP

Dataset	Method	Best			Average			S.D.		
		f1	f2	f3	f1	f2	f3	f1	f2	f3
Small1	HNSGA2LTS	**0**	**0**	**0**	**0.00**	**0.00**	**0.08**	**0.00**	**0.00**	**0.34**
	GSNSGA	0	0	0	1.33	6.74	9.91	0.94	3.90	4.66
Small2	HNSGA2LTS	**0**	**0**	**0**	**0.00**	**0.00**	**0.64**	**0.00**	**0.00**	**0.89**
	GSNSGA	0	0	0	1.63	5.75	5.90	1.35	4.12	3.51
Small3	HNSGA2LTS	**0**	**0**	**0**	**0.00**	**0.00**	**0.20**	**0.00**	**0.00**	**0.53**
	GSNSGA	0	0	0	0.65	2.06	7.38	0.76	2.17	5.51
Small4	HNSGA2LTS	**0**	**0**	**0**	**0.00**	**0.00**	**0.20**	**0.00**	**0.00**	**0.45**
	GSNSGA	0	0	0	1.02	1.14	20.46	0.93	1.48	8.86
Small5	HNSGA2LTS	**0**	**0**	**0**	**0.00**	**0.00**	**0.00**	**0.00**	**0.00**	**0.00**
	GSNSGA	0	0	0	1.52	2.04	15.10	1.52	2.22	6.84
Medium1	HNSGA2LTS	30	**71**	**0**	72.47	**110.90**	**5.82**	17.89	**17.29**	**2.75**
	GSNSGA	**3**	95	15	**8.74**	138.76	32.52	**3.75**	23.25	11.55
Medium2	HNSGA2LTS	34	**70**	**0**	74.47	**111.49**	**5.31**	17.54	**17.61**	**2.75**
	GSNSGA	**7**	94	3	**13.00**	176.60	21.00	**2.73**	24.68	6.73
Medium3	HNSGA2LTS	48	**90**	**0**	106.62	**135.28**	**6.97**	20.25	**16.48**	**2.71**
	GSNSGA	**1**	95	5	**6.90**	145.81	16.69	**2.33**	20.29	7.88
Medium4	HNSGA2LTS	16	65	**0**	59.74	101.28	**5.30**	16.24	**15.33**	**2.38**
	GSNSGA	**0**	**38**	2	**7.15**	**88.81**	22.38	**5.36**	20.29	15.44
Medium5	HNSGA2LTS	42	**59**	**1**	124.47	**108.72**	**8.95**	34.96	**19.78**	**3.52**
	GSNSGA	**5**	94	15	**23.15**	150.40	43.15	**9.72**	25.71	9.72
Large1	HNSGA2LTS	425	377	**22**	587.97	446.32	**37.95**	69.22	**44.87**	**6.59**
	GSNSGA	**30**	**221**	89	**39.72**	**345.94**	124.64	**5.46**	73.62	21.10

can perform better than GSNSGA on the average value of objective function f2 and f3 for dataset medium1–3 and medium5, while GSNSGA gave the better results on the objective function f1 and f3 for dataset medium4 and large. From the results in this section, it is noticeable that the local and tabu search were generally able to help our HNSGA2LTS to find better solutions.

4.3 Comparison with Other Algorithms on Single-Objective PECTP

Finally, we also compare the results of HNSGA2LTS with the other well-known, single-objective results by aggregating three objective function values into one objective function value. Here is the list of all algorithms for this section.

- **HNSGA2LTS**: The hybrid approach proposed in this paper.
- **HGALTS**: Hybrid genetic algorithm with local search and tabu search approach by *Blind Authors* [7] in 2015.
- **MMAS**: Max-min ant system by Socha et al. [8] in 2002.
- **EALS**: Evolutionary algorithm with local search by Rossi-Doria et al. [10] in 2003.
- **TSHH**: Tabu-search hyperheuristic by Burke et al. [14] in 2003.
- **FMHO**: Fuzzy multiple heuristic ordering by Asmuni et al. [15] in 2005.
- **VNS**: Variable neighborhood search by Abdullah et al. [16] in 2005.
- **RII**: Random iterative improvement by Abdullah et al. [17] in 2007.
- **GHH**: Graph-based hyper-heuristic by Burke et al. [18] in 2007.
- **HEA**: Hybrid evolutionary approach by Abdullah et al. [12] in 2007.
- **GAWLS**: Genetic algorithm with local search by Abdullah and Turabieh [19] in 2008.
- **GSGA**: Guided search genetic algorithm by Jat and Yang [20] in 2009.
- **MHSA**: Modified harmony search algorithm by Al-Betar and Khader [21] in 2012.

Table 4 shows the comparison results of HNSGA2LTS with 12 comparator methods in the list above. The numbers in the table represent the number of soft constraint violations, and the best result among all methods for each dataset is highlighted in bold font (lowest is the best). It can be seen from Table 4 that HNSGA2LTS was able to achieve the optimal solution on all small instances. HNSGA2LTS can beat the MMAS, EALS, TSHH, VNS and GAWLS on all medium instances, while it gave the best results on problem instances medium1–4 over all compared algorithms. What is more, the proposed method was able to provide a feasible solution on all problem instances, whereas four comparator methods were unable to find a feasible solution for the large instance. Moreover, comparing with the results of HGALTS, HNSGA2LTS was able to outperform HGALTS on four instances: medium1–4, and it ties HGALTS on all small

Table 4 Comparison of methods on different problem instances

Algorithm	Small1	Small2	Small3	Small4	Small5	Medium1	Medium2	Medium3	Medium4	Medium5	Large1
HNSGA2LTS (Best)	**0**	**0**	**0**	**0**	**0**	**127**	**122**	**172**	**110**	160	904
HGALTS (Best)	0	0	0	0	0	137	132	194	114	160	789
MMAS (Med)	1	3	1	1	0	195	184	248	164.5	219.5	851.5
EALS (Best)	0	3	0	0	0	280	188	249	247	232	100 % Inf
TSHH (Best)	1	2	0	1	0	146	173	267	169	303	80 % Inf
FMHO (Best)	10	9	7	17	7	243	325	249	285	132	1138
VNS (Best)	0	0	0	0	0	317	313	357	247	292	100 % Inf
RII (Best)	0	0	0	0	0	242	161	265	181	151	100 % Inf
GHH (Best)	6	7	3	3	4	372	419	359	348	171	1068
HEA (Best)	0	0	0	0	0	221	147	246	165	135	529
GAWLS (Best)	2	4	2	0	4	254	258	251	321	276	1027
GSGA (Best)	0	0	0	0	0	240	160	242	158	124	801
MHSA (Best)	0	0	0	0	0	168	160	176	144	**71**	**417**

instances and medium5. From the results, it can be seen that appropriate hybridization of LS techniques and TS heuristic can bring about the better results for solving the PECTP compared to other algorithms from the literature.

5 Conclusion and Future Work

This paper presents a hybrid approach, which combines a NSGA-II with two LS techniques and a TS heuristic, denoted HNSGA2LTS, to solve the multi-objective PECTP. The HNSGA2LTS was tested its performance on a set of benchmark, called MN instances. Experimental results indicated that HNSGA2LTS was able to produce promising results on all problem instances. When NSGA-II was integrated with the new LS (LS2) and TS then it gave improved results on this benchmark set, and the improvement was noticeable comparing with other algorithms on both multi-objective and single objective approaches. The two reasons why new LS and TS can bring about improved results are as following. The new LS takes a role to improve the quality of the solution by reducing the first soft constraint violation after applying the first LS (LS1), while TS is responsible for improving the quality of solutions after LS phrases as it can escape from the local optima. Moreover, when solving PECTP using multi-objective approach, the final result is a set of non-dominated solutions, and this give the opportunity for users to select the most preferable solution from the set of non-dominated solutions rather than limiting to a final single solution as in single objective approach. This shows that hybridization between explorative search ability of NSGA-II and exploitative search abilities of new LS and TS has the potential to produce improved results for the multi-objective PECTP on MN instances.

References

1. Even, S., Itai, A., Shamir, A.: On the complexity of time table and multi-commodity flow problems. SIAM J. Comput. **5**, 691–703 (1976)
2. Schaerf, A.: A survey of automated timetabling. Artif. Intell. Rev. **13**(2), 87–127 (1999)
3. Jat, S.N., Yang, S.: A guided search non-dominated sorting genetic algorithm for the multi-objective university course timetabling problem. In: Evolutionary Computation in Combinatorial Optimization, pp. 1–13 (2011)
4. Burke, E.K., Silva, J.D.L., Soubeiga, E.: Multi-objective hyper-heuristic approaches for space allocation and timetabling. In: Metaheuristics: Progress as Real Problem Solvers, vol. 32, pp. 129–158 (2005)
5. Datta, D., Deb, K., Fonseca, C.M.: Multi-objective evolutionary algorithm for university class timetabling problem. In: Evolutionary scheduling, pp. 197–236. Springer (2007)
6. Abdullah, S., Turabieh, H., McCollum, B., McMullan, P.: A multi-objective post enrolment course timetabling problems: a new case study. In: 2010 IEEE Congress on Evolutionary Computation (CEC), pp. 1–7. IEEE (2010)

7. Jaengchuea, S., Lohpetch, D.: A hybrid genetic algorithm with local search and tabu search approaches for solving the post enrolment based course timetabling problem: outperforming guided search genetic algorithm. In: 2015 7th International Conference on Information Technology and Electrical Engineering (ICITEE 2015), Chiang Mai, Thailand, pp. 29–34 (2015)
8. Socha, K., Knowles, J., Sampels, M.: A max-min ant system for the university course timetabling problem. In: Ant Algorithms, pp. 1–13 (2002)
9. Deb, K., Agrawal, S., Pratap, A., Meyarivan, T.: A fast elitist non-dominated sorting genetic algorithm for multi-objective optimization: NSGA-II. Lecture Notes in Computer Science, vol. 1917, pp. 849–858 (2000)
10. Rossi-Doria, O., Sampels, M., Birattari, M., Chiarandini, M., Dorigo, M., Gambardella, L.M., Knowles, J., Manfrin, M., Mastrolilli, M., Paechter, B.: A comparison of the performance of different metaheuristics on the timetabling problem. Pract. Theor. Autom. Timetabling IV. **2740**, 329–351 (2003)
11. Jat, S.N., Yang, S.: A hybrid genetic algorithm and tabu search approach for post enrolment course timetabling. J. Sched. **14**, 617–637 (2011)
12. Abdullah, S., Burke, E.K., McCollum, B.: A hybrid evolutionary approach to the university course timetabling problem. In: IEEE Congress on Evolutionary Computation, pp. 1764–1768. IEEE (2007)
13. Socha, K.: Metaheuristics for the Timetabling Problem. Facult́e des Sciences Appliques for the Diplome d'Etudies Approfondies (DEA), vol. Ph.D. thesis. Universit́e Libre de Bruxelles (2003)
14. Burke, E.K., Kendall, G., Soubeiga, E.: A tabu-search hyperheuristic for timetabling and rostering. J. Heuristics **9**, 451–470 (2003)
15. Asmuni, H., Burke, E.K., Garibaldi, J.M.: Fuzzy multiple heuristic ordering for course timetabling. In: Proceedings of the 5th United Kingdom Workshop on Computational Intelligence (UKCI 2005), pp. 302–309. Citeseer (2005)
16. Abdullah, S., Burke, E.K., Mccollum, B.: An investigation of variable neighbourhood search for university course timetabling. In: The 2nd Multidisciplinary International Conference on Scheduling: Theory and Applications (MISTA), pp. 413–427 (2005)
17. Abdullah, S., Burke, E.K., McCollum, B.: Using a randomised iterative improvement algorithm with composite neighbourhood structures for the university course timetabling problem. Metaheuristics, 153–169 (2007)
18. Burke, E.K., McCollum, B., Meisels, A., Petrovic, S., Qu, R.: A graph-based hyper-heuristic for educational timetabling problems. Eur. J. Oper. Res. **176**, 177–192 (2007)
19. Abdullah, S., Turabieh, H.: Generating university course timetable using genetic algorithms and local search. In: Third International Conference on Convergence and Hybrid Information Technology, vol. 1, pp. 254–260. IEEE (2008)
20. Jat, S.N., Yang, S.: A guided search genetic algorithm for the university course timetabling problem. In: The 4th Multidisciplinary International Scheduling Conference: Theory and Applications (MISTA 2009), Dublin, Ireland, pp. 180–191 (2009)
21. Al-Betar, M.A., Khader, A.T.: A harmony search algorithm for university course timetabling. Ann. Oper. Res. **194**, 3–31 (2012)

Part IV
Optimisation of Complex Networks

Conceptual Framework: The Adaptive Biometrics Authentication for Accessing Cloud Computing Services Using iPhone

Sorapak Pukdesree and Paniti Netinant

Abstract Presently, cloud computing technology is widely applied in many domains. The trends of cloud computing are forecasted to be worth around 250 billion dollars in 2017. In Thailand, cloud computing situations also correspondence to other countries. The survey report from Forester in 2012, showed that growth rate of utilization of cloud computing trended to increase continuously. Thai government also foresees the importance of cloud computing technology to support digital economy campaign by setting up Government Cloud Service project. However there are some issues which to be concerned when deploy cloud computing especially security on cloud computing. Currently, mobile devices are very more widely used in our society as we have seen the large number of delivered mobile devices in each year. Therefore we proposed adaptive biometric authentication for accessing cloud computing services using iPhone. The adaptive biometric authentication, which has advantages of fast, flexible and more secure, uses sequential authentications instead of multi-modal approaches. The current status of this research is on developing phase.

Keywords Biometric authentication · Cloud computing · iPhone · Security · iOS

S. Pukdesree (✉)
School of Science and Technology, Bangkok University,
Bangkok, Thailand
e-mail: sorapuck.p@bu.ac.th

S. Pukdesree · P. Netinant
College of Information and Communication Technology,
Rangsit University, Pathum Thani, Thailand
e-mail: netinant1@gmail.com

P. Meesad et al. (eds.), *Recent Advances in Information and Communication Technology 2016*, Advances in Intelligent Systems and Computing 463,
DOI 10.1007/978-3-319-40415-8_20

1 Introduction

Presently, cloud computing technology is widely applied in many domains. The trends of cloud computing are forecasted to be worth around 250 billion dollars in 2017 [1]. In 2016, one-fourth of typical Software can be accessed as services of cloud computing (SaaS) which would be worth from 18.2 billion dollars to 45.6 billion dollars in 2017. Furthermore, there was a forecast that enterprise businesses will spend more than 10 % of the budgets on cloud computing services [2].

In Thailand, cloud computing situations also correspondence to other countries. The survey report [3] showed that growth rate of utilization of cloud computing trends to increase continuously. Approximately 83 % of survey correspondences have already planned or adapted to use cloud computing technology. The correspondences have agreed that cloud computing will help the organizations to have more potential than competitors. Approximately 68 % of survey correspondences have also agreed that if the organizations would not use cloud computing, the organizations would not work as fast as competitors.

Thai government also foresees the importance of cloud computing technology to support digital economy campaign by setting up Government Cloud Service project named G-Cloud under controlling of Electronic Government Agency [4]. The survey report of IT market in 2013 showed that the growth rate was 15.6 % with worth 293,239 million baht. If we consider in more details, we found that the all of IT budget of Thai government was around 20.2 % with worth 59,818 million baht. Therefore, Thai government would like to reduce and duplicate of overall IT budgets but still has effectiveness to manage IT system for all sectors of government.

Currently, mobile devices are very more widely used in our society. The IDC reported of second quarter of 2015 [5, 6], there were 341.5 million smart phones delivered to end users, which was growth around 13 %. Samsung was the first rank of the market share with 21.4 % and Apple was the second rank with 13.9 %. The report also showed that the growth rate of smart phones trended to increase continuously according to 4G telecommunication network.

There are several information systems available in our diary life which typically use knowledge-based authentication such as password, therefore users have to remember all passwords which are difficult to remember all of those. If users write passwords on papers or save on to smart phones or computers which may be stolen or hacked by hackers. Everyone would get risk when using network or internet including cloud computing. One of the most important issues of cloud computing is security. Therefore researcher would like to propose adaptive biometric authentication for accessing services on cloud computing using smart phones. The proposed adaptive biometric authentication uses sequence authentications instead of multi-modal approaches which has advantages such as fast, flexible and security level. The current status of this research is on developing phase.

2 Related Work

Atmakuri [7] had proposed new biometric authentication framework using face authentication which should be better than previous. The main research objective was to study current several biometric authentication techniques. The researcher suggested that selection of biometric authentication technique should consider the available resources within mobile devices. The researcher also compared the proposed framework and other techniques by considering accuracy, cost, long term stability and security level.

Rotika [8] had studied and compared totally 13 biometric authentication techniques, then researcher had assessed all of these techniques by using seven factors proposed by Jain et al. [9] including universality, uniqueness, performance, collectability, permanence, acceptability and circumvention. Furthermore researcher suggested that using only one biometric authentication might not enough for biometric authentication because each biometric authentication has difference either advantages or disadvantages. And also each of biometric sensors may have difference approach to detect the biometric features. Therefore researcher had suggested using more than one biometric authentication which can improve security level. This research used biometric authentication using both face and hand authentications. The results showed that this approach had 99.82 % of accuracy. The researcher used six assessment factors including enrollment, renewal, machine requirements and public perception.

Romo [10] had proposed framework for biometric authentication on mobile devices named physical smart key which composed of five components including RFID manager, encryption layer, policy manager, hardware key and device id. The researcher represented the results of this study with 10 biometric authentication techniques. The researcher had assessed these techniques by using 3 factors including usability, deploy ability and security. The results of this research showed that the proposed framework had highest score with 84 points. However the researcher also mention about the proposed framework that if user had loss the physical smart key or the mobile devices was stolen, therefore the devices can also be accessed by other people.

Ahn et al. [11] had suggested that authentication was very importance for accessing cloud computing. They also had suggested that there still have problems for users to enter identity repetitiously for accessing cloud computing that are waste time and inconvenience for users. Therefore researchers had proposed framework which used concept of provisioning in cloud computing environment for authentication. This framework would analyze user information and authentication using user profile, therefore this could be solved the inconvenience way of authentication repetitiously. However this approach had some security issues due to storing user profile on the server, user profile would get in risk.

Sharmaand Mittal [12] had mentioned that cloud computing was the results of the advancement of the current technology. It was a combination of automated computing, client-server model, grid computing, mainframe computer, utility-computing peer-to-peer, and game on cloud. Cloud computing technology has several advantages for either personnel use or organizations, however there still has some issues especially security on cloud computing utilization. The researchers had compared several authentication techniques for verification of cloud users and cloud broker. The results of this research showed that the three appropriate authentication techniques were authentication using Kerberos, authentication using key distribution center and authentication using public key infrastructure.

Peer and Bule [13] had some opinions which the amounts of biometric information from several organizations are increasing rapidly; therefore the organizations have to plan and analyze either the capacity of the storage or the capacity of computation. The researcher has suggested that cloud computing technology could be utilized the solve problems of capacity of storage and computation dynamically. However using cloud computing technology should be concerned of problems and obstructions of changing technology, standards of cloud computing and the available approaches to solve the problems. The research was implemented by using finger printing biometric authentication for accessing learning management system on cloud computing.

Fathima et al. [14] have mentioned that biometric authentication systems have obtained more attention due to the security issues. The main objective of this study is to develop a multi-modal, multi-sensor based person authentication system (PAS) using joint directors of laboratories (JDL) fusion model. This study also explores the need for multi sensors, multiple recognition algorithms and multiple fusion level by using face, fingerprint and iris biometrics. The researchers have showed the advantage of using multiple modalities for authentication with using decision fusion scheme. The results of this research showed that the proposed multisensory PAS overcomes the weaknesses of each individual sensor and gives better detection rate.

From the literature reviews, we have found that security issue is one of an importance issue which should be concerned carefully. Even though cloud computing provides easy and convenience way to access from internet, the unauthorized people may also access to cloud computing which may cause of the unpredictable damages. Furthermore, accessing to cloud computing by using mobile devices is increase dramatically which also may cause of unauthorized accessing. Therefore authentication is one of appropriate approaches to prevent unauthorized accessing for cloud computing. The results of the literatures were the study or comparison of several authentication techniques which were to find the appropriate or the best approaches to solve the security issues. The authentications include knowledge-based authentication, device/token-based authentication or biometric authentication. Besides the study of Rotika was found that there still does not have a single perfectly approach of authentications, therefore composing of multiple authentications at same time as a hybrid should be used to improve security level of authentication.

3 Proposed Framework

The proposed of conceptual framework composes of three layers including client layer, application layer and database layer as Fig. 1. Client layer is an application which runs on client smart phones. Clients can use this application for either enrollment or verification. Clients will provide information such as account name, password, email, scanned face, voice and etc. The application will also display the result of authentication. Application layer is some applications which run on cloud computing platform. These applications will act as biometric authentication applications which were used in this research. For example, face authentication processes compose of face preprocessing, feature extractions, template creation, learning of face and face recognition. Database layer is database management system which also runs on cloud computing platform. DBMS is used to manage biometric information including add, delete, update or retrieval. For security enhancement, some critical information will be encrypted stored in database system.

This research puts either application layer or database layer on cloud computing because there are several advantages of cloud computing such as on-demand self-service, broad network access, resource pooling, rapid elasticity and measured service as described in Importance Characteristics of cloud computing by NIST [15]. The adaptive biometrics authentication approach is sequential authentications

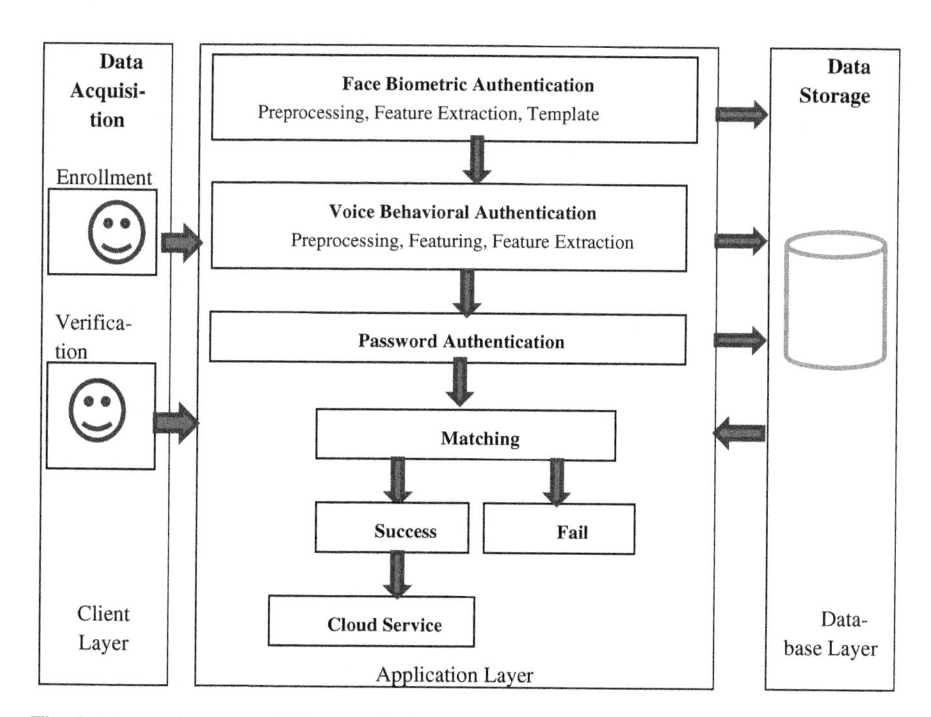

Fig. 1 Proposed conceptual framework of this research

which also uses more than one authentication which will provides fast, flexible and secure authentication. Authentication administrators can manage the authentication techniques from knowledge-based such as password, token-based such as magnetic-striped card or biometric such as face or voice. Authentication level may be one, two or more sequence of authentications. Authentication processes will work on cloud computing which provides not only flexibility for managing and extendibility of capacity of storage and computation.

4 Research Methodology

This research methodology composes of two major parts as following subsections. The first part is the research for develop the biometric authentication system which composes of client side application, applications on server side and database on another database server.

4.1 Development Research

The development research is the systematically processes to develop the adaptive biometric system by finding and discovering related information. There are three parts to be developed as described in conceptual framework including client side, server side and database management system. The mobile application will be developed on client side. The server side will be developed some applications for adaptive biometric authentications which work on cloud computing. Finally DBMS will be a open source DBMS will also work on cloud computing.
4.1.1 Populations are related documents or information.
4.1.2 Research tools are smart phones which using iOS or android operating system, internal or external cloud computing and developing tools.
4.1.3 Data will be collected by researchers by developing the proposed biometric authentication system.
4.1.4 Data analysis will use fundamental statistics such as frequently or mean for example.

4.2 Survey Research

This part aims to gather satisfactions from users who use the system by using online survey questionnaires which compose of four sections including general information, user interface and design, system performance, and usability.
4.2.1 Populations are Thai people. This research use simple sampling method which is around 100 students from the university.

4.2.2 Research tools are smart phones which using iOS or android operating system, internal or external cloud computing and developing tools.
4.2.3 Data will be collected by researchers by using survey questionnaires.
4.2.4 Data analysis will use fundamental statistics such as frequently or mean for example.

5 Conclusion

Security is one of the most importance issues which have to be concerned when deploying cloud computing technology. Therefore authentication should be used to verify the authorized users when connecting via internet. There are three majors authentication including knowledge-based, token-based and biometric authentication. Each of them may have some advantages or some drawbacks depending security level to be used. By using a single biometric authentication provides fast and easy approach, however this may not provide enough security level in this day. Therefore multi-modal is proposed to improve higher security level; however this approach is more complicate, slower and higher costs [14, 16]. Furthermore, there may some limitation of available resources or sensors on iPhone. Therefore we would like to propose adaptive biometric authentication which will provide fast, flexible and high security level. This research has two majors including development research and survey research.

References

1. Forecast: Public Cloud Services, Worldwide, 2011–2017, 4Q13. Gartner (2013)
2. Enterprise Cloud Adoption Survey 2014 Everest Global Inc., Everest Group (2014)
3. VMWare Cloud Index. Forester. http://www.vmware.com/ap/cloudindex
4. Government Cloud Service. https://www.ega.or.th/th/content/890/273/
5. Worldwide Smartphone Growth Forecast to Slow from a Boil to a Simmer as Prices Drop and Markets Mature, According to IDC. IDC (2015). http://www.idc.com/getdoc.jsp?containerId=prUS25282214
6. Smartphone Vendor Market Share, 2015 Q2. IDC. http://www.idc.com/prodserv/smartphone-market-share.jsp
7. Atmakuri, S.M.: A study of authentication techniques for mobile cloud computing. In: ProQuest LLC (2015)
8. Rokita, J.: Multimodal biometric system based on face and hand images taken by a cell phone. ProQuest Dissertations and Theses (2008)
9. Jain, A.K., Hong, L., Pankanti, S., Bolle, R.: An identity-authentication system using fingerprints. Proc. IEEE **85**(9), 1365–1388 (1997)
10. Romo, J.T.: Towards seamless and secure mobile authentication. In: ProQuest LLC (2014)
11. Ahn, H., Chang, H., Jang, C., Choi, E.: User authentication platform using provisioning in cloud computing environment. In: Advanced Communication and Networking. Communications in Computer and Information Science, vol. 199, pp. 132–138 (2014)

12. Sharma, S., Mittal, U.: Comparative analysis of various authentication techniques on cloud computing. Int. J. Innov. Res. Sci. Eng. Technol. **2**(4) (2013)
13. Peer, P., Bule, J., Gros, J.Ž., Štruc, V.: Building cloud-based biometric services. Informatica **37**, 115–122 (2013)
14. Fathima, A.A., Vasuhi, S., Babu, N.T.N., Vaidehi, V., Treesasa, T.M.: Fusion framework for multimodal biometric person authentication system. IAENG Int. J. Comput. Sci. **41**(1), 18–31 (2014)
15. Mell, P., Grance, T.: The NIST Definition of Cloud Computing. National Institute of Standards and Technology Special Publication 800-145. U.S. Department of Commerce (2011)
16. Ching, H.C.: Performance evaluation of multimodal biometric systems using fusion techniques. In: ProQuest LLC (2015)

An Insurmountable and Fail-Secure Network Interface

Wolfgang A. Halang, Panchalee Sukjit, Maytiyanin Komkhao and Sunantha Sodsee

Abstract The security risks of network connectivity include malware intrusion into servers and sabotage such as denial-of-service attacks. It will be shown here how these security risks can, in contrast to employing firewalls, effectively be coped with by an adequate network interface whose architecture combines two malware-proof programmable controllers with specific software functions. By design and physical separation, the interface behaves in a fail-secure way, i.e. there is never open access from a network to a computer even if the interface fails itself.

Keywords Network interfacing · Malware prevention · Air gapping · Hardware-based security · Fail-secure behaviour · Security by design

1 Introduction

The main security threats to data communication over networks are malware intrusion, denial-of-service attacks and eavesdropping. As the latter problem can quite effectively be coped with by end-to-end encryption, particularly utilising one-time keys [2], it is not the topic of this contribution. So-called firewalls are the means predominantly employed in trying to prevent malware intrusion. Their software-

W.A. Halang (✉) · P. Sukjit
Chair of Computer Engineering, Fernuniversität in Hagen, Hagen, Germany
e-mail: wolfgang.halang@fernuni-hagen.de

P. Sukjit
e-mail: panchalee.sukjit@fernuni-hagen.de

M. Komkhao
Faculty of Science and Technology, Rajamangala University of Technology Phra Nakhon, Bangkok, Thailand
e-mail: maytiyanin.k@rmutp.ac.th

S. Sodsee
Faculty of Information Technology, King Mongkut's University of Technology North Bangkok, Bangkok, Thailand
e-mail: sunanthas@kmutnb.ac.th

P. Meesad et al. (eds.), *Recent Advances in Information and Communication Technology 2016*, Advances in Intelligent Systems and Computing 463,
DOI 10.1007/978-3-319-40415-8_21

based working principle is to search for malware code in the data packets passing them. Needless to say, they can only detect the code of already known malware, and must necessarily fail totally with regard to new forms of malware. Furthermore, this approach is woefully inefficient, since the databases of known malware must be updated permanently, and since the size of these databases grows exponentially resulting in ever increasing search effort. Denial-of-service attacks, finally, exploit badly programmed software and inadequate hardware configurations to be successful.

It is the purpose of this paper to show that the two security problems malware intrusion and denial-of-service attacks can effectively be solved with adequately designed network interfaces whose architecture combines both hardware devices and software features. Such interfaces do not necessarily require databases of known malware, and their processing effort depends on the functionality of the computers or servers they interface to networks, but not on the amount of malware being around. Denial-of-service attacks are counteracted by preventing to overload the respective capacities of the computers interfaced to networks. The decisive advantage of the architecture presented here is that accordingly constructed interfaces behave in a fail-secure way, i.e. there is never open access from a network to a computer even if the interface between them fails itself.

In the next section we shall review the state of technology of network interfaces physically separating computers from networks. Based on this, a novel interface architecture will be presented, which combines insurmountability for malware and denial-of-service attacks with fail-secure behaviour. The two programmable controllers as envisaged in this architecture are, by design, immune to malware. Before a conclusion on this paper's results is drawn, the software functionality is summarised that these two controllers should provide.

2 Interfaces Based on Air Gapping

As *the ultimate method to secure a network is to disconnect it* [3], it makes no sense to be content with deficient means such as firewalls. In 1998 an application for a German patent was filed, and corresponding patents were granted by several national patent offices in later years, for a design physically separating computers and network segments from each other [1]. In other words, this invention realises the concept of so-called air gapping. As shown in Fig. 1, the design mimics the function of locks as found in rivers and canals. Alternatingly, the interface, called lock-keeper, is connected via the first lock gate with the external network, or via the second lock gate with an internal computer or network segment. The lock gates are realised as switches. At any time not more than one switch may be closed in order to guarantee that there never exists a direct link between the communicating computers or network segments, respectively. An interfacing system implementing this concept is commercially available [3]. It is realised in hardware and provides the possibility to subject the data packets passing through to certain checks.

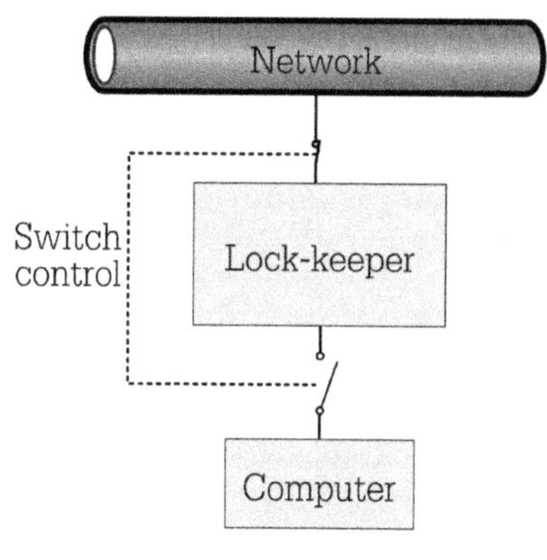

Fig. 1 Operation principle according to [1]

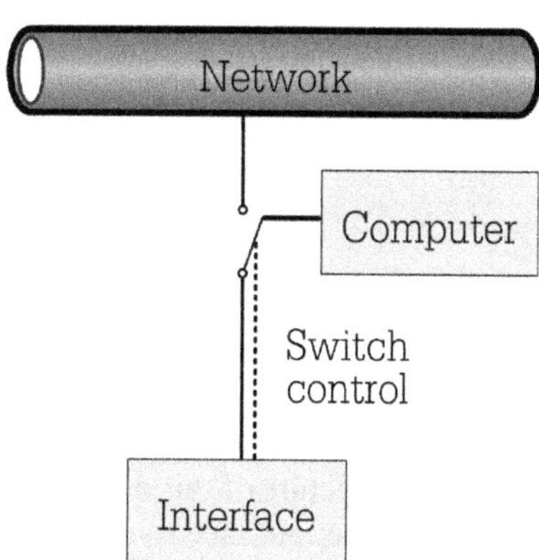

Fig. 2 Operation principle according to [4]

In 1992, i.e. already six years earlier, a more refined locking system based on the air-gap principle was designed and prototypically built in the course of a bachelor thesis supervised by this paper's first author [4]. As depicted in Fig. 2, instead of two switches it just needs a single toggle. Thus, there can never be a direct physical connection between the computer and the network segment interfaced—even if the toggle is incorrectly set by a failure condition. Thus, the locking system is fail-secure in contrast to the later design mentioned above, where a failure might cause both switches to be closed at the same time.

3 Architecture of an Insurmountable Network Interface

Based on this fail-secure realisation of the air-gap principle, we now design an insurmountable network interface to be attachable to any kind of computer or server, which incorporates the functionality of network controllers, and which is intended to replace them in order to enhance communication security.

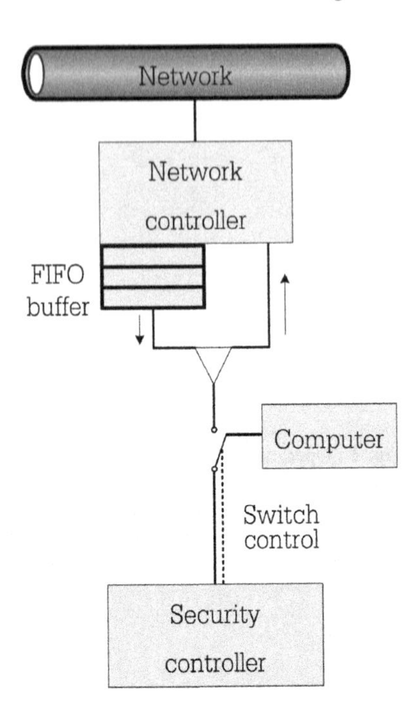

Fig. 3 Architecture of the network interface

The architecture of this device is schematically depicted in Fig. 3. The component connecting the interface to a network, which will be an Ethernet in most cases, is a classical network controller. This, in turn, connects via a toggle with a security controller. As this connection is not permanent, but the network controller may receive at unpredictable instants and possibly at a high rate data packets from the network, which may need to be forwarded to the computer interfaced, a first-in-first-out buffer of appropriate capacity for incoming data packets is placed between network controller and toggle. Functioning as described above, the latter is the decisive element in this design, ensuring in a fail-secure way that at no time there will be an immediate physical connection between network and computer. Incoming and outgoing data packets move via the toggle either between network and security controller or between security controller and computer.

The functions executed by both controllers are outlined in the next section. To allow for a wider range of functionality than the pure hardware solution [3], particularly the security controller is realised in form of a programmable device. This does not mean, however, that its software is stored in a read-write random access memory and, thus, becomes susceptible to malware. To render any known or still unknown malware ineffective, the controller's memory is, therefore, partitioned according to the Harvard architecture in separate program and data segments, the program segment is implemented as read-only memory, and it is architecturally safeguarded that instructions can be read from the program segment, only.

4 Controller Functionality

Being programmable devices, the two controllers of the network interface can be charged with a large variety of indispensable processing and security functions, but also with optional ones both relieving the computer interfaced from routine work, but also making up for deficiencies of the computer's operating system. Such functions are mentioned in this section in a non-exhaustive way.

First of all, the network controller must carry out the classical functions determined by its direct connection with a network, viz. receiving and transmitting data packets, performing the functions of the employed communication protocol's lower levels, and notifying the security controller of data packets received. Examples for further functions the network controller could optionally be charged with are discarding data packets arriving either from black-listed senders or from senders not white-listed. Thus, a great amount of unwanted data can be removed directly upon arrival, and the impact of denial-of-service attacks can already be reduced to a large extent.

Besides controlling the operation of the toggle, the security controller can be charged with any kind of access control tasks. Particularly when interfacing to a server, it may check in an application-specific way, if the content of arriving packets adheres to the data formats expected by the services provided. It is also possible to filter out data packets that cannot be properly authenticated. Many web services allocate a new data record for any open request while it is being processed. This is a point of attack for denial-of-service attacks, which try to place within short times more requests than servers can handle. Here the security controller can help by preventing a server to be overloaded. To this end, it needs to keep track of open requests and to discard requests when the server's capacity is exhausted.

5 Conclusion

The basic idea of the technical security measure described in this paper is by no means new, but it is consistently being disregarded for decades by mainstream academic computer science as well as by IT industry. Actually, when the thesis [4] was written in 1992, the author did not consider the approach to be really innovative—otherwise he would have patented it. Instead, he just built a prototype realising a straightforward concept. Thus, 24 or 18 (with regard to [3]) years have passed, during which the ignorance of the professional circles persisted. The reasons for this disregard are up to speculation. Maybe, it is just the inability to take notice of published results. But for academic circles it could also be the unwillingness to recognise certain problems as ultimately solved preventing further research with its possibility to publish more papers. The branch of the IT industry supplying malware detection and other security software would simply lose its complete business. The suppliers of computers want to sell new computers before the old ones cease functioning, and the customers of operating systems, which as software never wear out, need to be convinced by threats, errors and new features to buy new operating system versions. Hence, of each generation of new hardware and operating systems not more than only slight improvements with respect to security can be expected, in order to allow for many more generations to be marketed. And, finally, we should not forget those who are interested in spying out personal data and in invading and controlling our computers such as secret services, governments and big corporations.

References

1. Engel, T., Haffner, E.-G., Meinel, C.: Datenverbindung zwischen zwei Rechnern und Verfahren zur Datenübertragung zwischen zwei Rechnern (Data link between two computers and method for transmitting data between said computers). German Patent 198, 38, 253 (1998)
2. Halang, W.A., Komkhao, M., Sodsee, S.: Secure cloud computing. In: Boonkrong, S., Unger, H., Meesad, P. (eds.) IC2IT 2014. Advances in Intelligent Systems and Computing, vol. 265, pp. 305–314. Springer, Heidelberg (2014)
3. Hasso-Plattner-Institut. http://www.lock-keeper.org
4. Witte, M.: Eine virenresistente Netzschnittstelle (A Virus-resistent Network Interface). B.Eng. thesis. Bochum University of Applied Science (1992)

Finding an Optimal Parameter for Threshold Cryptography

Weena Janratchakool, Sirapat Boonkrong and Sucha Smanchat

Abstract Threshold cryptography is mostly used in securing the keys, which are the most important element in cryptographic systems. Threshold scheme protects the key by dividing and distributing the key among the entities in the system instead of storing it on a single server. In order to generate the key again, not all the shares are needed. However, there has been no research attempting to find a suitable threshold for any number of shares distributed. The objective of this paper is, therefore, to design and implement a guideline that can assist in choosing such value. The experiment was setup to collecting and observing the time used in distribution and reconstruction process to optimize proper threshold value.

Keywords Threshold cryptography · Distribution time · Reconstruction time

1 Introduction

Cryptography is a mathematical based tool used for providing security communication among the entities over the network. With technology's today and abilities of adversaries, cryptography has especially become a necessary tool to protect the privacy and confidentiality of data.

The process of cryptography is most associated with converting clear text into cipher text known as encryption and then back again called decryption. Cryptography can be categorized into two different types. The first is known as symmetric cryptography where the same key is used for both encryption and decryption.

W. Janratchakool (✉) · S. Boonkrong · S. Smanchat
Faculty of Information Technology, King Mongkut's University of Technology North Bangkok, Bangkok, Thailand
e-mail: yjijy@hotmail.com

S. Boonkrong
e-mail: sirapat_b@it.kmutnb.ac.th

S. Smanchat
e-mail: sucha.s@it.kmutnb.ac.th

P. Meesad et al. (eds.), *Recent Advances in Information and Communication Technology 2016*, Advances in Intelligent Systems and Computing 463,
DOI 10.1007/978-3-319-40415-8_22

The second is known asymmetric cryptography where a public key is used to encrypt data and private key is used to decrypt the cipher text. This is the type, specifically RSA [1] that all experiments in this paper will focus on.

The most important component is transforming text to cipher text and back is the key. This is agreed by Kerkhoff [2] whose principle states that, "a cryptosystem should be secure even if everything about the system is publicly known, except the key." This has put into context that the key is the foundation of the overall security of cryptography, which means that the protection of the key has also become an important issue. The original concept to secure a shared key is introduced by Shamir [3] and Blakley [4] using polynomial interpolation and hyperplane geometry respectively. These schemes have been used as blueprint of many key protection researches. For example, [5] introduced a method for using a monotone function to make a lightweight and simpler scheme. Moreover, [6, 7] provided a new paradigm to recombine keys which also raised an issue in key protection.

One of the methods that can reduce the risk of the key being compromised is threshold cryptography [8, 9]. The basic idea of threshold cryptography is that the key is divided into n shares before being distributed to the involved entities. In order to generate the key again, not all the shares are needed. Instead, an entity can combine only k shares (known as the threshold value) to reconstruct the key. In other word, even though the key is divided into n shares, only k out of n shares is needed to reconstruct the key.

This threshold cryptography scheme is an advance step to securing the key and to preventing the key from being compromised. This is because an adversary will need to attack k entities in order to obtain k shares to generate the key, rather than compromising one entity to obtain the key. This makes it more difficult for an attacker. As a result, threshold cryptography has been used in many applications such as in mobile ad hoc networks [10, 11] and cloud computing [12, 13].

However, there has been no research attempting to find a suitable threshold for any number of shares distributed. The objective of this paper is, therefore, to design and implement a method that can assist in finding such value.

The rest of the paper is organized as follows. Section 2 provides background knowledge on threshold cryptography as well as the work related to it. The experimental designs are explained in Sect. 3. The results of the experiments and discussion are given in Sect. 4. Section 5 provides an attempt to find a suitable threshold. Section 6 then concludes the paper.

2 Background Knowledge and Related Work

This section provides some background knowledge on threshold cryptography as well as the existing work related to a method for finding a threshold.

2.1 Overview of Threshold Cryptography

Threshold cryptography is a secret key sharing scheme for public key cryptography, denoted as (*n, k*)-threshold, introduced by Shamir [3]. It works by dividing the private key into sub *n* keys or shares according to number of entities. The combination of partial sub *k* keys distributed among them can be reconstructed to form the original private key. In other words, only a few subkeys or shares can be use to re-generate the entire private key.

Generally, the threshold cryptography scheme is divided in two phases: shared-key distribution and shared-key reconstruction.

2.1.1 Distribution Phase

This phase begins with a private key of any asymmetric cryptography algorithm. The private key will then be split into *n* shares, according to the number of parties involved in the system. This is done by using threshold cryptography polynomial algorithm. The shares are then distributed to each party. The process can be summarized as follows.

- Step 1: The asymmetric cryptography key pair (private and public key) is calculated.
- Step 2: The private key is split into (*n, k*)-threshold scheme based on a random polynomial of degree $k - 1$, where the coefficient a_0 is private key as in

$$f(x) = a_o + a_1 x + \cdots + a_{k-1} x^{k-1} \quad (1)$$

- Step 3: The new shared secrets are distributed to each entity.

2.1.2 Reconstruction Phase

If the private key needs to be reconstructed, *k* of *n* shares would be needed. Here, *k* is known as a threshold value or number of shares needed to reconstruct the private key. The reconstruction process is done by applying the Lagrange interpolation [14], shown in Eq. 2.

$$f(x) = \sum_{i=1}^{k} y_i \left(\prod_{1 \le j \le k, j \ne i} \frac{x - x_j}{x_i - x_j} \right) \quad (2)$$

On the whole, threshold cryptography will be applied on the private key. This means that the private key will be divided into shares and distributed to the same number of entities. Only some of those shares will then be combined so that the private key can be reconstructed. How many shares are needed will be answered in the next sections.

2.2 Related Work

Change point detection [15, 16] is a statistic analysis tool, which able to find a point when the change happened. Firstly, the data is collected and then transferred to a graph. Finally, a change point is operated based on the analysis of all the collected data. This is one of methods prefers to find a threshold value. However, this method is suitable for dealing with large data sets when needed result is closely to be real. This is definitely that this method is inappropriate with this circumstance.

It appears that the existing methods for finding a threshold value are not suitable and have not been extensively studied. It is, therefore, of our interest in trying to find such method that will be more suitable to threshold cryptography.

In order for threshold cryptography to perform efficiently in terms of speed, a suitable threshold must be found. We will attempt to provide a guideline in choosing the suitable threshold value in the next sections.

3 Experimental Designs

This section designs an experiment to that a suitable threshold value for any number of shares can be found. As stated earlier, the paper focuses on applying threshold cryptography to asymmetric cryptography, specifically the private key of the RSA algorithm. In other words, the private key will be split into shares, distributed and then reconstructed. The distribution and reconstruction processes will be analyzed closely in this paper so that a suitable threshold value can be found.

3.1 Distribution Phase

The experiment is designed for observing, collecting and analyzing the distribution time by varying the threshold value, k, while the number of parties, n, is constant.

Until now, 1024-bit private keys of RSA have been used in common. However, the policy of using key size is changed to 2048 bits, according to the recommendation [17] by NIST through the year 2030. Thus, the key size used in the experiment will be only the NIST recommended 2048-bit private keys.

The experiments in the distribution phase were carried by splitting the private key into n shares, where n was 256 shares, 128 shares and 64 shares. The time taken to distribute the shares was then recorded.

3.2 Reconstruction Phase

The experiments in the reconstruction phase were carried out as follows. First, after the private key was split into 256 shares, 128 shares and 64 shares in the previous section, the threshold, k, was chosen to begin with the value of 8. It was then incremented by 8 up to the maximum value of n or the number of shares. For example, if n were 256, the values of k used in the experiment would be 8, 16, 24, 32…256. The time taken to reconstruct the private key based on the different threshold values was then recorded.

3.3 Experimental Setup

The RSA key generation and all the threshold cryptography processes were implemented using JAVA via NetBeanIDE 8.1 Application. All the experiments were run on Mac Os with CPU 2.4 GHz Intel Core 2 Duo, RAM 8 Gb.

Nodes were simulated by the creation of various virtual machines that were located within the same physical machine of the specified specification. Therefore, the shares of the private key were distributed and the private key was reconstructed using the virtual machines in this manner. The results of the simulation will be presented in the next section.

4 Experimental Results

This section provides the results of the performance of the distribution phase and reconstruction phase of the 2048-bit private keys from the experiments and methods explained in the previous section.

4.1 Distribution Phase

The average times taken to distribute the private key shares where n equaled 256 shares, 128 shares and 64 shares were obtained. Figure 1 shows the results of the share distribution time for the 2048-bit private keys.

Figure 1 shows that when the number of shares is 256, the distribution time linearly increases from 301 to 484 ms as the threshold value increases from 8, 16, …, 64. Similarly, when the number of shares is 128 and 64, the distribution time gradually increases in an approximately linear fashion as the threshold value increases. That is, the time taken to distribute the shares when n is 128 increases from 135 to 283 ms when the threshold value goes from 8 to 64. When the number

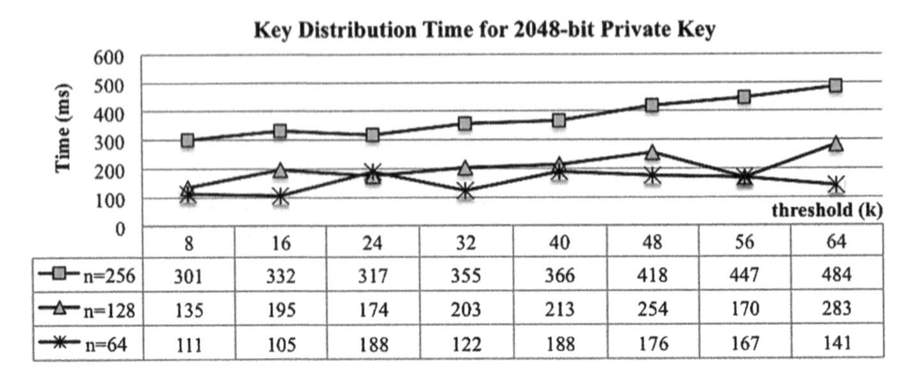

	8	16	24	32	40	48	56	64
n=256	301	332	317	355	366	418	447	484
n=128	135	195	174	203	213	254	170	283
n=64	111	105	188	122	188	176	167	141

Fig. 1 Key distribution time of 2048-bit private key when n = 64, 128 and 256 shares

of shares is 64, the distribution time goes from 111 to 141 ms as the threshold value increases.

Overall, it can be seen that on average the distribution time increases almost linearly when the threshold value increases, no matter how many shares the private keys are split into.

4.2 *Reconstruction Phase*

The average times taken to reconstruct the private keys using the threshold value, k, set in the previous section were obtained and are shown in Fig. 2.

Figure 3 shows that when the total number of shares is 256, the reconstruction time increases from 45 to 600 ms when the threshold value increases from 8 to 64.

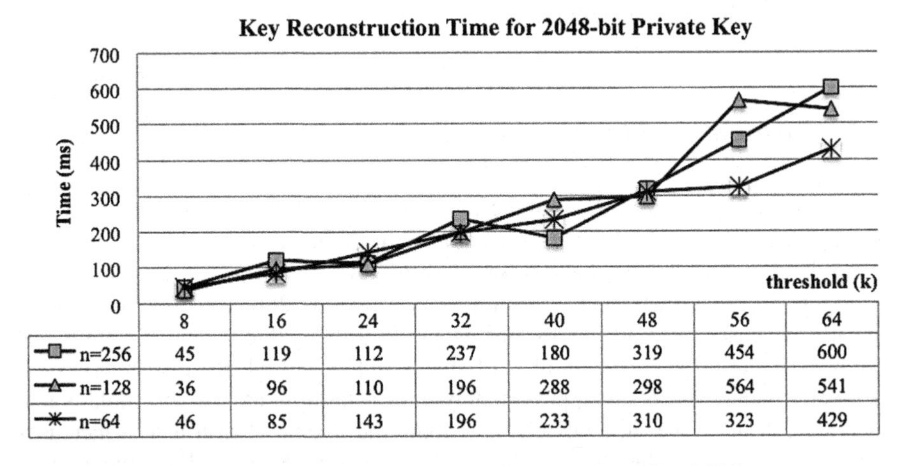

	8	16	24	32	40	48	56	64
n=256	45	119	112	237	180	319	454	600
n=128	36	96	110	196	288	298	564	541
n=64	46	85	143	196	233	310	323	429

Fig. 2 Key construction time 2048-bit private keys when n = 64, 128 and 256

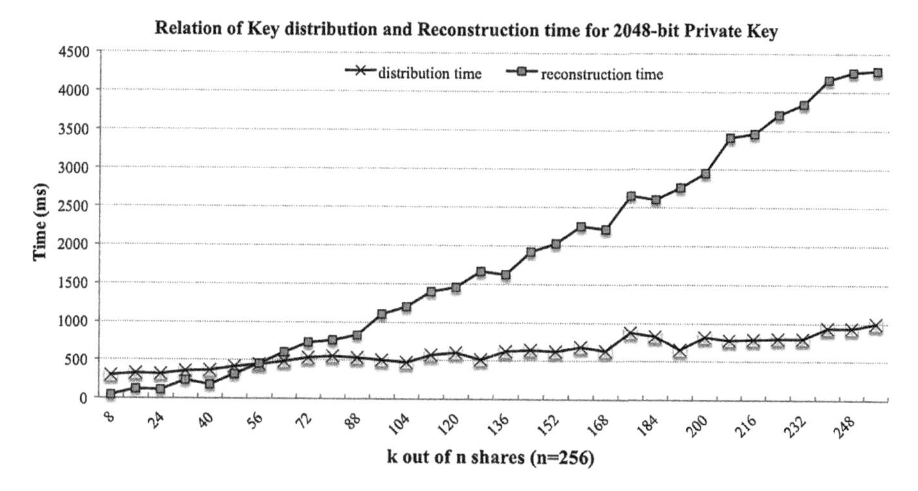

Fig. 3 Relation of key distribution time and reconstruction time of 2048-bit key size where number of shares, n, is 256

When the total number of shares is 128, the reconstruction time goes up linearly from 36 to 541 ms as the threshold value goes from 8 to 64. Finally, when the total number of shares is 64, the reconstruction time gradually increases in a linear fashion from 46 to 429 ms as the threshold value increases from 8 to 64.

It can be seen from the graphs in Fig. 3 that the time taken to reconstruct a 2048-bit private key increases in an approximately linear form when the threshold value increases. It can also be observed here that the rate of increase in the reconstruction time is greater than that of the key distribution time.

It appears that there is enough information to attempt to find a suitable threshold value for the threshold cryptography scheme. The method for achieving such value is presented in the next section.

5 Finding a Suitable Threshold

This section explains how a suitable threshold value for threshold cryptography can be found. At the end of this section, we should end up with a simple guideline for choosing a suitable threshold value, k, for a given number of shares, n.

From the results and graphs obtained in the previous section, the graph of the distribution phase and the graph of the reconstruction phase will be put together in order to find the intersection. This is the point where they graphs cross and is what we believe to be a suitable threshold value for that particular scenario.

Using the proposed method for finding a suitable threshold value, the following graphs, shown in Figs. 3, 4 and 5, are obtained.

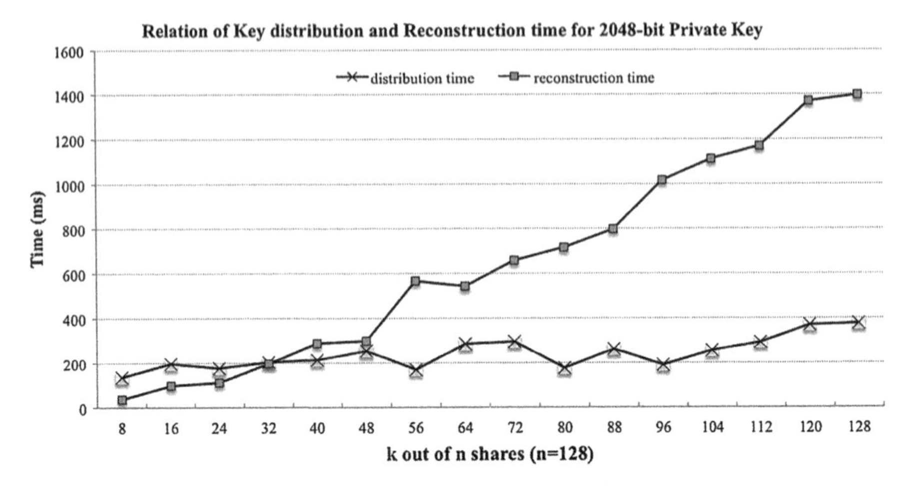

Fig. 4 Relation of key distribution time and reconstruction time of 2048-bit key size where number of shares, n, is 128

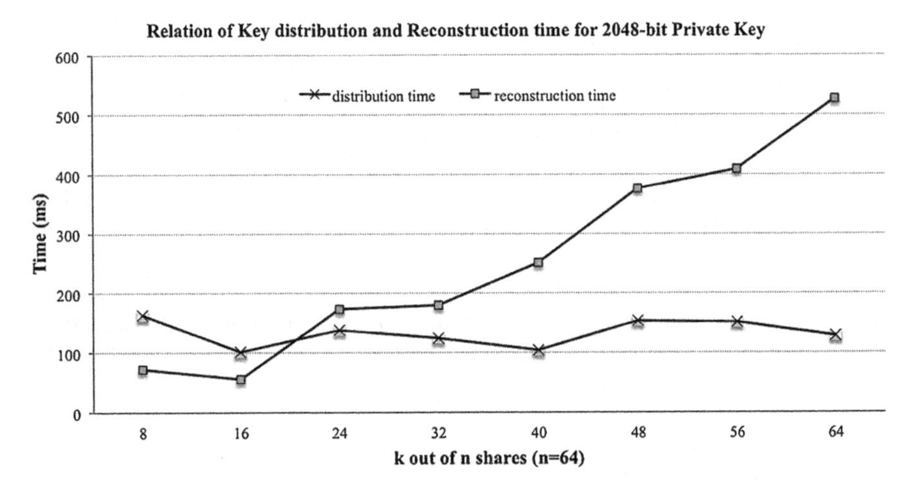

Fig. 5 Relation of key distribution time and reconstruction time of 2048-bit key size where number of shares, n, is 64

Figure 3 shows that when the private key is 2048 bits long and is split into 256 shares, the distribution line and reconstruction line cross at the point where *k* is around 56. That is, the threshold value, *k*, is approximately 21 % of all available shares, *n*.

Figure 4 shows that when the 2048-bit private key is divided into 128 shares, the intersection point of distribution line and reconstruction line at the point where the value of *k* 32. That is, the threshold value, *k*, is approximately 25 % of all available shares, *n*.

When the 2048-bit private key is divided into 64 shares, Fig. 5 shows that the line graphs of the distribution time and the reconstruction time cross at the point where the threshold value is 21. That is, the threshold value, k, is approximately 33 % of all available shares, n.

Using this simple method of finding an intersection between the key distribution time and the key reconstruction time, it can be seen that, on average, a suitable value of the threshold, k, is at approximately 26 % of all shares, n. It is, in this case, claimed suitable in terms of the time taken to both distribute and reconstruct. In other words, by using this value of k, the number of threshold is so low that it is easy for an adversary to obtain. At the same time, it is not so large that the time taken to reconstruct is long. It can also be explained that the threshold value, k, should not be too high since the graphs of the key reconstruction suggests that when k increases the time also increases in higher rate, compared with the key distribution time.

On a related note, Gharib and Belloulata [18] carried out a study that suggested that the security of threshold cryptography would decrease if the difference between the number of shares, n, and the threshold value, k, is large. In addition, it has to be said that one of the biggest protection for threshold cryptography is the fact that the shares are all geographically distributed over a set of machines. This makes it difficult for an adversary to find them. However, what we have done here is the simplification of the problem that only the efficiency of the working of the threshold scheme was taken into account. This is what this research is aiming for.

6 Conclusions

The security of any asymmetric cryptography partly relies on the fact that the private key is kept secure. In order to increase this security aspect, threshold cryptography has been proposed. This scheme works by dividing a private key into n shares before distributing them to all involved entities. However, when reconstructing the key only some threshold value k out of n shares is needed. The problem is that there has not been a method for finding a suitable k in terms of efficiency or the performance. This paper, therefore, presented a simple method for achieving such objective.

By carrying out simulated experiments in a virtual environment for finding the time taken to distribute the shares to all entities as well as the time taken to reconstruct the 2048-bit private keys using different threshold value, we were able to find a point where the two graphs intersect. This is the value k claimed to be suitable in terms of speed and efficiency. The results from our experiments show that the value k should be at approximately 26 % of all available shares, n. Having said that, this value does not yet take into account the geographic distribution of the machines holding the shares. Although this is the limitation of the results, it can be used as a simple guideline for choosing a suitable threshold value, k.

References

1. Rivest, R.L., Shamir, A., Adleman, L.: A method for obtaining digital signatures and public key cryptosystems. Commun. ACM **21**, 120–126 (1978)
2. Kerckhoffs, A.: La Cryptographie militaire. J. Des. Sci. Mil. **9**, 5–83 (1883)
3. Sharmir, A.: How to share a secret. Commun. ACM **22**, 612–613 (1979)
4. Blakley, G.R.: Safeguarding cryptographic keys, In: Merwin, R.E., Zanca, J.T., Smith, M. (eds.) Proceedings of the 1979 AFIPS National Computer Conference, AFIPS Conference Proceeding, vol. 48, pp. 313–317. AFIPS Press (1979)
5. Benaloh, J., Leichter, J.: Generalized secret sharing and monotone functions. In: Goldwasser, S. (ed.) Advances in Cryptology-CRYPTO'88. Lecture Notes in Computer Science, vol. 403, pp. 27–35. Springer (1990)
6. Bertilsson, M., Ingemarsson, I.: A construction of practical secret sharing schemes using linear block codes. In: Seberry, J., Zheng, Y. (eds.) Advances in Cryptology-AUSCRYPT'92. Lecture Notes in Computer Science, vol. 718, pp. 67–79. Springer (1993)
7. Capocelli, R.M., Santis, A.D., Gargano, L., Vaccaro, U.: On the size of shares for secret shairing schemes. J. Cryptol. **6**, 157–168 (1993)
8. Lysyanskaya, A., Peikert, C.: Adaptive security in the threshold setting: from cryptosystem to signature schemes. In: Boyd, C. (ed.) ASIACRYPT 2001. LNCS, vol. 2248, pp. 331–350. Springer, Heidelberg (2001)
9. Desmedt, Y., Frankel, Y.: Threshold Cryptosystems. In: Brassard, G. (ed.) CRYPTO 1989. LNCS, vol. 435, pp. 307–315. Springer, Heidelberg (1990)
10. Kaya, T., Lin, G., Yilmaz, G.N.: Secure multicast groups on ad hoc networks. In: SASN'03: Proceedings of the 1st ACM Work-shop on Security of Ad Hoc and Sensor Networks, pp. 94–102. ACM Press, New York (2003)
11. Khalili, A., Katz, J., Arbaugh, W.: Toward secure key distribution in truly ad hoc networks. In: Proceedings of IEEE Workshop on Security and Assurance in Ad-Hoc Networks, in Conjunction with the 2003 International Symposium on Applications and the Internet, pp. 342–346 (2003)
12. Mishra, M., Bhatele, M.: Improved cloud security approach with threshold cryptography. Int. J. Master Eng. Res. Technol. **2**, 119–126 (2015)
13. Bharill, S., Hamsapriya, T., Lalwani, P.: A secure key for cloud using threshold cryptography in Kerberos. Int. J. Comput. Appl. **79**, 35–41 (2003)
14. Desmedt, Y., Frankel, Y.: Threshold cryptosystems. In: Proceedings of Advances in Cryptology—Crypto'89. Lecture Notes in Computer Science, Santa Barbara, vol. 435, pp. 307–315 (1990)
15. Basseville, M., Nikiforov, I.V.: Detection of Abrupt Changes: Theory and Application. Prentice Hall (1993)
16. Brodsky, B.E., Darkhovsky, B.S.: Nonparametric Methods in Change-point Problems. Kluwer Academic Publishers (1993)
17. NIST SP 800-57.: Recommendation for Key Management Part1: General (Revised), NIST Special Publication, Special Publication 800-57 (2007)
18. Gharib, H., Belloulata, K.: Authentication architecture using threshold cryptography in kerberos for mobile ad hoc networks. Adv. Sci. Technol. Res. J. **8**, 12–18 (2014)

Effective Linear Transformation Matrices for Block Cipher Based on Bi-regular Matrix

The Dung Luong

Abstract Although the separation matrix with maximum distance (MDS matrix) has been widely used in the cipher and hash function, the implementation of the linear transformation based on MDS matrix in many current block ciphers is not effective because the number of occurrences of 1 in matrices is not much and the number of different elements in the matrices is quite large. In this paper, we propose a method to develop the effective MDS matrices based on the bi-regular matrix that maximizes the number of occurrences of 1 and minimizes the number of different elements in MDS matrices. By using the proposed method, we will construct the square MDS matrices 4×4, 8×8, 16×16 for using in the block ciphers.

Keywords MDS matrix · Bi-regular matrix · Block ciphers

1 Introduction

The replacement of permutation layer on Substitution-Permutation Network (SPN) by a diffusion linear transformation will improve the SAC property of the block ciphers, thereby increasing the resistance to the linear and differential attacks [1–3]. The MDS matrices play an important role in the design of block ciphers and hash functions to provide the level of high security that resists many strong attacks on block ciphers such as: linear attack and differential attack. The MDS linear transformation was first proposed by Vaudenay [4] and then used in block ciphers SHARK [6] and SQUARE [5]. The advantage of linear transformation layer is that it creates the minimum number of active S-boxes in two consecutive rounds of a linear approximation or in two consecutive rounds of a differential property to be $m + 1$ (where m is the number of S-boxes in a round of SPN), in theory, this is the possible largest value of the minimum number of active S-boxes in two consecutive rounds. In other words, the MDS matrix changes the number of branches in diffusion layer. Therefore

T.D. Luong (✉)
Academy of Cryptography Techniques, Hanoi, Vietnam
e-mail: thedungluong@gmail.com

P. Meesad et al. (eds.), *Recent Advances in Information and Communication Technology 2016*, Advances in Intelligent Systems and Computing 463,
DOI 10.1007/978-3-319-40415-8_23

it reaches the maximum value. For block ciphers, the level of security against strong attacks (such as linear attack, differential attack) depends on the number of branches of the diffusion layer. Therefore, if the diffusion layer of block ciphers uses the MDS matrix, the resistance of the block ciphers against linear and differential attacks will be the best quality. In fact, the MDS matrices are used for the diffusion layer of many block ciphers such as: AES [11], SHARK [6], KHAZAD, SQUARE [5], Anubis, ... MDS matrix is also used in the design of hash functions such as Maelstrom, and the family of lightweight hash functions in which they use the MDS matrix as a key component in the diffusion layer. The MDS matrix plays an important role, but there is not many systematic studies on how to find the effective matrix. Pascal Junod and Serge Vaudenay are the first two authors introducing to this problem [7] that is referred by many other studies such as in [1, 8, 9], so it shows that the construction of the efficient MDS matrices is an interested problem. However, there are not any studies to develop ideas of S. Vaudenay P. Junod. In [7], the authors define the MDS matrices as the multi-linear permutations to design the MDS matrix with a large number of elements 1 and minimize the number of different elements in the matrix to obtain the best implementation performance. Thus it will create the optimal matrix that minimizes the number of XOR operations, the temporary variables and lookup table are smallest. Based on the bi-regular array (or be called the bi-regular matrix), the authors have developed the effective matrices of 4×4 and 8×8 in $GF(2^8)$. However, the construction method in [7] only applies to particular cases without in general. In this paper, we also base on the idea of the effectiveness of the matrix in [7] by using the bi-regular matrix. However, we propose an algorithm to build a square bi-regular matrix that maximizes the number of elements 1 with any sizes, meanwhile minimizes different elements for the general case. These results will be an important contribution to construct the efficient MDS matrices for block ciphers. In Sect. 2, we introduce some knowledge about the MDS matrix and some effective MDS matrices. Section 3 presents an algorithm to build a square bi-regular matrix that optimizes the number of elements 1. Section 4, we provide estimations about the minimum number of different elements in any square bi-regular matrix. Section 5 is conclusion.

2 Related Work

2.1 MDS Matrix

MDS Code has been studied for a long time in the theory of error correcting codes [10]. If C is a linear code with parameters $[n, k, d]$, where n is the code word length, k is the length of the plaintext and d is the distance of the code C, then these parameters will satisfy the inequality $d \leq n - k + 1$ that is called Singleton limitation. MDS codes are linear codes $[n, k, d]$, where $d = n - k + 1$, it means that the code distance is optimal. C code is associated with the generation matrix (G) that it consists of

k rows and n columns with rank $k(k \leq n)$, when coding a message of length k, then implementing multiplication of the vector a and matrix G will obtain a codeword of length $n : c = aG$. We always write G in the normal form $G = [I/A]$, where I is the unit matrix of rank k, matrix A has k rows, $n - k$ columns. If C is MDS code then the matrix A is called MDS matrix. In [7], the authors presented the definition about MDS matrix and the linear multipermutation as following:

Definition 1 Let K be a finite field and p, q be two integers. Let $f : x \rightarrow M \times x$ be a mapping from K^p to K^q defined by the $q \times p$ matrix M. We say that it is a linear multipermutation (or an MDS matrix) if the set of all pairs $(x, M \times x)$ is an MDS code, i.e. a linear code of dimension p, length $p + q$ and minimal distance $q + 1$.

In [10] the authors give a theorem about the MDS matrix as follow:

Theorem 1 *An (n, k, d) code with generation matrix $G = \lfloor I_{k \times k} A_{k \times (n-k)} \rfloor$ is an MDS code if and only if every square submatrix of A is non singular and the matrix A is called an MDS matrix.*

2.2 The Effective MDS Matrices

In [7], the goal of authors is to build the effective MDS matrices that maximize the number of occurrences of 1 and the minimize the number of different elements. To achieve this goal, the authors provide a definition of the two values v_1 and c as follows:

Definition 2 Let $K *$ be a set including a distinguished one denoted 1. Let M be a $q \times p$ matrix whose entries lie in $K *$.

1. We let $v_1(M)$ denote the number of (i, j) pairs such that $M_{i,j}$ is equal to 1. We call it the number of occurrences of 1.
2. We let $c(M)$ be the cardinality of $M_{i,j}; i = 1, \ldots, q; j = 1, \ldots, p$. We call it the number of entries.
3. If $v_1(M) > 0$ we let $c_l(M) = c(M) - 1$. Otherwise we let $c_l(M) = c(M)$. We call it the number of nontrivial entries

In [7], the authors also define a bi-regular array, which is useful for formulating MDS matrix.

Definition 3 Let $K *$ be a set including a distinguished one denoted 1.

1. We say that a 2×2 array with entries in $K *$ is bi-regular if at least one row and one column have two different entries.
2. We say that a $q \times p$ array with entries in $K *$ is bi-regular if all 2×2 sub-arrays are bi-regular.
3. An array which is not bi-regular called bi-singular.

4. Two arrays are equivalent if we can obtain the second by performing a finite sequence of simple operations on the first one. Simple operations are per mutation of rows, columns, transpose, and permutation of $K*$ elements in which 1 is a fixed point.

Note that an MDS matrix must be a bi-regular array but the reverse is not true. Equivalent arrays have the same metrics of v_1 and c. In [7], the authors also give some results about the optimal values of v_1 and c for a bi-regular matrix with size $q \times p$ as follows:

1. $v_1^{q,p} = v_1^{p,q}$ and $c_{q,p} = c_{p,q}$ since we can transpose bi-regular arrays
2. We have $v_1^{1,p} = p, c^{1,p} = 1$ for any $p \geq 1$.
3. $v_1^{q,p}$ and $c^{q,p}$ increase when p and q increase.
4. we have $v_1^{4,4} = 9, v_1^{6,6} = 16, v_1^{8,8} = 24, c^{4,4} = 3, c^{6,6} = 4, c^{8,8} = 5$.

Lemma 1 *If p is a prime power, for any integers $\alpha > 1$ and $q \leq \frac{p^{(\alpha-1)}(p^\alpha-1)}{(p-1)}$, we have $v_1^{(q,p^\alpha)} \geq q \times p$.*

From this lemma we have: for $\alpha > 1$and $q = p = n$ then $v_1^{(n,n)} \geq n\sqrt{n}$.

Lemma 2 *For any k we have $c^{2k-1,2k-1} \leq k$*

Note that the high values of v_1 and low values of c for the elements with small Hamming weight is our goal to build effective MDS matrices. In [7] authors also mentions the concept of optimal matrix, in which a matrix M of size $q \times p$ is optimal if it gets two optimal values v_1 and c, that means $v_1(M) = v_1^{(q,p)}$ and $c(M) = c^{(q,p)}$. However, Building the optimal matrix is very difficult, because it is multi-objective problem, we have to achieve optimal values for both v_1 and c. Therefore, it only obtain optimal partial matrices or matrices with the high v_1 and the low c that is not optimal.

3 Building the Square Bi-regular Matrix with Optimal v_1

In this section, we give some remarks as well as stating and proving some propositions, then proposing algorithms for building a square bi-regular matrix with optimal v_1. Firstly, we have some remarks as follows.

Remark 1 We can swap rows and columns of the bi-regular matrix that the number of elements 1 is constant. So without loss of generality we can assume the columns of the matrix are arranged so that the number of elements 1 in columns decreases from left to right.

Remark 2 If on a column has k elements 1 (such as column i) then the maximum number of elements 1 in the remaining columns decreases: $(k-1)(n-1)$.

Fig. 1 8 × 8 bi-regular array example

1	1	x	1	x	x	x	x
1	x	1	x	x	x	1	x
1	x	x	x	1	x	x	x
1	x	x	x	x	1	x	1

For the bi-regular arrays, on the remaining column, we can only select a location containing the element 1 coincides with a certain row that contains the element 1 of column i and the remaining $k-1$ positions will be eliminated (it is not able to fill 1) on the rows containing the numbers of 1 of the column i. This means that in each column, the number of elements 1 can fill to able to be reduced $k-1$. Thus, the number of elements 1 can be filled in the matrix to able to reduce $(k-1)(n-1)$. For example (see Fig. 1), in the following example we have a 8×8 matrix with first column containing four locations of 1. Meanwhile, from the second column to the eighth column we can only enter the elements 1 into at most one location (in the rows that contains elements 1 of the first column) to satisfy the bi-regular property, the remaining positions will be eliminated. Therefore, if the first column contains four positions of 1 then the number of positions in the matrix will be eliminated is $(4-1)(8-1) = 21$. Consider the following propositions:

Proposition 1 *We have* $v_1^{(n,n)} \geq 3n - 3$

Proof Because $v_1^{(n,n)}$ is the maximum number of elements 1, therefore in order to point out that Proposition 1 is correct we only need to show a building way that $v_1 = 3n - 3$. We can build a $n \times n$ bi-regular matrix as follow: first column has $n-1$ elements 1, other columns have 2 elements 1. Therefore $v_1 = n - 1 + (n-1) \cdot 2 = 3n - 3$.

For example (Fig. 2), we fill the elements 1 in a 6×6 matrix as below. The number of elements 1 obtains $3 \cdot 6 - 3 = 15$.

Fig. 2 6 × 6 bi-regular array example

1	1				
1		1			
1			1		
1				1	
1					1
	1	1	1	1	1

Proposition 2 *We have* $v_1^{(n,n)} \geq n\sqrt{n}$.

From Lemma 1 [7].

Proposition 3 s_1 *is the number of elements 1 at the first column of* $n \times n$ *bi-regular matrix. Then* $\sqrt{n} \leq s_1 \frac{(n+1)\sqrt{n}+1}{\sqrt{n}+1}$.

Proof From Remark 2, we see that if column 1 have s_1 elements 1, the number of elements 1 of the matrix is reduced $(s_1 - 1)(n - 1)$. Thus, the maximum number of elements 1 in the matrix is $n^2 - (s_1 - 1)(n - 1)$. From Proposition 2, we have:

$$n^2 - (s1 - 1)(n - 1) \geq n\sqrt{n} \quad \text{or} \tag{1}$$

$$n^2 - n\sqrt{n} \geq (s_1 - 1)(n - 1) \quad \text{or} \tag{2}$$

$$s_1 \leq \frac{n^2 - n\sqrt{n}}{n-1} + 1 = \frac{n\sqrt{n}(\sqrt{n}-1) + (\sqrt{n}-1)(\sqrt{n}+1)}{(\sqrt{n}-1)(\sqrt{n}+1)} \quad \text{or} \tag{3}$$

$$s_1 \leq \frac{(n+1)\sqrt{n}+1}{\sqrt{n}+1} \tag{4}$$

In other hand, we have $v_1^{(n,n)} \geq n\sqrt{n}$ and the first column contains the most elements 1 (under the assumption of Remark 1) therefore $ns_1 \geq n\sqrt{n}$ or $s_1 \geq \sqrt{n}$.

From the above remarks and propositions, we give an algorithm to build a $n \times n$ square bi-regular matrix with the optimal v_1.

3.1 Algorithm for Constructing a $n \times n$ Square Bi-regular Matrix Which Is Able to Be Optimal for v_1

INPUT: An $n \times n$ empty matrix.
OUTPUT: A $n \times n$ matrix has the greatest number of elements 1 and satisfies bi-regular property.

The steps of the algorithm.

Step 1. Choose a permutation of n elements $1, 2, \ldots, n$, each element represents a location of the row that contains element 1 for each column.

Step 2. On each column add an element 1 such that the matrix ensures bi-regular property. It is easy to see that the matrix contains the most elements 1 when the row added also is a permutation of n elements. Therefore, in this step, we also choose a permutation of n elements $1, 2, \ldots, n$ to fill 1 in the columns and remove the positions that violate bi-regular property.

Iteration 2.1. Repeat Step 2 until no more 1 will be added, we will obtain a bi-regular matrix with optimal v_1

Note that, the obtained matrix depends on the first permutation that we choose in Step 1.

Proving the correctness of the algorithm. If in Step 1, in columns we choose the locations containing 1 to form a permutation (row) of n elements $1, 2, \ldots, n$ then base on the Remark 2, the number of elements 1 in columns will reduce to 0. That means the number of locations eliminated by violating bi-regular property is 0. Therefore, in Step 2, the ability to choose the location for the element 1 in the columns is as much as possible.

Conversely, suppose that in Step 1 we choose at least two columns that contain the elements 1 (in the same row), then following Remark 2, the number of locations containing 1 at those two columns will be reduced at least $(2-1)(n-1) = n-1$ locations. In other words, there are at least $n-1$ positions eliminated in these two columns. Thus, the ability to choose the locations for the element 1 in this case is less than the above case.

Therefore, the way to choose locations of 1 in columns to form a permutation of n elements $1, 2, \ldots, n$ is the best choice, that we have the maximum number of elements 1 for the matrix.

Assess the complexity of algorithm. Due to Proposition 3, the number of steps required in the above algorithm does not exceed $\frac{(n+1)\sqrt{n+1}}{\sqrt{n+1}}$. Each step at each column needs $(n-1)^2$ choices to remove the locations that contain the elements 1, at the same time we have n column in a matrix so we need n choices just like that. Therefore, the complexity does not exceed $O(n \times n^2 \times n) = O(n^4)$. Here, we will illustrate the algorithm on the 8×8 and 16×16 matrix.

3.2 Example

Example 1 With a 8×8 matrix

Step 1. Choose a permutation of the 8 elements 1, 2, …, 8 such as (3, 5, 6, 4, 2, 1, 8, 7) to fill 1 in the columns (Fig. 3).

Step 2. Choosing the cell 4 of column 1, cell 6 of column 2, cell 7 of column 3, cell 5 of column 4, cell 3 of column 5, cell 2 of column 6, cell 1 of column 7, and cell 8 of column 8 to obtain a permutation (4, 6, 7, 5, 4, 2, 1, 8) and remove the cells do not satisfy the bi-regular property (marked as x) (Fig. 4).

Fig. 3 Example 1, Step 1

	1	2	3	4	5	6	7	8
1						1		
2					1			
3	1							
4				1				
5		1						
6			1					
7								1
8							1	

Fig. 4 Example 1, Step 2

	1	2	3	4	5	6	7	8
1					x	1	1	
2	x				1	1		
3	1				1	x		
4	1			1	x			
5	x	1	x	1				
6		1	1	x				
7		x	1				x	1
8			x			x	1	1

Fig. 5 Example 1, Step 3

	1	2	3	4	5	6	7	8
1	x	x	1	x	x	1	1	x
2	x	x	x	x	1	1	x	1
3	1	x	x	x	1	x	1	x
4	1	x	x	1	x	1	x	x
5	x	1	x	1	1	x	x	x
6	1	1	1	x	x	x	x	x
7	x	x	1	1	x	x	x	1
8	x	1	x	x	x	x	1	1

Step 3. Just like the Step 2, choose a permutation followed by (6, 8, 1, 7, 5, 4, 3, 2) to fill the elements 1 and remove the cells that do not satisfy the bi-regular property (Fig. 5).

After this step we can see that no more cells can be filled, therefore we had a plan for a 8×8 bi-regular matrix with optimal v_1 ($v_1 = 24$).

4 Assessment of Value c of the Square Bi-regular Matrix

In this section, we give some remarks and propositions to make an overall assessment for the value c of a $n \times n$ square matrix.

Proposition 4 *Suppose k is the number of different elements in any two consecutive rows of a $M_{(nn)}$ bi-regular matrix on $GF(2^p)$, the set of k elements is $A = a_1, a_2, \dots a_k$, then we have:*

$$k^2 - k + 1 \geq n \quad or \quad k \geq \frac{1 + \sqrt{4n - 3}}{2} \tag{5}$$

Proof Indeed, There are k^2 with the formulation (a, b) can have k elements in A. The pairs are:

$$\begin{bmatrix} (a_1, a_1) & (a_1, a_2) & \dots & (a_1, a_k) \\ (a_2, a_1) & (a_2, a_2) & \dots & (a_2, a_k) \\ \dots\dots & \dots\dots & \dots\dots & \dots\dots \\ (a_k, a_1) & (a_k, a_2) & \dots & (a_k, a_k) \end{bmatrix}$$

Because M is bi-regular matrix, so above k elements must be filled in two consecutive rows of the matrix that satisfies the bi-regular property. Therefore, in above k^2 pairs, we have to remove $k - 1$ pairs in $(a_1, a_1), (a_2, a_2), \dots, (a_k, a_k)$ to fill in these two rows. According to the bi-regular property, there is no sub-array as

$$\begin{bmatrix} a & b \\ a & b \end{bmatrix}$$

therefore we can only take a pair with form (a, a) in above k pairs. So, we have $k^2 - k + 1$ columns (of two considered rows) to be able to fill k elements in A.

At the same time, to ensure the bi-regular property we need $k^2 - k + 1 \geq n$, or $k \geq \frac{1+\sqrt{4n-3}}{2}$. Because, if in the reverse case, with k different elements, we can build an array $2 \times m$ of M satisfying the bi-regular property with $m < n$, and the pairs in $n - m$ remaining columns will be identical to m pairs of previous columns. This does not satisfy the bi-regular property.

Example 2 If $k = 3$, set $A = a, b, c$ and the bi-regular matrix with size $n = 8$ then we have $3^2 = 9$ pairs generated from A and only have $3^2 - 3 + 1 = 7$ columns that can be filled the elements of A that satisfy the bi-regular property, as follows (here only hold a identical pair (a, a): Thus, at the last column, if only taking values in the set A to fill then two values in the last column will coincide with two values of the previous 7 column. It will break the bi-regular property.

a	a	b	c	c	b	a
a	b	a	a	b	c	c

So there must be $k^2 - k + 1 \geq 8$, or $k \geq \frac{1+\sqrt{4.8-3}}{2}$ or $k \geq 4$.

Corollary 1 *We have* $c^{(n,n)} \geq \frac{1+\sqrt{4n-3}}{2}$, *with* $n \geq 2$.

Remark 3 We have $c^{2k-2,2k-2} \leq k$

Indeed, from Lemma 2 [7], we have: $c^{2k-1,2k-1} \leq k$. In addition, $c^{q,p}$ increases when p and q increase.

Therefore $c^{2k-2,2k-2} \leq c^{2k-1,2k-1} \leq k$.

Remark 4 From Remark 3, it is easy to deduce that with every $n \times n$ square matrix, we have: $c^{n,n} \leq [n/2] + 1$

Proposition 5 *We have:*

$$\frac{1+\sqrt{4n-3}}{2} \leq c^{n,n} \leq [n/2] + 1 \quad with \quad c^{n,n} \in N \tag{6}$$

Proof From Corollary 1 and Remark 4, we have something to prove. So far, we obtain a general assessment of the value c of $n \times n$ bi-regular matrix, that is:

$$\frac{1+\sqrt{4n-3}}{2} \leq c^{n,n} \leq [n/2] + 1 \quad with \quad c^{n,n} \in N$$

Example 3 We assesses the value c for some specific cases as follows:

- With $n = 4$ then $3 \leq c \leq 3$, so $c = 3$.
- With $n = 5$ then $3 \leq c \leq 3$, so $c = 3$.
- With $n = 6$, $3 \leq c \leq 4$.
- With $n = 7$, $3 \leq c \leq 4$.
- With $n = 8$, $4 \leq c \leq 5$.
- With $n = 16$, $5 \leq c \leq 9$.

5 Conclusions

In this paper, we propose a method to build square MDS matrices with any size based on bi-regular that maximizes the number of elements 1. When the number of elements 1 is maximized the performance of multiplication in implementation of permutation layer will improve. We also give an assessment for the number of minimum different elements in the general case of square bi-regular matrix. That will reduce the size of look up table when design the permutation layer for block cipher.This is an important result for designing permutation layer of block ciphers.

References

1. Heys, H.M., Tavares, S.E.: The design of substitution-permutation networks resistant to differential and linear cryptanalysis. In: Proceedings of 2nd ACM Conference on Computer and Communications Security, pp. 148–155, Fairfax, Virginia (1994)
2. Heys, H.M., Tavares, S.E.: The design of product ciphers resistant to dierential and linear cryptanalysis. J. Crypt. **9**(1), 1–19 (1996)
3. Heys, H.M., Tavares, S.E.: Avalanche characteristics of substitution-permutation encryption networks. IEEE Trans. Comput. **44**, 1131–1139 (1995)
4. Vaudenay, S.: On the need for multipermutations: cryptanalysis of MD4 and SAFER. In: Proceedings of Fast Software Encryption (2). LNCS 1008, pp. 286–297. Springer (1995)
5. Daemen, J., Knudsen, L., Rijmen, V.: The block cipher Square, Fast Software Encryption (FSE'97). In: LNCS 1267, pp. 149–165. Springer (1997)
6. Rijmen,V., Daemen, J., Preneel, B., Bosselaers, A., Win, E.D.: The cipher SHARK, fast software encryption. In: LNCS 1039, pp. 99–112. Springer (1996)
7. Junod, P., Vaudenay, S.: Perfect diffusion primitives for block cipher Building efficient MDS matrices in Selected Areas in Cryptology (SAC 2004). In: LNCS 3357, pp. 84–99. Springer (2004)
8. Abrahao, E.: A new involutory MDS matrix for the AES. Int. J. Netw. Secur. **9**(2), 109–116 (2009)
9. Elumalai, R., Reddy, A.R.: Improving diffusion power of AES Rijndael with 8×8 MDS Matrix. Int. J. Sci. Eng. Res. **2**(3) (2011)
10. MacWilliams, F.J., Sloane, N.J.A.: The Theory of Error-Correcting Codes. North-Holland Publishing Company, Amsterdam (1977)
11. Daemen, J., Rijmen,V.: AES Proposal: Rijndael (Version 2). NIST AES (1997)

A High Radix Hierarchical Interconnection Network for Network-on-Chip

Mohammed N.M. Ali, M.M. Hafizur Rahman, Rizal Mohd Nor and Tengku Mohd Bin Tengku Sembok

Abstract Architecture of the interconnection network has a great influence on the speed of the multi-core processor design. The main aims of any new architecture are to avoid the latency, and to decrease the cost. In this paper, we proposed new hierarchical interconnection network, in order to build fast parallel computing system. We have evaluated the static network performance of the proposed network such as: node degree, diameter, cost, arc connectivity, bisection width, and wiring complexity. The proposed topology achieved low cost and small diameter comparing to 2D-mesh, and 2D-torus topologies. As well as, it gives good results in the other static parameters. Hence, the proposed network is good solution to improve the performance, and decrease the cost of the interconnection networks for the future generation parallel computing systems.

Keywords Network-on-chip · Interconnection network · Hierarchical interconnection network · Static network performance

1 Introduction

Architecture of modern computer systems should meet the advancement in technology, therefore, most of IT companies developed the architecture of network on chip (NoC), in order to improve the performance, and decrease the cost of these

M.N.M. Ali (✉) · M.M.H. Rahman · R.M. Nor
Department of Computer Science, KICT IIUM, Kuala Lumpur, Malaysia
e-mail: moh.ali.exe@gmail.com

M.M.H. Rahman
e-mail: hafizur@iium.edu.my

R.M. Nor
e-mail: rizalmohdnor@iium.edu.my

T.M.B.T. Sembok
Cyber Security Center, UPNM, Kuala Lumpur, Malaysia
e-mail: tmtsembok@gmail.com

P. Meesad et al. (eds.), *Recent Advances in Information and Communication Technology 2016*, Advances in Intelligent Systems and Computing 463,
DOI 10.1007/978-3-319-40415-8_24

networks. Network on chip composed of number of network clients: processors, DSPs digital signal processing, memories, peripheral controllers, gateways to networks on other chips, and custom logic, each client is placed in rectangular tile and communicates with all other clients by using the network resources [1]. Furthermore, the routers in NoC are similar in their performance to those routers in local area network, making their decisions based on routing algorithms. However, choosing the routing algorithm depends on the structure of NoC [2, 3]. It's clear that, NoC is the basic architecture of parallel computing systems which provide efficient solutions for many of difficult problems in a reasonable time [4].

Interconnection networks of a multiprocessor computing system are a critical factor in determining the performance of the modern computing devices [5], due to their role in connecting processors and memories. In addition, the topology of these networks plays main role in determining the network diameter, whereas small diameter indicates low latency and good dynamic communication performance, leading to accelerate the speed of the system [6, 7]. New types of interconnection networks have been revealed to meet the advancement in signaling technology based on high-radix routers. High speed signaling technology, as well as the high number of signals available to a router chip caused increasing in the bandwidth; this additional bandwidth can be most effective by increasing the radix or the degree of the router, in addition building high radix routers with thin channels more efficient than using routers with low radix and wide channels [8, 9].

On the other hand, advances in VLSI allowed placing billions of fast transistors in small sized chip, which leads to multi-core technology as dominant organization for future microprocessor chips [10]. The development in the chips size passed in many stages from 10 μm in 1971 until reached 14 nm in 2014, and by referring to Moore's law, the number of transistors in one chip expected to be doubled in every two years, therefore the expected next proposed chip size will be 10 nm in 2016.

While the signaling technology provided us high bandwidth, it's be the motivation of the researchers to look forward in developing the parallel computing systems to reap the benefits from this technology in serving the world. Therefore many of new interconnection networks have been proposed in the recent years, thrusting for the optimal one.

In this paper we propose new architecture of hierarchical interconnection network. This new design of interconnection network guarantees high performance by reducing the network diameter, which will ease the packets movement from their source nodes to the destination nodes; hence it will help in reducing the latency, and improving the dynamic communication performance of the network on chip.

The remainder of the paper is organized as follows. In Sect. 2 we will describe the structure of our proposed hierarchical interconnection network, and then we will evaluate the static network performance in Sect. 3, and finally in Sect. 4 will be the conclusion of this paper.

2 Architecture of the Proposed Network

The proposed network is a new hierarchical interconnection network and it composed of multiple levels to build up a reliable system with multiple numbers of basic modules (BMs). The basic module (BM) of the new system consists of six nodes; these nodes are completely connected by using electrical wires. As well as, every node has a direct connection link to each other nodes from the same group, which will reduce significantly the number of hops between the source and the destination nodes.

Figure 1 depicts the design of the BM of the proposed new system, and illustrates the communication between the different nodes from the same group; in addition, it shows the direct connection paths between these nodes. Hence, each node has a direct connection to five nodes; leading to every node will be connected in five pairs. The relation between (N_1) and the other nodes can be represented as (N_1, N_2), (N_1, N_3), (N_1, N_4), (N_1, N_5), (N_1, N_6). Likewise, we can represent the relation between any node and the other nodes.

Routers in every node choose the shortest and direct path to transfer the packet from the source node to its destination node. And it is interesting to observe that, in the proposed system the packet can reach the destination node by passing only one hop, which will play main role in decreasing the diameter and in turns the latency for message transfer between the nodes in the network. Finally it will improve the performance of the new system. Figure 1 represents the BM of the proposed system which indicates to Level-0 network. In addition it is the BM of Level-1 of the proposed network, which is depicted in Fig. 2.

Figure 2 illustrates Level-1 of the proposed system. In this system there are 2^m number of groups, where m is an integer number. Each group represents the BM of the Level-1 Network, as well as, every group composed of six nodes which are connecting to each other based on the description of Fig. 1. These groups connected by using global wires to create the level-1 of the proposed network, furthermore,

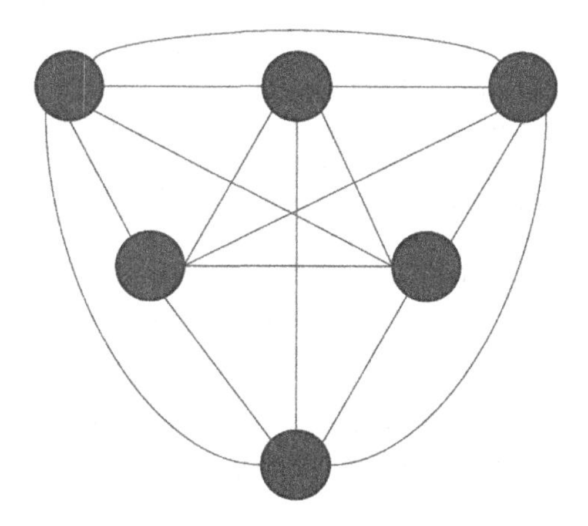

Fig. 1 Basic module of the proposed network

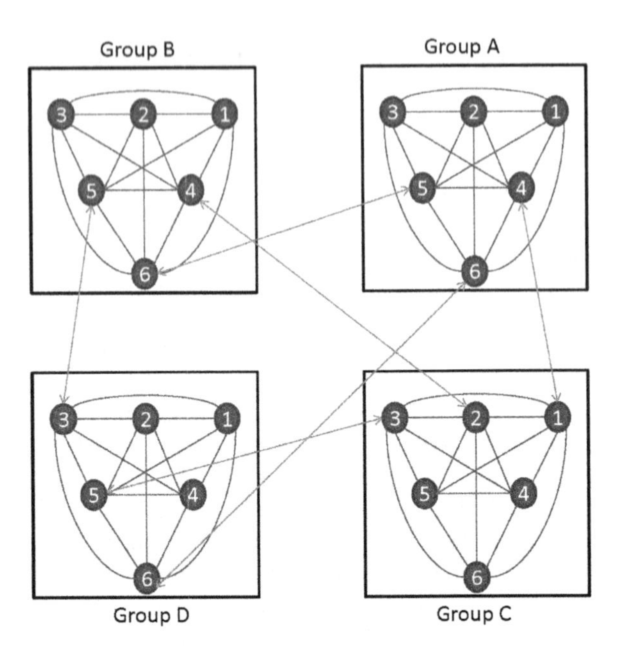

Fig. 2 Level-1 of the proposed network

these wires could be either electrical or optical. Level-1 is the BM of the higher level of the system which is Level-2.

The idea beyond connecting the nodes in Level-1 based on shifting the binary numbers of each node, the default shifting direction is the left in case of one direction link. As we see from the figure above, each group has six nodes, Group A has $(A_1, A_2, A_3, A_4, A_5, A_6)$, group B has $(B_1, B_2, B_3, B_4, B_5, B_6)$, group C has $(C_1, C_2, C_3, C_4, C_5, C_6)$, and group D has $(D_1, D_2, D_3, D_4, D_5, D_6)$ nodes. To connect a node from one group to a node from another group by using one direction link we follow that: if A_i $(1 \leq i \leq 6)$ wants to connect to Bi, one digit from the binary number of Bi will be shifted to the left, and it will be connected to the equal number of A_i. As well as, if A_i wants to connect to C_i we will shift the binary number of C_i two digits to the left. On the other hand, we will shift the binary number of D_i three digits to the left in case we want to connect A_i to D_i. Likewise, B_i connected to C_i by shifting the binary number of C_i one digit to the left, and it can be connected to D_i by shifting the binary number of D_i two digits to the left.

Shifting in the proposed system depends on the distance between the connected groups, the distance between A and B is one hop, therefore, we shift B_i one digit to the left. As well as, the distance between A and C is two hops, so the shifting in C_i will be two digits to the left, same with A and D the distance between them is three hops, therefore, we will shift D_i three digits to the left. Likewise, the communication between these pairs: (B_i, C_i), (C_i, D_i), or (B_i, D_i). Figure 3 depicts the distance, and the communication between the groups A, B, C, and D.

It's clear from Fig. 2 that, the connection between every two nodes from each group is bidirectional connection; we achieved that by shifting the binary digits of

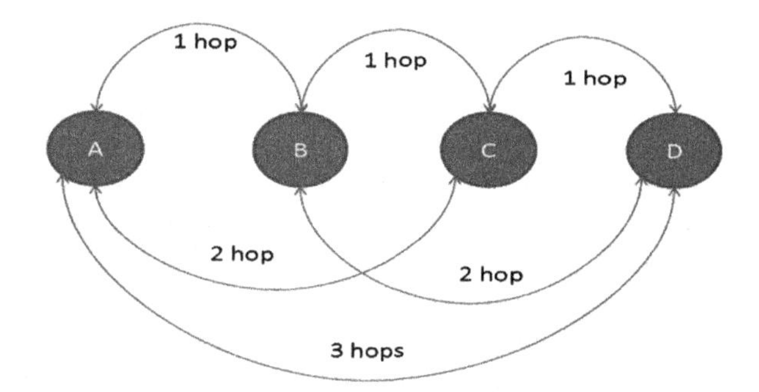

Fig. 3 Distances between groups

each node to the right or to the left, depending on the distance between their groups. As we explained earlier, the default direction to connect two nodes is the shifting to the left, however, if two nodes want to use the same global link to be connected, as shown in Fig. 2. Therefore, we connect the two nodes in bi- directional path by shifting the binary digits of one node to the left. In contrast, shifting the binary digits of the other node in the opposite direction, which will be to the right. Figure 4 shows an example of the bidirectional connection between two nodes.

By shifting the binary numbers of B_i we will obtain new binary number for every digit equal to another binary number in A_i, and these two equal numbers will be connected. We can explain the shifting idea in the following example, if we need to connect nodes from group A, with nodes from groups B, C, and D. As Fig. 2 shows the following pairs are connected to each other: (A_5, B_6), (A_4, C_1), and (A_6, D_6). The binary number of node 5 in group A is 101, and the binary number of node 6 in group B is 110, therefore, when we shift the binary digits of 6 one move to the left it will be 101 which is equal to the binary number of 5 in group A. As well as, the binary number of node 4 in group A is 100, and the binary number of node 1 in group C is 001, and due to the distance between A and C which is two hops, we will shift the binary number of node 1 in group C two digits to the left, therefore it will be equal the binary number of node 4 in group A. Moreover, if we need to connect node 6 in group A to another node in group D, and as we know the binary number of node 6 in group A is 110, therefore, if we shift the binary digits of each node in group D three digits to the left, we will find node 6 in group A can be connected to node number 6 in group D. By applying the shifting mechanism to the other nodes

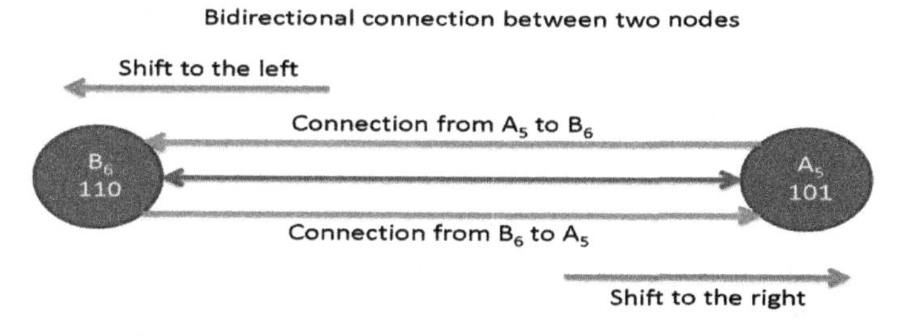

Fig. 4 Bidirectional connection between nodes

in each group we will find each node in every group has a node in the other groups to connect with. As we explained earlier the left is the default shifting direction in the case of the single paths. However, in our system we used only one bi-directional link to connect the nodes from different groups. In addition, the connection between these nodes based on the shifting in two directions one to the left, and the other one to the right. The benefit of using the bidirectional connection is to reduce the cost by decreasing the number of the global wires. In contrast, increasing the number of the global links to connect the groups, it could be useful to guarantee the diversity of the system communication; also it will be useful to control the high bandwidth, as well as, to avoid the congestion. Figure 2 illustrated the connection between some nodes from each group; we used these nodes as an example to make the idea clear. However each node in every group could be connected to another node from different group, therefore we can connect A_4 to B_2, or A_3 to B_5. Likewise, each node from each group. Decision of connecting the groups by using more than one global path, it would be left to the environment and the nature of the system, depending on the bandwidth, place, number of users, and the size of the data.

The main motivation and objective of this research is to propose a new hierarchical interconnection network for the future generation massively parallel computers. Four level of hierarchy will be used. These include chip-level hierarchy, board-level hierarchy, cabinet-level hierarchy, and system level hierarchy. According to the available VLSI technology, 24 processing nodes can be interconnected in a two levels of hierarchy as proposed in this paper.

3 Static Network Performance

The underlying interconnection network has main role in improving the performance of massively parallel computing systems. Several of topological properties can be used to evaluate and to compare the different networks; technology and physical implementation of the system have a significant impact on the network performance. These topological properties can be derived from the graph theoretic model of the topology. In this section we will use some of the static parameters such as degree, diameter, arc connectivity, bisection width, cost, and wiring complexity- to compare networks such as 2D-mesh, and 2D-torus to our proposed network. Table 1 shows the comparison between these networks, whereas the number of

Table 1 Static networks performance of different interconnection networks

Network	Node degree	Diameter	Cost	Arc connectivity	Bisection width	Wiring complexity
2D-Mesh	4	8	32	2	5	40
2D-Torus	4	5	20	4	10	50
Proposed network	6	3	18	3	4	66

nodes in mesh, and tours is 25 nodes; however the proposed network has 24 nodes. It is quite difficult to make different networks with same number of nodes because every network has different architecture. To make the performance comparison as fair as possible, we have compared the network performance of the proposed network consist of 24 nodes with (5×5) mesh and torus networks.

3.1 Node Degree

The basic module of the proposed network has six nodes; every node has direct link to the other five nodes at the same group, in addition, every node can be connected by direct link to the other nodes in different groups. Node degree is the maximum number of the physical links emanating from the node. The number of physical links of the Level-0 of the proposed network is 5 Links; however the number of the physical links of Level-1 is 6 links. As Table 1 shows the node degree of our proposed network is a little bit higher than the other topologies; however, as we mentioned before in this paper, high speed signaling technology, as well as the high number of signals available to a router chip caused increasing in the bandwidth; therefore, it's more efficient to use higher degree nodes with thin links than using nodes with low degree.

3.2 Diameter

In the interconnection networks the real distance between two nodes could be measured by knowing the length of the link which connects them, however, the logical distance between two nodes is the maximum number of hops that must be traversed to send packet from the source node to the destination node by choosing the shortest path, and according to the routing algorithm, which refers to the definition of the network diameter.

The performance of the interconnection networks depends widely on the length of the network diameter, and as we mentioned in this paper the small diameter improves significantly the performance of the network. According to Table 1, the diameter of the proposed network is low compared to 2D-mesh, and 2D-torus. Therefore, the proposed network will provide the system less latency and less message congestion, which leads to high performance network.

3.3 Cost

Network diameter and node degree has crucial impact on inter-node distance, message traffic density, and fault tolerance. The relation between cost and performance of the system can be measured as the result of (diameter × node degree). High node degree is very expensive, while the network with large diameter has a very low message passing bandwidth [11]. The comparison between the costs of the different networks plotted in Table 1 shows that the new proposed network cost is lower than the other networks. Therefore, the proposed network could be a solution in decreasing the cost of the interconnection networks.

3.4 Arc Connectivity

Reliability of interconnection network is measured by the Arc connectivity, which is known as the minimum number of links that must be cut to partition the network into two disjoint parts. Network performance impact positively by the influence of the high arc connectivity, which helps in avoiding links congestion, and also it improves the fault tolerance. Also we can measure the static fault tolerance performance from the ratio between arc connectivity and the degree of the node. In addition, if the arc connectivity and the degree of the node are the same, then the network will be reliable and robust. From Table 1 it is interesting to see that, arc connectivity of the proposed network comes in the middle between 2D-mesh and 2D-torus; whereas it little bit higher than mesh and little lower than torus network.

3.5 Bisection Width

Bisection width is the minimum number of links that must be cut to split the network into two equal parts. Partitioning the network in two equal halves will solve many of problems by separating the input data between these two halves. Data in each part will be manipulated in parallel and independently, and then the results will be merged from each part to obtain ideal solution.

Moderate bisection width preferable more than the low and the large one, due to low bisection width implies low bandwidth between the two halves; in addition, it will cause congestion in the middle of the network. In contrast, large bisection width refers to extra chip wires, which will affect negatively the design of VLSI of the interconnection network.

Table 1 show that bisection width of Level-1 of the proposed network is lower than that of 2D-mesh, and 2D-torus networks. Referring to Sect. 2 in this paper, we proved that Level-1 of the new system is flexible, and capable to increase the number of the global links between the groups; which will lead to higher bisection width. Therefore, the proposed system will be able to control the bisection width of the interconnection network depending on the system purpose.

3.6 *Wiring Complexity*

Wiring complexity indicates to the total number of links needed to connect each node in the interconnection network to the other nodes. As well as, it depends on the node degree and it affects the cost and the complexity of the interconnection network.

A (5 × 5) 2D-mesh and 2D-torus networks have $\{N_x \times (N_y - 1) + (N_x - 1) \times N_y\} = 5 \times (5 - 1) + (5 - 1) \times 5 = 40$ and $(2 \times N_x \times N_y = 2 \times 5 \times 5) = 50$ links, respectively. N_i represents the number of switches in the *i*th dimension. From Table 1, it is clear that the wiring complexity of the proposed system is little bit higher than that of the 2D-mesh, and 2D-torus. The nodes in the BM of the system are connected to each other by 15 electrical wires. As well as, Level-1 of this interconnection network composed of four BMs, and these BMs are connected to each other by using 6 global links. Therefore, the number of the electrical wires in level-1 is 60 link, and by adding 6 global electrical or fiber optic links, the total will be 66 link. Hence, wiring complexity of the proposed network will be 66. The only drawback of the proposed network is a bit high wiring complexity. However, this extra wire yields optimum diameter and less cost.

4 Conclusion

A new architecture of hierarchical interconnection network has been proposed for the network-on-chip. The architecture of the proposed new system and its connectivity has been discussed in details. We have evaluated the static network performance of the proposed network. And to show the superiority, the performance of this network is compared with popular mesh and torus networks. It is shown that the proposed network results small diameter, low cost, low bisection width, and moderate arc connectivity compared to mesh and torus networks. On the other hand, the only drawback is that the wiring complexity of the proposed network is a little bit higher than that of those networks.

This paper focused on the architecture of a new hierarchical interconnection network for NoC system. Issues for further exploration include (1) 3D NoC implementation and (2) dynamic communication performance evaluation by the dimension order routing algorithm.

Acknowledgments This research is supported by the project FRGS13-065-0306, Ministry of Education, Government of Malaysia. The authors would like to thank the anonymous reviewers for their constructive comments and suggestions on the paper which have helped to improve the quality of the paper.

References

1. Dally, W., Towles, B.: Route packets, not wires: on-chip interconnection networks. In: 38th Conference on Design Automation, pp. 684–689. New York (2001)
2. Sarkar, D.: Cost and time-cost effectiveness of multiprocessing. IEEE Trans. Parallel Distrib. Syst. **4**, 704–712 (1993)
3. Kim, J., Balfour, J., Dally, W.: Flattened butterfly topology for on-chip networks. In: 40th annual IEEE/ACM International Symposium on Micro-architecture (Micro-40), pp. 172–182. Chicago (2007)
4. Awal, M., Rahman, M.M.H., Akhand, M.: A new hierarchical interconnection network for future generation parallel computer. In: 16th International Conference on Computer and Information Technology (ICCIT), pp. 314–319. Khulna, Bangladesh (2014)
5. Kim, J., Dally, W., Scott, S., Abts, D.: Technology-driven, highly-scalable dragonfly topology. In: 35th International Symposium on Computer Architecture, IEEE Computer Society, pp. 77–88. Washington, DC USA (2008)
6. Rahman, M.M.H., Inoguchi, Y., Faisal, F., Kundu, M.: Symmetric and folded tori connected torus network. J. Netw. **6**, 26–35 (2011)
7. Rahman, M.M.H., Jiang, X., Masud, M., Horiguchi, S.: Network performance of pruned hierarchical torus network. In: 6th IFIP International Conference on Network and Parallel Computing, pp. 9–15. Gold Coast, Australia (2009)
8. Kim, J., Dally, W., Towles, B., Gupta, A.: Microarchitecture of a high-radix router. In: 32nd annual International Symposium on Computer Architecture, pp. 420–431. IEEE Computer Society, Washington, DC, USA (2005)
9. Villar, J., Andujar, F.J., Sanchez, J., Alfaro, F., Duato, J.: C-switches: increasing switch radix with current integration scale. In: 13th IEEE International Conference on High Performance Computing and Communications, HPCC 2011, pp. 40–49. Banff, Alberta, Canada (2011)
10. Qiao, B., Shi, F., Ji, W.: THIN: A new hierarchical interconnection network-on-chip for SOC. In: 7th International Conference, ICA3PP, pp. 446–457. Hangzhou, China (2007)
11. Rahman, M.M.H., Inoguchi, Y., Sato, Y., Horiguchi, S.: TTN: a high performance hierarchical interconnection network for massively parallel computers. J. IEICE Trans. Inf. Syst. **E92D**(5), 1062–1078 (2009)

Prioritized Probabilistic Caching Algorithm in Content Centric Networks

Warit Sirichotedumrong, Wuttipong Kumwilaisak, Saran Tarnoi and Nattanun Thatphithakkul

Abstract We propose a new prioritized probabilistic caching scheme for content-centric networks (CCN) to selectively cache more important data more than the others by using heterogeneous caching probabilities. The proposed caching algorithm is used along with Least Recently Used (LRU) cache replacement policy. The caching algorithm allows each CCN node to cache data based on popularity and priority of the data. The popularity of data is determined based on the number of corresponding interest packets sent by content requesters. The priority of data is dictated by the influence of the data to the quality of content reconstructed from the data. With these popularity and priority, the caching probability can be computed by weighing summation. We conduct computer simulations to evaluate the performance of the proposed caching algorithm. Results show that the proposed caching algorithm outperforms a universal caching scheme in terms of content quality at a requester and caching performance such as cache-hit percentage and server load reduction.

Keywords Prioritized probabilistic caching scheme · Content-centric networks · Caching probability · Universal caching scheme

---

W. Sirichotedumrong (✉) · W. Kumwilaisak
King Mongkut's University of Technology Thonburi, Bangkok, Thailand
e-mail: warit.siri@mail.kmutt.ac.th

W. Kumwilaisak
e-mail: wuttipong.kum@kmutt.ac.th

S. Tarnoi
National Institute of Informatics, Tokyo, Japan
e-mail: saran@nii.ac.jp

N. Thatphithakkul
National Electronics and Computer Technology Center, Khlong Luang, Pathum Thani, Thailand
e-mail: nattanun.thatphithakkul@nectec.or.th

P. Meesad et al. (eds.), *Recent Advances in Information and Communication Technology 2016*, Advances in Intelligent Systems and Computing 463,
DOI 10.1007/978-3-319-40415-8_25

1 Introduction

Content-centric networking (CCN) was proposed to be one of the promising Future Internet architectures [1]. In CCN, the current Internet architecture, which was designed for end-to-end communications, is replaced by a new architecture that focuses on the content rather than the location of the content. CCN utilizes prefix names to identify content objects. The content forwarding mechanism in CCN is operated by routers, which are called "CCN routers". There are three main tables in a CCN router: the FIB (Forwarding Information Base); PIT (Pending Interest Table); and CS (Content Store). CS is an important part of CCN because it allows CCN routers to cache content data, then CCN router can disseminate cached content data to content requesters. Caching content data in CS can reduce the network traffic, round-trip hop distance, and round-trip time in CCN. With CS, a CCN node can respond to an "interest packet", which carries a content request sent by a content requester, with a "data packet", which contains content data that corresponds to the content request. Nevertheless, the capacity of CS is limited by memory devices. Only a small fraction of the content available in networks can be cached in CS. Therefore, caching management, which consists of a caching scheme and a cache replacement policy, is of essential.

According to forecasting in Cisco Visual Networking Index (Cisco VNI) [2], the global consumer Internet video traffic will be increasing from 64 to 80 % during 2014–2019. It shows that the video contents are widespread and increasing on the internet. A video content is divided into many video frames, and a video frame can be encoded by using two different coding techniques: intra-frame coding and inter-frame coding [3]. The results of the frame coding are I-frame, B-frame, and P-frame. Because of using a reference frame for predicting other frames in video compression, it has to be ensured that the reference frame is not corrupted during transmission. In a viewpoint of disseminating video as content in CCN, each video frame can be considered a content data. Data packets contain multiple coded frames which are called "data layers". According to the characteristic of video compression, reference frames should obtain higher priority compared to those assigned to other frames. Unequally prioritizing data layers is an important issue in video transmission on practical networks.

Various caching strategies were proposed [1, 4–9]. By default, CCN employs a universal caching scheme, which allows each CCN nodes to cache every content object that does not exist in CS [1]. The popularity-based caching strategies were proposed in [4, 5]. These caching strategies let CCN nodes cache content data based on popularity of the contents. The probabilistic caching strategies were presented in [6–8]. These strategies use caching probability to determine which content is cached. A cache replacement strategy for scalable video streaming (SVC) in CCN was presented in [9]. However, the previous work does not consider the unequal priority of data when caching contents in CCN, which is of great importance especially in multimedia applications. Therefore, we propose a new caching scheme that takes both popularity and priority of data into the consideration. The proposed caching scheme selectively caches more important content data more than the others by using caching probabilities. The caching probability is calculated from both data popularity and priority. We evaluate the performance of our proposed caching scheme by computer simulations via normalized information values reflecting the quality of reproduced content at requesters, cache-hit ratio, and server load comparing with the universal caching scheme.

The rest of this paper is organized as follows. Sections 2 and 3 describe our caching system model and the method in calculating priority data caching probability, respectively. Section 4 presents simulation results and discussion for various study cases. Finally, we summarize this work in Sect. 5.

2 System Model

The network used to study the prioritized probabilistic caching algorithm is shown in Fig. 1. This network consists of M CCN routers, a content provider, and a content requester. We assume that the content provider, CCN routers, and the requester are connected by bottleneck-free links with sufficiently large bandwidth. Cached content data can reside in a CS until the CS evicts the content data. When a data packet arrives at the CCN router, CS makes a decision whether it will cache content data carried by the data packet based-on a caching scheme. If the memory of CS is full, and the caching decision is positive, the CCN router evicts the least-recently used content data from its CS to make a room for new content data.

2.1 Content Request Behavior

A content requester utilizes Zipf-mandelbrot distribution as a content request model and sends interest packets for multiple data layers to request for desired content. When the content provider receives an interest packet, multiple data layers of the requested content will be sent back to the content requester. For example, if the content requester desires a video frame, "/video/", the content requester will send n interest packets for n data layers: "/video/l0", "/video/l1", ..., "/video/ln".

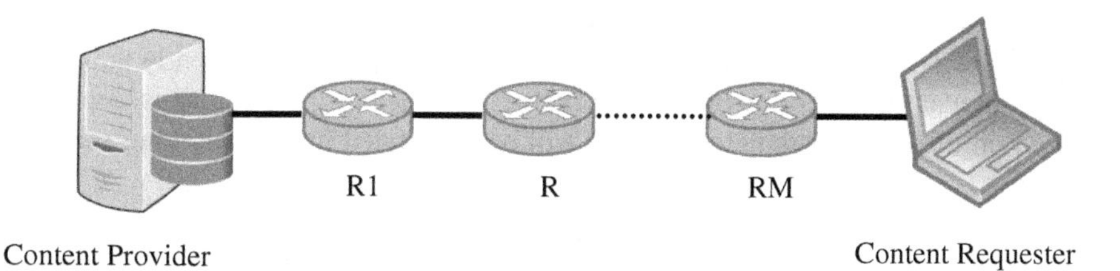

Fig. 1 The network topology

2.2 *Round-Trip Time*

A data packet is not considered useful for decoding if its round-trip time is larger than T, where T is the maximum acceptable delay time for video decoder. Even though the link capacity in our network is sufficient for satisfying every interest packets, T affects the quality of reproduced video, which is described in the next section.

3 Prioritized Probabilistic Caching Scheme

It is widely known that in-network caching at CCN routers plays an important role in the CCN network architecture. When the CCN routers receive interest packets sent from receivers, they can immediately send the corresponding data packets to the receivers if requested data exists in CS, instead of retrieving the data from an original content provider. This can shorten the time spent by receivers to obtain the requested data and reduce the server load of the content provider.

In this section, we propose the prioritized probabilistic data caching scheme for CCN, which considers both the importance and popularity of data packets. The content data in a data packet is randomly cached by CCN routers with a caching probability. We only focus on the caching decisions in CS and use Least-Recently used (LRU) as a cache replacement policy. CS at each CCN router stores content data based on its importance and popularity under storage constraints. The importance of content data is defined based on its contribution to the quality of reconstructed content on the receiver side. For example, in video data transmission, video data corresponding to an I-frame is more important than video data corresponding to P- and B-frames. The loss of data packets from I-frame can decrease reconstructed video quality significantly. Here, we categorize video data into multiple layers such as a base layer and enhancement layers. When the content requester requests video data, it will send multiple interest packets for multiple data layers to CCN routers, which in turn forward these interest packets to the content provider, to obtain video content with the best quality under constraints on delay time and network resources. At least, a base layer of the video content must be received at the requester so that the requester reconstructs the video with base quality. The popularity of content data is measured via the frequency content requesters request the content data. For example, content data corresponding to well-known movies tend to be requested from customers more frequently than those of unpopular ones. In our proposed scheme, the caching probability is computed based on these two parameters, which will be elaborated in the following sections.

3.1 Least Recently Used (LRU)

LRU maintains the most recently used data at the top of a cache. When there is new data entering the cache, the data will be placed at the top of the cache, pushing the rest of data in the cache down to the bottom. If a cache limit has been reached, data that are least recently requested will be cleared out from the cache for the incoming data. When data in the cache matches a request, the data will be moved to the top of the cache.

3.2 Caching Decision Factors

To decide whether a CCN router caches or does not cache content data in incoming data packets, popularity and priority are taken into consideration. These two parameters are utilized to compute caching probability for each data packet. Define the popularity of caching content carried by a data packet, α, via a ratio between a number of interest packets corresponding to the data packet and the total interest packets. Define the priority value γ of each data packet based on its contribution to the reconstructed data quality. For example, in video applications, the importance of data packet can be based on the error propagation effect to the reconstructed video quality when these data packets are lost. However, in this paper, we assign the priority value to each data packet based on the data layer which its content belongs to. The more importance of data packet, the higher priority value it obtains. The example of information value assignment can be shown in Table 1.

3.3 Caching Probability

The proposed caching scheme will cache content data carried by data packets based on their caching probabilities, which integrate the effects on both priority and popularity of data packets.

Define $P(n)$ as a caching probability of the nth data layer, which can be expressed as

$$P_i(n) = a \cdot \alpha_i + b \cdot \gamma_n, \tag{1}$$

where α_i is the popularity of the ith content, and γ_n is the priority values of the nth data layer, $a, b \in [0, 1]$, and $a + b = 1$.

Table 1 The example of priority values of data layers

Layer	Base layer	First layer	Second layer	Third layer
Priority value(γ)	1.0	0.3	0.2	0.1

Note that the caching probabilities for data packets are dynamic along the time due to the change of popularity values, whereas the priority values are static as shown in Table 1. In general, these caching probabilities will be calculated in advance by individual CCN routers. When a data packet arrives at a CCN router, the CCN router will randomly select a positive real number between zero and one. If the random number is less than the caching probability for the underlying data packet, content data inside the data packet will be cached into the CS. Otherwise, it will not be cached.

For the CCN router, there are total N caching probabilities corresponding to N data layers. The CCN router may possess different $P_i(n)$, when the popularity of content data inside a data packet is taken into account with the priority of the data packet. This is because different CCN routers can have different numbers of interest packets. For example, consider Fig. 1. Suppose that the first CCN router (R1) receives 100 interest packets, which consist of 70 interest packets for the base layer and 30 interest packets for the first enhancement layer. Therefore, the popularity value (α_0) of the base layer at R1 is equal to 0.7, and the popularity value (α_1) of the first enhancement layer at R1 is equal to 0.3. Moreover, suppose that the second CCN router (R2) receives 150 interest packets, which consist of 100 interest packets for the base layer and 50 interest packets for the first enhancement layer. The popularity values of the base layer and the first enhancement layer at R2 are approximately equal to 0.67 and 0.33 respectively. If only the priority of data packet is considered without the popularity of data packet, all CCN routers will have the same set of caching probabilities because a number of interest packets will not be taken into consideration and the priority of data layer is static.

4 Simulation Results

We conduct our simulations to evaluate the performance and characteristics of the prioritized probabilistic caching scheme by using ndnSIM-2.0 [10], which is an NS-3 based Named-Data-Networking (NDN) simulator. The CS in each CCN router uses LRU as a cache replacement policy and our proposed prioritized probabilistic caching scheme.

4.1 Simulation Set-Up

The network in this simulation consists of a CCN router (R1), a content provider, and a content requester. The CCN router connects to a content provider, and a content requester, as shown in Fig. 1. The capacity of CS in our simulation is set to 1% of the content population. The link capacity and propagation delay of the simulation are 10 Mbps and 100 ms, respectively. The total simulation time is equal to 1,000 s.

We use Zipf-Mandelbrot distribution as a content request model for the requester. The simulation consists of various cases with different popularity and priority coefficients. First, the caching probability depends on only the popularity of the data layer, which is defined by setting $a = 1, b = 0$. The caching probability in the second scenario only depends on the priority of the data layer, which is defined by setting $a = 0, b = 1$. Equally weighing between popularity and priority values is the third scenario. In this case, CCN routers consider both popularity and priority values of the data packet. This case is defined by setting $a = 0.5, b = 0.5$. We also conduct the simulation with other popularity and priority coefficients to find the set of coefficient that gives maximum normalized information value. Furthermore, the simulations are performed with a different number of layers in a set of $\{1, 2, 3\}$ for each scenario. We define "/prefix0" and "/prefix1" as names of data and assign suffixes to represent the layers of the data. For example, "/prefix0/l0" stands for the base layer of prefix0, "/prefix0/l1" presents the first enhancement layer of prefix0, "/prefix0/ln" denotes the nth enhancement layer of prefix0, and "/prefix1/ln" stands for the nth enhancement layer of prefix1. For all data requested by a requester, a content provider will send all data layers back to the requester. For simplicity, we assume that all data layers have the same size.

4.2 Evaluation Metrics

We evaluate our simulations using three metrics, which are described as follows.

Normalized Information Value. Since we focus on the priority of data, normalized information value (V) is used for evaluating the quality of reconstructed data and for obtaining caching performance. Higher normalized information value implies that the requester obtains better content quality. In the simulation, different weights are assigned to different layers. Base layer (l0) has the largest weight, and other enhancement layers have lower weight, as shown in Table 2. Specifically, the normalized information value is calculated by

$$V = \frac{w_0 \cdot N_0 + \sum_{i=1}^{n-1} (w_i \cdot N_i)}{\sum_{j=0}^{n-1} (w_j \cdot K_j)}, \quad (2)$$

where w_0 is the weight of the base layer and w_i is the weight of the ith enhancement layer. K is a number of the total received data. Let T be the maximum delay that data packet is still usable at receivers. N_0 is the number of data packets carrying base layer data that has round-trip time less than T. N_i is the number of data packets that carry the data of the ith enhancement layer and satisfy all of the following conditions: (1) the requester already receives all of the lower layers; and (2) the round-trip time is less than T. According to the real-time media end-to-end maximum delay in [11], T is equal to 250 ms to evaluate the data quality of real-time multimedia.

Table 2 The weight of each layer for normalized information values calculation

Layer	Base layer	First layer	Second layer
Weight	1	0.3	0.2

Cache-hit Ratio: Cache hit-ratio is a ratio between cache-hits and the sum of cache-hits and cache-misses. It is widely used for evaluating the performance of caching schemes. A high cache-hit ratio shows that the caching scheme has high performance. In this simulation, we utilize average cache-hit ratio for comparing caching performance among different scenarios.

Server Load: Server load is one of the evaluations metric that describes the performance and the contribution of in-network caching. The server load is equal to the total number of requests that a content provider receives from content requesters. The low server load means that CS can perform well and there are many cache-hits at CCN routers in a network. The high server load could imply that the content provider would need to be upgraded to more capable hardware for providing the acceptable quality of services to the requesters.

Proportion of Data Layers in CS: As we split data into multiple layers, there are many layers of data that are cached by CSs of CCN routers. For studying characteristics and properties of prioritized probabilistic caching scheme, we calculate an average proportion of cached data layers in the CCN router, R1. An average proportion of cached data layers also explain the reasons of increasing or decreasing normalized information values.

4.3 Experimental Results and Discussions

According to the network topology in Fig. 1, the content requester requests for two contents, which are named as "/prefix0" and "/prefix1", with different content request frequencies. Figure 2 shows the normalized information value while "/prefix0" and "/prefix1" are requested with the equal and unequal request frequencies. In the case of unequal request frequency, "/prefix0" is requested at a rate of 140 times per second, and "/prefix1" is requested at a rate of 60 times per second. From the results, every case of prioritized priority caching strategy offers better a normalized information value than the universal caching scheme. Considering the case of equal content request frequency, the prioritized probabilistic caching scheme gives the best normalized information value when popularity and priority coefficients are equal to 0.5 and 0.5, respectively. In the case of unequal request frequency, the normalized information value will be high when the weighting coefficient corresponding to the popularity is high.

As shown in Fig. 3a, b, universal caching scheme equally treats all data layers. On the other hands, the prioritized probabilistic data caching strategies cache data of the base layer more than that of the first enhancement layer, ensuring that the content requester will perceive at least the content quality of the base layer.

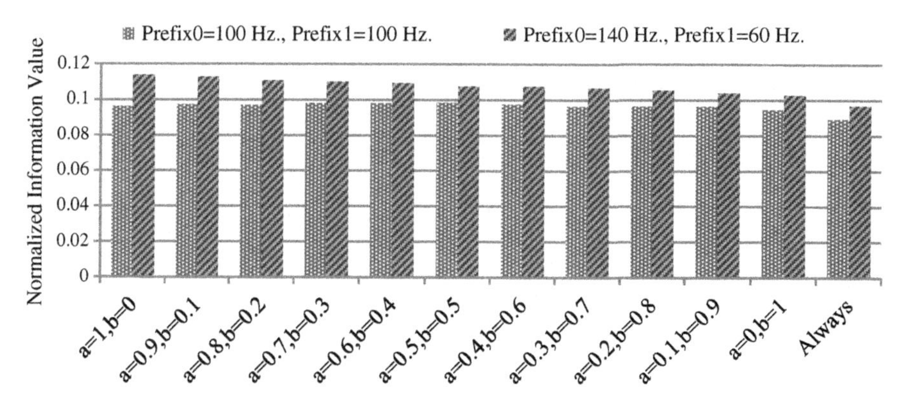

Fig. 2 Normalized information value

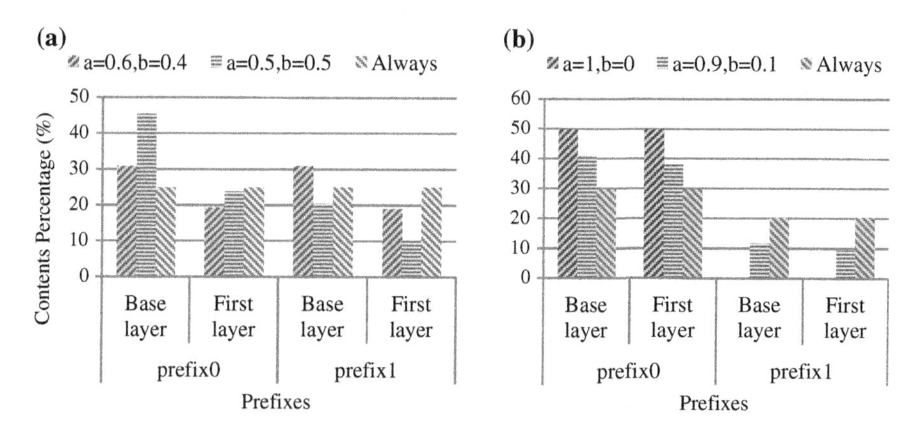

Fig. 3 Average data layers proportion of each prefix in the CS. **a** Equally content request frequency. **b** Unequally content request frequency

In the case of equal request frequency, when $a = 0.5, b = 0.5$ and $a = 0.6, b = 0.4$ our caching scheme provides the promising normalized information value, as shown in Fig. 2. However, both of them differently give attention to the data layers. In Fig. 3a, when $a = 0.5, b = 0.5$ our caching scheme prefers base layer of "/prefix0" to the others, and when $a = 0.6, b = 0.4$ the caching algorithm equally takes care base layer of "/prefix0" and "/prefix1". Therefore, setting $a = 0.6, b = 0.4$ gives the optimal popularity and priority coefficients when the receiver request content equally. Moreover, Fig. 3b also shows that considering only content popularity rarely caches "/prefix1", which is a less popular content. Although only considering the popularity offers the best information value, "/prefix1" is rarely cached and nearly ignored by CS. Prioritized caching strategy with $a = 0.9, b = 0.1$ is more suitable if the receiver requests contents with different frequency because CCN router does not ignore "/prefix1" and takes care of "/prefix0" more than "/prefix1".

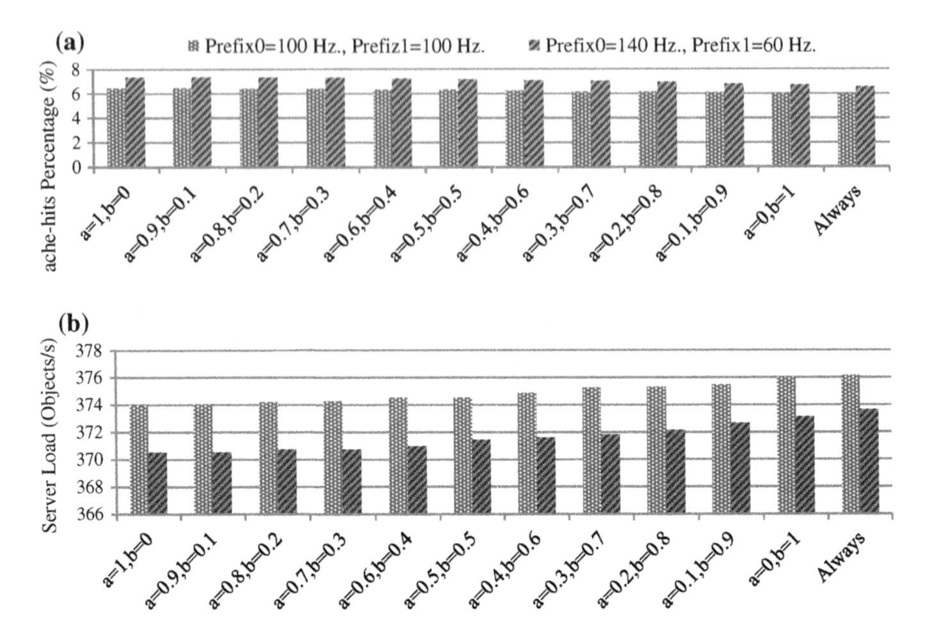

Fig. 4 Cache-hit percentage and server load. **a** Average cache-hit percentage at CCN router. **b** Average server load

Other than normalized information value, caching performance is evaluated by cache-hit percentage and server load. As shown in the Fig. 4a, b, prioritized probabilistic data caching strategies provide better performance than universal caching scheme in terms of cache-hit percentage and server load. Interestingly, from our experimental results, a set of coefficients giving better cache hit performance may not lead to a better perceived content quality (lower normalized information value).

From the normalized information values and caching performance evaluation, setting $a = 0.6$, $b = 0.4$ are the optimal coefficients when the requester equally requests the two videos. If the receiver requests prefer a video to the other video, $a = 0.9, b = 0.1$ are the optimal coefficients.

5 Conclusion

We proposed a new prioritized probabilistic caching scheme which randomly caches data layers via different caching probabilities in CCN architecture. The caching probability of this caching scheme depends on two factors: popularity values and priority values of data layers. The computer simulation results showed that the proposed caching scheme gave better performance than the universal caching scheme in terms of reconstructed data quality at requesters, cache-hit percentage, and server load reduction. With the optimal popularity and priority coefficients, our prioritized probabilistic caching scheme offers the best normalized information values.

Acknowledgements This research is supported in part by Thailand Graduate Institute of Science and Technology (TGIST) National Science and Technology Development Agency (NSTDA) with contract number TGIST 01-57-022.

References

1. Jacobson, V., Smetters, D.K., Thornton, J.D., Plass, M.F., Briggs, N.H., Braynard, R.L.: Networking named content. In: Proceedings of ACM CoNEXT, pp. 1–12. ACM, Rome (2009)
2. Cisco Visual Networking Index: Forecast and Methodology, 2014–2019 White Paper. http://www.cisco.com
3. Kumwilaisak, W.: Image and Video Communication Systems: Representation, Compression, and Networks. Jarunsanitwong Press, Bangkok (2015)
4. Bernardini, C., Silverston, T., Festor, O.: MPC: popularity-based caching strategy for content centric networks. In: IEEE International Conference on Communication (ICC), pp. 3619–3623. IEEE, Budapest (2013)
5. Cho, K., Lee, M., Park, K., Kwon, T.T., Choi, Y., Pack, S.: WAVE: popularity-based and collaborative in-network caching for content-oriented networks. In: Proceedings of INFOCOM WKSHPS, pp. 316–321. IEEE, Orlando, FL (2012)
6. Psaras, I., Chai, W.K., Pavlou, G.: In-network cache management and resource allocation for information-centric networks. In: IEEE Transactions on Parallel and Distributed Systems, vol. 25, no. 11, pp. 2920–2931. IEEE (2013)
7. Psaras, I., Chai, W.K., Pavlou, G.: Probabilistic in-network caching for information-centric networks. In: Proceedings of the Second Edition of the ICN Workshop on Information-centric Networking, pp. 55–60. ACM, Helsinki (2012)
8. Tarnoi, S., Suppakitpaisarn, V., Ji, Y.: Adaptive Probabilistic Caching for Information-Centric Networking. IEICE Technical Report (2015)
9. Lee, J., Lim, K., Yoo, C.: Cache replacement strategies for scalable video streaming in CCN. In: 19th Asia-Pacific Conference on Communications (APCC), pp. 184–189. IEEE, Denpasar (2013)
10. Mastorakis, S., Afanasyev, A., Moiseenko, I., Zhang, L.: ndnSIM 2.0: A new version of the NDN simulator for NS-3. NDN, Technical Report NDN-0028 (2015)
11. Perea, R.M.: Internet Multimedia Communications Using SIP. Morgan-Kaufmann, San Francisco (2008)

KSBS Solution of Power Allocation in Multi-user Multi-relay Networks Using Stackelberg Game

Supenporn Somjit, Pattarawit Polpinit and Chatchai Khunboa

Abstract This paper considers a multi-user multi-relay network. The relay power allocation among the users and pricing problem are studied. We model the interaction between the relay and users as a Stackelberg game, where the relay is the leader who gets paid for helping users forward signal, and user is the followers who pay to receive relay service. For the relay power allocation, a bargaining game is deployed to model the negotiation among the users on relay power allocation. We propose the Kailai–Smorodinsky Bargaining Solution (KSBS) of bargaining game in order to formulate the relay power allocation. Simulation results are shown the proposed KSBS-based scheme has better network sum-rate than even power allocation and achieves close-to-sum-rate-optimal solution.

Keywords Kailai–Smorodinsky bargaining solution (KSBS) · Multi-user multi-relay network power allocation · Stackelberg game

1 Introduction

Recently, cooperative communications have been adopted in order to improve the performance of communication in wireless network [1]. The basic idea is to have some nodes help each other's transmission to achieve diversity and increase data rate. The cooperative strategy aim is to optimize the global network performance. Two main relaying schemes used cooperative strategies are amplify-and-forward (AF) and decode-and-forward (DF) [2]. For multi-user multi-relay network, the

S. Somjit (✉) · P. Polpinit · C. Khunboa
Faculty of Engineering, Department of Computer Engineering,
Khon Kaen University, Khon Kaen, Thailand
e-mail: s.supenporn@kkumail.com; s.supenporn@gmail.com

P. Polpinit
e-mail: polpinit@kku.ac.th

C. Khunboa
e-mail: chatchai@kku.ac.th

P. Meesad et al. (eds.), *Recent Advances in Information and Communication Technology 2016*, Advances in Intelligent Systems and Computing 463,
DOI 10.1007/978-3-319-40415-8_26

performances in cooperative communications depend on resource allocation among the users. Recently, numerous works have investigated the relay power allocation as shown in [3–13].

Game theory was recently used in resource allocation in cooperative communication to optimize the global network performance. In [6, 7], Stackelberg game model was proposed to perform power allocation in a relay network. In [8–10], Nash bargaining game was applied in relay network in order to analyze the relay power allocation among the user. They consider the balance between the network sum-rate and the user fairness. The work in [11] studies the relay power allocation and pricing problem, uses the KSBS of bargaining game to formulate the relay power allocation. The relay power allocation problem is considered to maximize the network sum-rate. The power allocation in multi-user multi-relay was studied in [12, 13].

However, most existing work focuses on a game theoretic approach to power allocation in multi-user single-relay network. This observation motivates this research to study an even more general problem of the relay power allocation and pricing problems in multi-user multi-relay network. The problem will be investigated KSBS-based power allocation and Stackelberg game.

The reminder of paper is organized as follows: Sect. 2 describes the multi-user multi-relay network. The relay power allocation and pricing solution are proposed and studied in Sect. 3. Simulation results are shown in Sect. 4. Section 5 contains the conclusion.

2 System Model

We consider multi-user multi-relay network as shown in Fig. 1. There are N users and R Relay where each user communicating with its destinations of which one or more relay will be used. Define $i \in N$ and $r \in R$, we denote the transmit power of User i as Q_i, the total power constraint of all relay as P and Relay r uses power $P_{i,r}$ to help User i's transmission.

In this paper, the dimension of transmission is $N \times R$. We define the transmission on this network as an N-dimensional model, which equivalent to that of a single-relay network with $N \times R$ transmissions purchasing power from one relay with power constraint P. We define User i's transmission as transmission t wants to buy power $P^t_{i,r}$ from the Relay r where $t \in N \times R$.

The amplify-and-forward (AF) cooperation protocol is employed in the system. Denote channel gain from User i's transmission to relay r as $f^t_{i,r}$, the channel gain from relay r to Destination i as $g^t_{i,r}$ and channel gain of direct link between the User i to Destination i as h_i. We consider Reylieigh flat-fading channel and assume the relay has global and perfect knowledge of channel state information (CSI).

The half-duplex two-step AF protocol is used. In the first step, User i transmits $(\sqrt{Q_i}S_i)$ to each relay node r and Destination i, where S_i is the information symbol

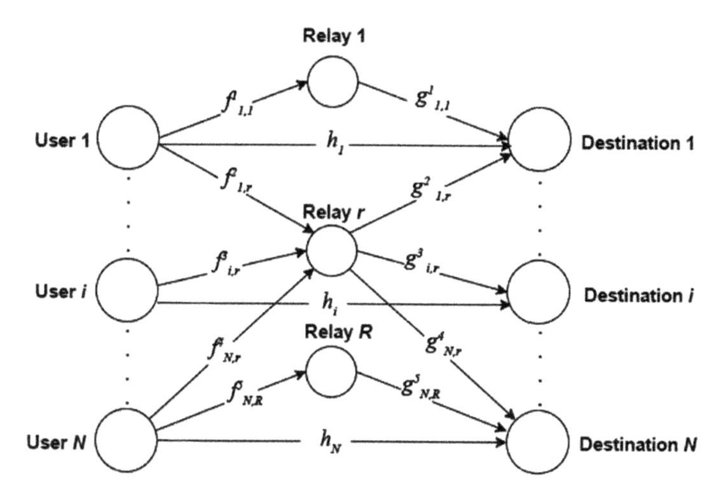

Fig. 1 Multi-user multi-relay network

normalized as $E(|S_i|^2) = 1$ where E stands for the average. The received signal at Relay r and Destination i are

$$y_{i,r}^t = \sqrt{Q_i} S_i f_{i,r}^t + n_{i,r}^t \text{ and } y_i = \sqrt{Q_i} S_i h_i + n_i \quad (1)$$

where $n_{i,r}^t$ and n_i are the noise at the Relay r and the Destination i in the first step.

The second step, Relay r amplifies $y_{i,r}^t$ and forwards it to Destination i with transmitted power $P_{i,r}^t$. The received signal at Destination i is

$$y_{r,i}^t = \sqrt{\frac{Q_i P_{i,r}^t}{Q_i |f_{i,r}^t|^2 + 1}} s_i f_{i,r}^t g_{i,r}^t + \sqrt{\frac{P_{i,r}^t}{Q_i |f_{i,r}^t|^2 + 1}} g_{i,r} n_{i,r} + n_{r,i} \quad (2)$$

where $n_{r,i}^t$ is the noise at the Destination i in the second step.

We assume all noise in different channels are i.i.d. additive circularly symmetric complex Gaussian with zero-mean unit variance. After combining the direct path and relay path the signal to noise ratio (SNR) of User i at the Destination i is

$$SNR_i = \sum_{r=1}^{R} \frac{Q_i P_{i,r}^t |f_{i,r}^t g_{i,r}^t|^2}{P_{i,r}^t |g_{i,r}^t|^2 + Q_i |f_{i,r}^t|^2 + 1} + Q_i |h_i|^2 \quad (3)$$

If User i is not helped by any relay and uses direct transmission only the SNR at the Destination i is

$$SNR_i = Q_i |h_i|^2 \quad (4)$$

3 Relay Power Allocation and Pricing Solution

In this section, the Stackelberg game is employed to model the interaction between the user and the relay. We use bargaining game to analyze the conflict and interaction among users and formulate the relay power allocation problem.

3.1 Stackelberg Game

The stackelberg game, one player take action first as a leader and the other player observe the leader's action and act accordingly as the follower [14].

We consider relay as the leader who set the price of power in helping the users. The relay will set the price to gain the maximum revenue. The revenue of relay is $\sum_{i=1}^{N}\sum_{r=1}^{R}\lambda P_{i,r}^{t}(\lambda)$ where λ is the normalized unit price of the relay power and $P_{i,r}^{t}$ is the power of relay r in helping transmission t.

We consider the user as the followers who react in unit price of relay power. The user will buy relay power to gain the maximum utility and user can buy relay power from multi-relay. The bargaining game is used to model the cooperative interaction among users. We formulate the utility function from SNR of User i (3) we can define User i's utility function as

$$u_i = \sum_{r=1}^{R} \frac{Q_i P_{i,r}^{t} \left| f_{i,r}^{t} g_{i,r}^{t} \right|^2}{P_{i,r}^{t} \left| g_{i,r}^{t} \right|^2 + Q_i \left| f_{i,r}^{t} \right|^2 + 1} + Q_i |h_i|^2 - \sum_{r=1}^{R} \lambda P_{i,r}^{t} \tag{5}$$

The first two terms of (5) correspond to the effective received SNR of User i, and the last term represents the User i's total cost in purchasing power from relays.

If user does not purchase any relay power and use the direct transmission only, its utility is

$$u_{i,0} = Q_i |h_i|^2 \tag{6}$$

which $u_{i,0}$ is the minimum utility that User i expects when it does not purchase power from any relays.

From the received utility of User i can be calculated the network sum-rate follows:

$$U_{sum} = \sum_{i=1}^{N} \log_2(1 + u_i) \tag{7}$$

3.2 Relay Power Allocation

The bargaining game is used to model interaction among independent users. The user game is to find the relay power allocation among the user for a given unit power price λ. We look for KSBS to formulate relay power allocation, which guarantees fairness in the sense of equal penalty. The KSBS-based power allocation problem is equivalent to that of a single-relay network allocate power constraint P to $N \times R$ transmissions.

To find KSBS-based power allocation, we first calculate ideal utility of User i's transmission by relay r when User i's transmission expect power $P_{i,r}^{t,I}$ from relay r, $u_{i,r}^{t,I}$ of given λ to maximize its utility and all relay use the same unit relay power is λ. Ideal utility of User i's transmission by Relay r when User i's transmission expect power $P_{i,r}^{t,I}$ from Relay r can be found using Lemma 1 as follows [11].

Lemma 1 [11] *Define the price above which transmission t of User i will not purchase any relay power from Relay r as*

$$\bar{\lambda}_{i,r}^{t} \triangleq \frac{Q_i \left| f_{i,r}^{t} g_{i,r}^{t} \right|^2}{Q_i \left| f_{i,r}^{t} \right|^2 + 1} \tag{8}$$

Given the unit price of all relay power λ, the ideal power of transmission t demand from Relay r to maximize its utility $u_{i,r}$ is

$$P_{i,r}^{t,I}(\lambda) = \begin{cases} 0, & \text{if } \lambda \geq \bar{\lambda}_{i,r}^{t} \\ \frac{Q_i \left| f_{i,r}^{t} \right|^2}{\sqrt{\bar{\lambda}_{i,r}^{t}}} \left(\frac{1}{\sqrt{\lambda}} - \frac{1}{\sqrt{\bar{\lambda}_{i,r}^{t}}} \right), & \text{if } \bar{\lambda}_{i,r}^{t} > \lambda > \bar{\lambda}_{i,r}^{t} \left(\frac{\bar{\lambda}_{i,r}^{t} P}{Q_i \left| f_{i,r}^{t} \right|^2} + 1 \right)^{-2} \\ P, & \text{if } \lambda \leq \bar{\lambda}_{i,r}^{t} \left(\frac{\bar{\lambda}_{i,r}^{t} P}{Q_i \left| f_{i,r}^{t} \right|^2} + 1 \right)^{-2} \end{cases} \tag{9}$$

The ideal utility of transmission t demand from Relay r to maximize network sum-rate is

$$u_{i,r}^{t,I}(\lambda) = \begin{cases} u_{i,0}, & \text{if } \lambda \geq \bar{\lambda}_{i,r}^{t} \\ Q_i \left| f_{i,r}^{t} \right|^2 \left(1 - \sqrt{\lambda / \bar{\lambda}_{i,r}^{t}} \right)^2 + u_{i,0}, & \text{if } \bar{\lambda}_{i,r}^{t} > \lambda > \bar{\lambda}_{i,r}^{t} \left(\frac{\bar{\lambda}_{i,r}^{t} P}{Q_i \left| f_{i,r}^{t} \right|^2} + 1 \right)^{-2} \\ \frac{\bar{\lambda}_{i,r} P}{\left(Q_i \left| f_{i,r}^{t} \right|^2 \right)^{-1} \bar{\lambda}_{i,r}^{t} P + 1} - \lambda P + u_{i,0}, & \text{if } \lambda \leq \bar{\lambda}_{i,r}^{t} \left(\frac{\bar{\lambda}_{i,r}^{t} P}{Q_i \left| f_{i,r}^{t} \right|^2} + 1 \right)^{-2} \end{cases} \tag{10}$$

From Lemma 1, in case 1 (9), the ideal power demand when the price is too high transmission t will not buy any power from the Relay r. When the price is too low, in case 3 (9), transmission t wants to buy total power constraint of all relay. When the price range in case 2 (9) transmission t will buy part of relay power from Relay r that give ideal balance between its SNR and its payment to maximize its utility.

To find the KSBS-based power allocation of each transmission t, First the number of transmission who enter the bargaining game is determined. This is achieved by sorting $\bar{\lambda}_{i,r}^{t}$ of User i's transmission in descending order. Then let L be the number transmission who receive $P_{i,r}^{t}$ from Relay r and $\lambda \leq \bar{\lambda}_{i,r}^{t}$, where $\bar{\lambda}_{i,r}^{L} > \lambda > \bar{\lambda}_{i,r}^{L+1}$. Therefore $\bar{\lambda}_{i,r}^{t}$ of transmission t are sorted in descending order as follows

$$\bar{\lambda}_{i,r}^{1} \geq \bar{\lambda}_{i,r}^{2} \geq \cdots \geq \bar{\lambda}_{i,r}^{L} \geq \bar{\lambda}_{i,r}^{L+1} \geq \cdots \geq \bar{\lambda}_{i,r}^{N \times R} \tag{11}$$

With the given price λ and all relay in network use the same price λ, for transmission t who want to receive $P_{i,r}^{t}$ from Relay r but $\lambda \geq \bar{\lambda}_{i,r}^{t}$ as shown in Lemma 1, they do not enter the game, thus their ideal power is 0.

The first L transmission who helped by Relay r will participate in the bargaining game and buy relay power. To find the KSBS-based power allocation of L transmission is equivalent to the following problem, whose proof can be found in [15]

$$\max_{P_{i,r}} k \qquad s.t. \frac{\dfrac{\bar{\lambda}_{i,r}^{t} P_{i,r}^{t}}{\left(Q_i \left|f_{i,r}^{t}\right|^2\right)^{-1} - \bar{\lambda}_{i,r}^{t} P_{i,r}^{t} + 1} - \lambda P_{i,r}^{t}}{u_{i,r}^{t,I} - u_{i,0}} = k \tag{12}$$

where k is constant independent of transmission t who helped by Relay r.

The constraint in (12) is due to the sum of relay power of transmission t not exceed total relay power of all relay. Moreover the other constraint is $0 < P_{i,r}^{t} \leq P_{i,r}^{t,I}(\lambda)$ for ensuring the user cannot request relay power larger than ideal power of transmission t demanding from Relay r.

Solution of Eq. (12) call the KSBS-based power allocation and only first L transmission t who helped by relay r and $\bar{\lambda}_{i,r}^{t}$ larger than the unit price of relay power λ participate in bargaining game. The remaining transmission t whose $\bar{\lambda}_{i,r}^{t}$ less than the unit price of relay power λ does not request any relay power. The problem KSBS-based power allocation under consideration of relay power constraints, for given unit price of relay power λ, total power demand by the transmission t exceeds the relay power constraint. The KSBS-based power allocation will allocate all relay power to the transmission fairly, as shown in the following Lemma 2, whose proof can be found in [10].

Lemma 2 [11] *For a fix λ, let the ideal power allocation of transmission t demand relay power from Relay r be $P_{i,r}^{t,I}(\lambda)$, the KSBS-based power allocation be $P_{i,r}^{t,K}(\lambda)$. When sum of ideal power of L transmission less than total relay power P, we have*

$P_{i,r}^{t,K}(\lambda) = P_{i,r}^{t,I}(\lambda)$; when sum of ideal power of L transmission larger than total relay power P, we can find $P_{i,r}^{t,K}(\lambda)$ from (12) and total $P_{i,r}^{K}(\lambda)$ of L transmission equal to P, where $P_{i,r}^{K}(\lambda)$ is the relay power from Relay r allocate to transmission t based on the KSBS for the given price λ.

3.3 Optimal Relay Power Price

The relay modeled as the service provider who sets the unit power price for relay service and maximize relay revenue. When the unit price of the relay power is λ and by using the KSBS-based power allocation in Sect. 2, the revenue of the relay is

$$\sum_{i=1}^{N} \sum_{r=1}^{R} \lambda P_{i,r}^{t,K}(\lambda) \tag{13}$$

The relay pricing problem can be formulated as

$$\max_{\lambda} \sum_{i=1}^{N} \sum_{r=1}^{R} \lambda P_{i,r}^{t,K}(\lambda) \tag{14}$$

To find the optimal relay power price, we first show the following Lemma 3 [11].

Lemma 3 [11] *The optimal price is inside the interval $[\bar{\lambda}_{lb}, \bar{\lambda}_{i,r}^{1})$, where $\bar{\lambda}_{lb}$ price at which the total ideal power demands of transmission is P. To find $\bar{\lambda}_{lb}$ in Lemma 3, we need to solve the following equation:*

$$\sum_{i=1}^{N} \sum_{r=1}^{R} \lambda P_{i,r}^{t,I}(\lambda) = \Phi(\bar{\lambda}_{lb}) = P \tag{15}$$

Note that $\Phi(\bar{\lambda}_{lb})$ in (15) monotonically decreases from ∞ to 0 as $\bar{\lambda}_{lb}$ increases from 0 to $\bar{\lambda}_{lb}$. To find the value of $\bar{\lambda}_{lb}$, we can first find the M transmission such that $\Phi\left(\bar{\lambda}_{i,r}^{M}\right) < P$ and $\Phi\left(\bar{\lambda}_{i,r}^{M+1}\right) > P$. Thus $\bar{\lambda}_{lb} \in [\bar{\lambda}_{i,r}^{M}, \bar{\lambda}_{i,r}^{M+1}]$, where M is number of transmission of user i who receive $P_{i,r}^{t}$ from relay r and sum of M transmission's ideal power not exceed P. We can calculate $\bar{\lambda}_{lb}$ using following equation:

$$\bar{\lambda}_{lb} = \left(\sum_{t=1}^{M} \frac{Q_i |f_{i,r}^{t}|^2}{\sqrt{\bar{\lambda}_{i,r}^{t}}} \right)^2 \left(P + \sum_{t=1}^{M} \frac{Q_i |f_{i,r}^{t}|^2}{\bar{\lambda}_{i,r}^{t}} \right)^{-2} \tag{16}$$

From ordering of users based on their $\bar{\lambda}_{i,r}^{t}$ when we sort in descending order and M is index such that $\bar{\lambda}_{i,r}^{M} \geq \bar{\lambda}_{lb} \geq \bar{\lambda}_{i,r}^{M+1}$.

We define $\gamma_{i,r}^{t} \triangleq \bar{\lambda}_{i,r}^{t}$ for $t = 1, \ldots, M$ and $\bar{\lambda}_{i,r}^{M+1} \triangleq \bar{\lambda}_{lb}$ and define $\tau_t = [\bar{\lambda}_{i,r}^{t+1}, \bar{\lambda}_{i,r}^{t}]$ for $t = 1, \ldots, M$ as the price range where t transmission purchase the relay power. The price range $[\bar{\lambda}_{lb}, \bar{\lambda}_{i,r}^{1}]$ can be divided into the following M intervals

$$\begin{aligned}[\bar{\lambda}_{lb}, \bar{\lambda}_{i,r}^{1}] &= [\bar{\lambda}_{lb}, \bar{\lambda}_{i,r}^{M}] \cup [\bar{\lambda}_{i,r}^{M}, \bar{\lambda}_{i,r}^{M+1}] \cup \ldots \cup [\bar{\lambda}_{i,r}^{3}, \bar{\lambda}_{i,r}^{2}] \cup [\bar{\lambda}_{i,r}^{2}, \bar{\lambda}_{i,r}^{1}] \\ &\triangleq \tau_M \cup \tau_{M-1} \ldots \cup \tau_2 \cup \tau_1\end{aligned} \tag{17}$$

From Lemma 2, $P_{i,r}^{t,K}(\lambda) = P_{i,r}^{t,I}(\lambda)$ we can rewrite the price optimization problem in (12) as

$$\max_{t=1,2,\ldots M} \max_{\lambda \in \Gamma_t} \sum_{j=1}^{t} \lambda P_{i,r}^{t,I}(\lambda) \tag{18}$$

In (18), we have decomposed the optimization problem into M subproblems, where the tth subproblem is to find the optimization problem in (18) are solved in the following theorem, the proof of which can be found in [11].

Theorem 1 [11] *Define* $c_{i,r}^{t}$ *following equation*

$$c_{i,r}^{t} \triangleq \left(\frac{\sum_{j=1}^{t} Q_j \left| f_{j,r}^{t} \right|^2 / \sqrt{\bar{\lambda}_{j,r}^{t}}}{2 \sum_{j=1}^{t} Q_j \left| g_{j,r}^{t} \right|^2 / \bar{\lambda}_{j,r}^{t}} \right)^2 \tag{19}$$

The solution to subproblem i is

$$\lambda_{i,r}^{t} \triangleq \begin{cases} c_{i,r}^{t}, \text{ if } c_{i,r}^{t} < \gamma_{i,r}^{t+1} \\ \gamma_{i,r}^{t}, \text{ if } c_{i,r}^{t} > \gamma_{i,r}^{t} \\ c_{i,r}^{t}, \text{ if } \gamma_{i,r}^{t+1} \leq c_{i,r}^{t} \leq \gamma_{i,r}^{t+1} \end{cases} \tag{20}$$

With the subproblems solved, to find the optimal relay power price following equation is used:

$$\lambda^{*} = \arg\max_{\lambda_{i,r}^{t}} \left\{ \left(\sum_{j=1}^{t} \frac{Q_j \left| f_{j,r}^{t} \right|^2}{\sqrt{\bar{\lambda}_{j,r}^{t}}} \right) \sqrt{\lambda_{i,r}^{t}} - \sum_{j=1}^{t} \frac{Q_j \left| f_{j,r}^{t} \right|^2}{\bar{\lambda}_{j,r}^{t}} \lambda_{i,r}^{t} \right\} \tag{21}$$

where λ^{*} *is optimal relay power price*

We can find the optimal price for the relay power price by solving the M subproblem in (18) and then find the optimal price among the M subproblem solutions that result in the maximum relay revenue.

4 Simulation Result

To evaluate the performances of the KSBS-based power allocation, we performed simulations for multi-user multi-relay network with 7 users. Consider the Rayleigh flat-fading model, the channel gains $f_{i,r}^{t}$, h_i and $g_{i,r}^{t}$ are modeled as independent and identically distributed (i.i.d.) random variables following the distribution $\mathcal{CN}(0,1)$. The transmit power of the user are set to be 10 dB and the relay power is 15 dB. We set the relay power price to be optimal according to Theorem 1.

The simulation result shown the comparison among the network sum-rate of KSBS compare to power allocation solution and the even power allocation. The sum-rate optimal power allocation solution is relay power allocation among users that maximize network sum-rate. With the even power solution, the relay allocate to each transmission t is $P/(N \times R)$.

Figure 2 shows the network sum-rate of KSBS-based power allocation compare to the sum-rate optimal power allocation solution and the even power allocation. We can observe from Fig. 2 that, the proposed KSBS-based scheme has better network sum-rate than even power allocation and achieves close-to- sum-rate-optimal solution.

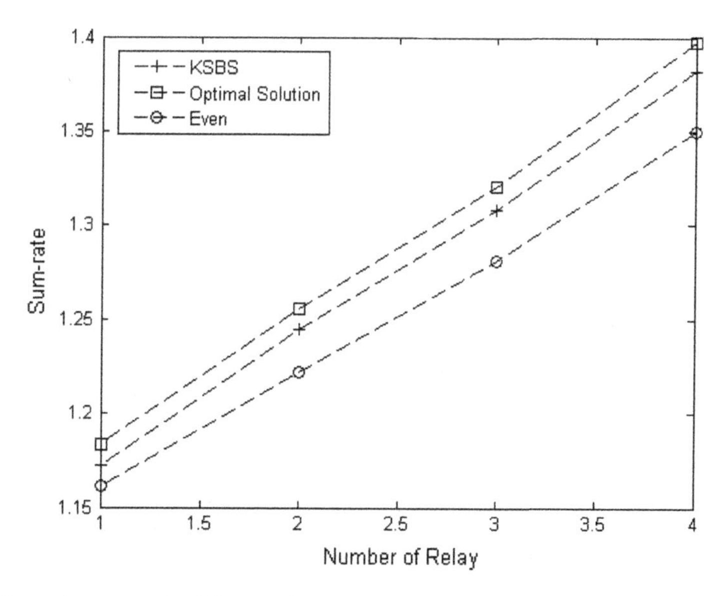

Fig. 2 Network sum-rate in different relay power allocation scheme

5 Conclusion

In this paper, we consider a multi-user multi-relay network and use game theory to model and analyze the user and relay. We model the interaction between the relay and users as Stackelberg game model. The relay power allocation among users is model as a cooperative bargaining game. We propose the KSBS of bargaining game for faired allocation of relay. Based on the KSBS relay power allocation, we can find the optimal relay power price. From the simulation results, we find that the proposed KSBS-based scheme has better network sum-rate than even power allocation and achieves close-to- sum-rate-optimal solution.

References

1. Nosratinia, A., Hunter, T.E., Hedayat, A.: Cooperative communication in wireless networks. IEEE Commun. Mag. **42**(10), 74–80 (2004)
2. Laneman, L.N., Tse, D.N.C., Wornell, G.W.: Cooperative diversity in wireless networks: efficient protocols and outage behavior. IEEE Trans. Inf. Theory **52**(12), 3062–3080 (2004)
3. Phan, K.T., Le, L.B., Vorobyov, S.A., Le-Ngoc, T.: Centralized and distributed power allocation in multi-user wireless relay networks. IEEE Int. Conf. Commun. (2009)
4. Phan, K.T., Le-Ngoc, T., Vorobyov, S.A., Tellambura, C.: Power allocation in wireless multi-user relay networks. IEEE Trans. Wireless Commun. **8**(5), 2535–2545 (2009)
5. Shen, Y., Feng, G., Yang, B., Guan, X.: Fair resource allocation and admission control in wireless multiuser amplify-and-forward relay networks. IEEE Trans. Veh. Technol. **61**(3), 1383–1397 (2012)
6. Wang, B., Han, Z., Liu, K.J.R.: Distributed relay selection and power control for multiuser cooperative communication networks using Stackelberg game. IEEE Trans. Mobile Comput. **8** (7), 975–990 (2009)
7. Al-Tous, H., Barhumi, I.: Resource allocation for AF cooperative communication using Stackelberg game. In: IEEE Trans. Signal Process. Commun. 1–6 (2012)
8. Cao, Q., Jing, Y., Zhao, H.V.: Power bargaining in multi-source relay networks. Proc. IEEE Commun. Conf. 3905–3909 (2012)
9. Cao, Q., Jing, Y., Zhao, H.V.: Power allocation in multi-user wireless relay networks through bargaining. IEEE Trans. Wireless Commun. **12**(60), 2870–2882 (2013)
10. Zhang, G., Yang, K., Chen, H.H.: Resource allocation for wireless cooperative networks: a unified cooperative bargaining game theoretic framework. IEEE Wireless Commun. Mag. **19** (2), 38–43 (2012)
11. Cao, Q., Zhao, H.V., Jing, Y.: Power allocation and pricing in multiuser relay networks using Stackelberg and bargaining games. IEEE Trans. Veh. Technol. **61**(7), 3177–3190 (2012)
12. Shen, Y., Feng, G., Yang, B., Guan, X.: Distributed fair resource allocation in wireless multi-user multi-relay networks with heterogeneous rate constraints. In: Proceedings of the IEEE ISCIT, pp. 333–338. Hangzhou (2011)
13. Rashid, U., Kha, H.H., Tuan, H.D., Nguyen, H.H.: Joint design of source power allocation and relay beamforming in multi-user wireless networks. IEEE Trans. Signal Process. Commun. 1–4 (2012)
14. Fudenberg, D., Tirole, J.: Game Theory. MIT Press, Cambridge, MA (1991)
15. Kalai, E., Smorodinsky, M.: Other solutions to Nash's bargaining problem. Econometrica **43** (3), 513–518 (1975)

Energy-Distance Aware Clustering Scheme (E-DACS) for Wireless Sensor Networks

Abdullah Said Alkalbani, Teddy Mantoro and Satyanarayana Degala

Abstract In this paper, popular clustering protocols for Wireless Sensor Networks (WSNs) are investigated. There are many issues and challenges that effect network lifetime. One of these critical issues is high power consumption which reduces network lifetime. The aim of this paper is to propose Energy-Distance Aware Clustering (E-DACS) approach that considers sensor energy level and distance during cluster heads selection to increase the throughput. In this method, cluster head selected among other nodes in the cluster by optimal criteria that calculate the distance to each node, distance to base station, and residual energy. Performance, network efficiency and maximizing network lifetime are measured by simulation, analysis and comparison with other well-known mechanisms.

Keywords WSNs · Network efficiency · Power consumption · Clustering protocol · Energy aware clustering · Cluster head

1 Introduction

Due to recent developments in networks of low-cost, low-power, multifunctional sensors have led to increased attention in this area. In the last few years, WSNs and related applications have gained significant momentum, due mainly to the fact that the technology is maturing and moving out of the purely research-driven environment into commercial interests [1].

A.S. Alkalbani (✉) · S. Degala
Electrical & Computer Engineering Department, College of Engineering, University of Buraimi, Al-Buraimi, Oman
e-mail: abdullah.s@uob.edu.om

S. Degala
e-mail: degala.s@uob.edu.om

T. Mantoro
Faculty of Science and Technology, USBI-Sampoerna University, Jakarta, Indonesia
e-mail: teddy@ieee.org

P. Meesad et al. (eds.), *Recent Advances in Information and Communication Technology 2016*, Advances in Intelligent Systems and Computing 463,
DOI 10.1007/978-3-319-40415-8_27

WSNs use distributed sensors to monitor various conditions of remote locations such as smoke, sound, vibration, pressure, temperature, motion, and pollution. A WSN is configured autonomously by sensor nodes equipped with sensing, computing and wireless communication capabilities [2]. The main purpose of sensor nodes with the wireless techniques is to collect useful data and transmit these back to the base station for possible use. Data transmission throughput is very important due to nodes limitation. Every node is capable of sensing, data processing, and communication, and operates on its limited amount of battery energy consumed mostly in transmission and reception at its radio transceiver [3]. Hence, the energy consumption of sensor nodes needs to be seriously considered to achieve longer surveillance [4]. A fundamental part of the WSNs configuration is how the sensors reciprocate data among the network till they reach base station; a specialized routing protocol is required to manage transmission band. The design of such routing protocol must take into consideration the limited resources and security objectives. Numerous protocols have been designed in this area but few of them were constructed to care for the limited resources, clustering and routing issues at the same time [5]. The main problem for this research is high energy consumption when transmitting data from source to base station and sensors rapidly dying, which render isolated parts of WSN useless.

In this work, CH selection not depends on considering energy level only like previous studies but also the distances between proposed node, the base station and other members of cluster.

In this research, the nodes are assumed static. Hence, the contributions are suitable for networks which are either static or whose topology changes slowly enough such that there is enough time for optimally balancing the traffic in the periods between successive topology changes.

The remaining part of the paper is organized as follows: In Sect. 2, the related work in this area is given. Section 3 describes the research framework and mathematical models. In Sect. 4, mechanism steps are described. In Sect. 5, extensive experiments by simulation are conducted to prove the efficiency and quality of the proposed model. Section 6 presents the research contribution, results discussion, and analysis. Conclusion as well as the challenges, encountered and future directions for research is given in Sect. 7.

2 Related Work

In the last decade, deep research showed the collaboration between sensors in data collection and processing, and management of the sensing activity and data transmission. Due to the limitations of sensor nodes in power and communication bandwidth, innovative mechanisms were proposed to minimize energy consumption that decreases WSNs lifetime and efficient use of the limited bandwidth. Many challenges are faced in construction and management of power efficient and secure routing protocols for WSNs due to the limitations and constraints such as a large number of sensors deployed, self-organizing requirement, and energy limits [6].

Researchers in [7] described well-defined clustering protocols such as Low-Energy Adaptive Clustering Hierarchy (LEACH), Hybrid Energy-Efficient Distributed clustering (HEED), Two-Level Hierarchy LEACH (TL-LEACH), Energy Efficient Clustering Scheme (EECS), Energy-Efficient Uneven Clustering (EEUC), Algorithm for Cluster Establishment (ACE), Base-Station Controlled Dynamic Clustering Protocol (BCDCP). Figure 1 presents well-known clustering based energy mechanisms, and Table 1 shows a comparison between famous techniques based on some factors [8].

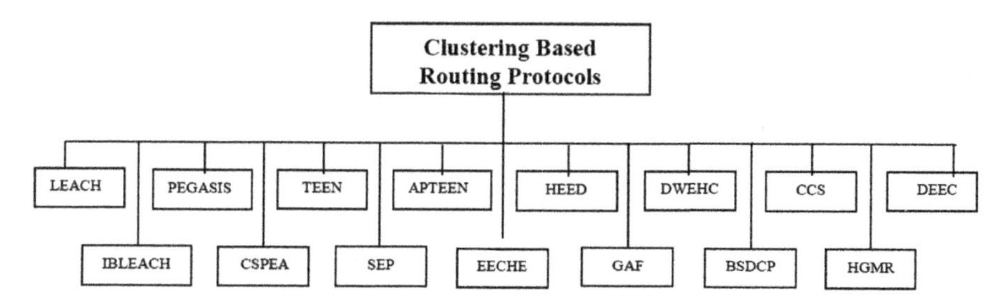

Fig. 1 Clustering based energy efficient routing protocols

Table 1 Comparison of clustering protocols in WSNs

Protocol	Cluster stability	Delivery delay	Energy efficiency
LEACH	Medium	Very small	Very poor
TEEN	High	Small	Very high
APTEEN	Very low	Small	Medium
GAF	Medium	Poor	Medium
CSPEA	Medium	Medium	High
PEGASIS	Flat	Very large	Poor
SEP	Medium	Very small	Medium
HGMR	High	Medium	Poor
DEEC	High	Very small	High
DWEHC	High	Medium	Very high
IBLEACH	High	Very small	Very high
CCS	Low	Large	Poor
EECHE	Medium	Small	Very good
HEED	High	Medium	Medium
BCDCP	High	Small	Very poor

One of the researches that studied the area of energy aware mechanisms for WSNs is An Optimal Power Conservation Cluster based Routing Algorithm using Fuzzy Verdict Mechanism for Wireless Sensor Networks. In this mechanism, Fuzzy Verdict Mechanism Cluster Protocol (FVMCP) is used to construct knowledge based system effectively for the eligibility of the sensor node to be elected as a Cluster Header (CH) [9].

In [10], researcher analyzed and evaluated two clustering-based mechanisms for WSNs. These mechanisms are Single-hop Energy-Efficient Clustering Protocol (S-EECP) and Multi-hop Energy-Efficient Clustering Protocol (M-EECP). In S-EECP, the cluster heads (CHs) are selected by a weighted probability based on the ratio between residual energy of each node and an average energy of the network. The nodes with high initial energy and residual energy will have more chances to be elected as CHs than nodes with low energy whereas, in M-EECP, the elected CHs communicate the data packets to the base station via multi-hop communication approach.

The Design of a Distributed Energy-Efficient Clustering (DEEC) algorithm for heterogeneous wireless sensor networks is proposed to increase the scalability and lifetime of the network [11]. In this mechanism, the cluster-heads are elected by a probability based on the ratio between residual energy of each node and the average energy of the network. The epochs of being cluster-heads for nodes are different according to their initial and residual energy. Researchers in [12] proposed Enhanced Developed Distributed Energy Efficient Clustering scheme (EDDEEC) for heterogeneous WSNs. This technique is based on changing dynamically and with more efficiency the Cluster Head (CH) election probability.

Our research proposes clustering mechanism based on some parameters such as residual energy, distance from neighbors and base station. Simulation results evaluated and compared with mechanisms discussed in [12].

3 Research and Mathematical Models

3.1 Research Model

One method to prolong WSNs lifetime is optimizing energy-distance aware cluster heads selection method. This research optimizes such scheme by considering residual energy and distances from cluster members and the base station.

As shown in Fig. 2, this mechanism improves network lifetime by applying energy aware in selecting CH. CH is the sensor node, which has a computing power higher than the other nodes available in the network. There may be more than one CH in a network. The CH performs actions like managing sensor nodes under its range and providing the information to the base station. The base station, in turn gives a routing command to CH based on the nodes energy level and its position. This information is then passed to sensor node which needs routing action.

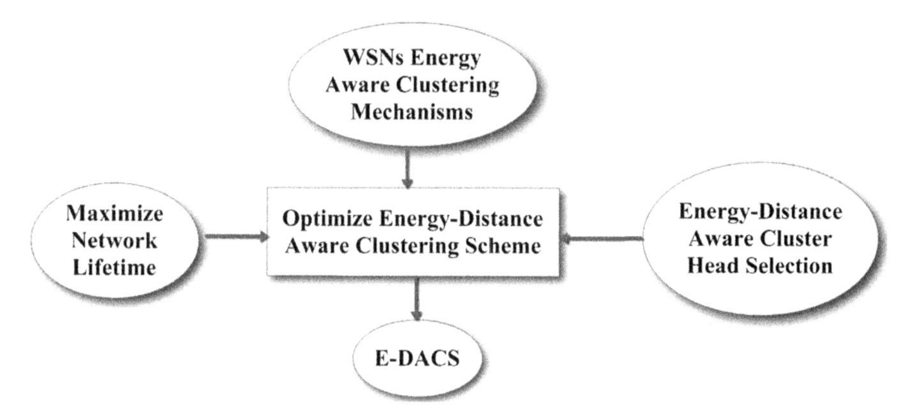

Fig. 2 Research model for E-DACS

3.2 *Mathematical Model*

In this section, the mathematical model for nodes and all network energy consumption during routing is presented. The energy consumed for a node (i) is the total energy used when receiving data and energy used for transmission as follows:

$$E_i = E_{transmit} + E_{receive} \tag{1}$$

In details, the equation that represents energy consumed when the sensor receives a message of size k is

$$E_{receive} = E_{elec} * k \tag{2}$$

where E_{elec} is electronics energy, and the energy consumed on sending a message of size k is

$$E_{transmit} = E_{elec} * k + E_{amp} * r^2 * k \tag{3}$$

where r is the sensor sensing range and E_{amp} is amplifier consumed energy [13]. So, the total energy consumed for the network is

$$E_{total} = \sum_{i=1}^{n} E_i \tag{4}$$

In this research we assume that the N nodes are distributed uniformly in an M * M region and the base station is located in the center of the field for simplicity. The distances between proposed CH and other cluster members calculated as follows:

$$D_{node-cluster} = \frac{M}{\sqrt{2\pi k}} \tag{5}$$

where k is number of clusters [12]. Finally, the average distance between CH and base station estimated by Eq. (6)

$$D_{cluster-base\ station} = 0.765\left(\frac{M}{2}\right) \tag{6}$$

where 0.765 is constant represents distribution of solar radiation.

4 Energy-Distance Aware Clustering Scheme

There are many challenges remains face data routing such as high energy consumption during data transmission and security of the routing path. E-DACS considers energy consumption issue. The following phases describe E-DACS scheme; network construction and clustering. There are many challenges remains face data routings such as high energy consumption during data transmission and security of the routing path. E-DACS considers energy consumption issue. The following paragraphs describe E-DACS scheme; network construction and clustering.

Once the sensors are deployed in a specific area, the network starts to construct itself. Cluster Heads (CHs) are elected for each cluster depending on the node's residual energy level, distance from cluster members and distance to base station, because the cluster head is the node that consumes the most energy in each cluster. CH selection must be done periodically in each cluster once current CH energy level becomes less than any other node in its group. The following steps describe in detail network construction after deployment and how the cluster was created:

```
1. Initialize node (ni)in initial state and broadcast
   energy level
2. If current cluster (cc) energy level > ni energy
   level then
    send acknowledgement to ni to be as member
   else   Calculate optimization as following
     Optimal (cc_Energy level, Distance_cc-cluster , Distance_cc-base station,
                 ni_Energy level, Distance_ni-cluster , Distance_ni-base station)
   If optimal is cc send acknowledgement to ni to be as
   member
   else send acknowledgement to ni to be as Cluster
   Head
3. Repeat for all members every threshold time
```

Figure 3 represents these steps.

Figure 4 shows the structure of WSN and Fig. 5 displays the communication between network parts.

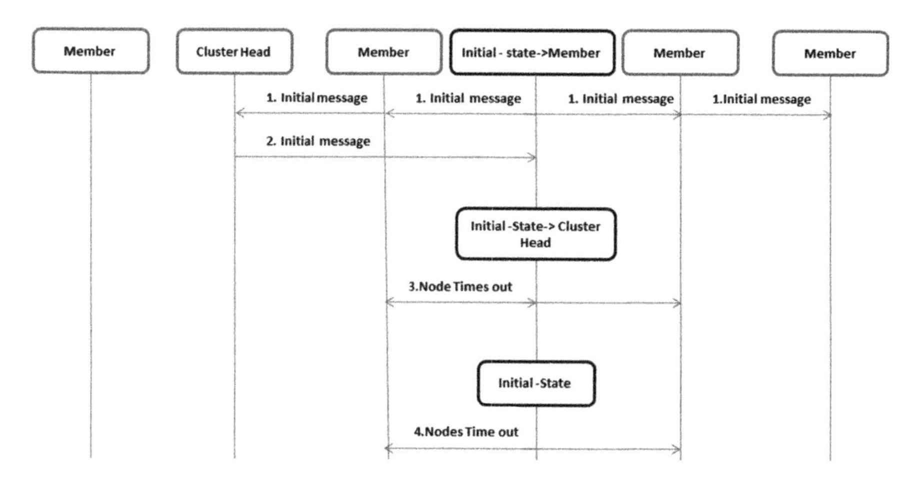

Fig. 3 WSN clusters construction steps

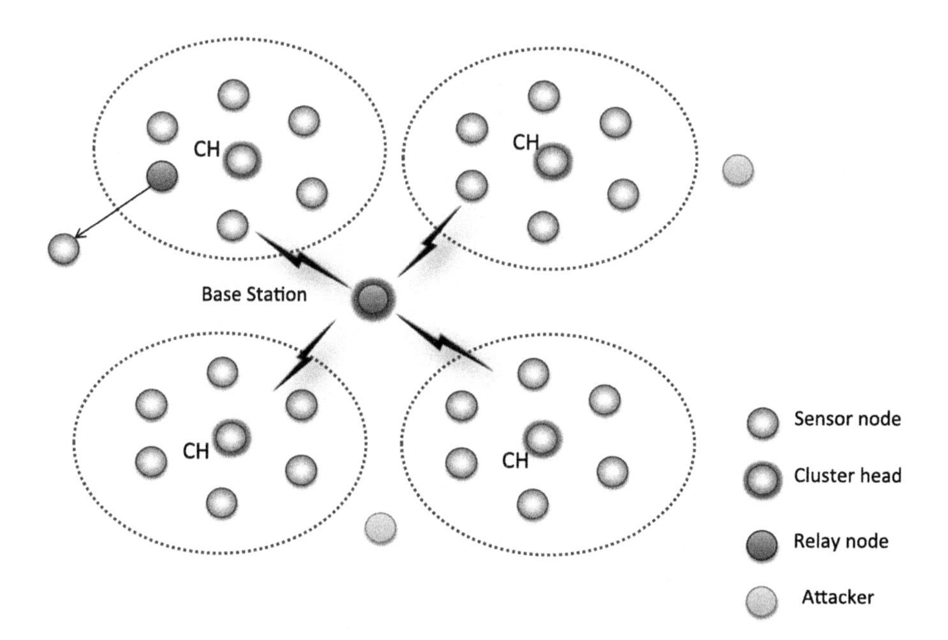

Fig. 4 WSN structure

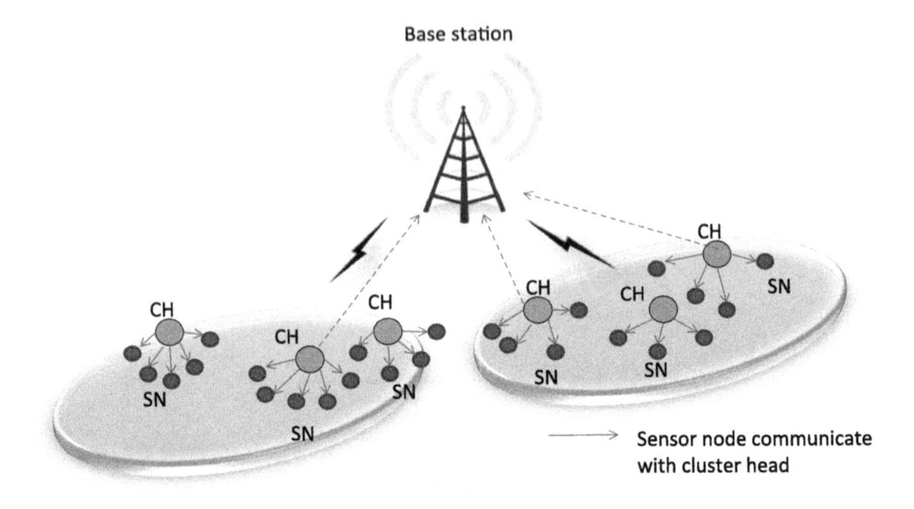

Fig. 5 WSN communication

5 Experiments and Simulation Results

5.1 Simulation Tool and Definitions of Parameter

In this research, NS2 of version 2.35 is used for WSNs simulation running under Ubuntu 12.04 operating system in Samsung machine. Experiments carried out consisting of 100 nodes randomly distributed over an area of 100 square units. Another assumption in this simulation is that every node only knows its neighbors within its Radio Frequency (RF) range. Each sensor node is assumed to have an initial energy of 100 J. A node is considered dead if its energy level is 0 J. Simulation parameters and default values used in the experiments are summarized in Table 2.

Table 2 Simulation and network parameters

Parameter	Value	Comments
N	100	Number of nodes
E_0	100 J	Node initial energy
rx power	1.0 rp	Receive power
tx power	1.0 tp	Transmit power
Data packet	50 bytes	Max packet in Ifq
max Packet	50,000 bytes	Max packet limit
I_t	10.0 s	Interval set for packet transfer
B	512, 1024, 1500, 2000, 2500 bytes	Broadcast packets
P_{opt}	0.05	Optimal probability

In addition, this research assumed that all sensors are homogeneous, so each sensor has the same communication range. Also, sensors can only communicate with the neighbors within communication range due to limited power. Multi-hop is required to communicate with farther ones.

5.2 *Experiments Results*

Since one of the essential constraints that effect on WSNs is battery limits and high energy consumption during transmission and reception, a dynamic WSN is simulated in our experiments. In these networks some sensors go into an idle state for a while if they do not receive any request from neighbors within a specific period of time. A sensor during idle state does not receive or transmit any data. After a certain timeout, they wake up again.

In this research, experiments used WSNs that contain 100 sensors and deployed throughout 100 times 100 square network areas. The main objective of these experiments is to measure the energy consumed during network life and the average number of alive sensors during network lifetime. Also a number of packets transmitted are calculated to measure network efficiency. Simulation results are summarized in Table 3. Sections 5.3 and 5.4 describe in detail these results with graphical representation.

5.3 *Energy Consumption*

Long lifetime requirement of different applications and limited energy storage capability of sensor nodes has increased the need for reducing power consumption upon nodes. To increase sensors lifetime this mechanism is energy efficient.

In our simulation, initially, we set the energy level for each node equivalent to 100 J. Then, from the result after we had run our simulation, the total energy used to transmit data did not exceed 100 J. Energy consumption reaches its minimum value in experiments when 5000 rounds were used as shown in Fig. 6. Results show that energy conservation is achieved using proposed mechanism.

Table 3 Simulation results

No. of rounds	Total energy consumed (J)	Packets size (Byte) sent to base station	Alive nodes during network lifetime
1000	91	1	100
2000	80	2	99
3000	72	3	97
4000	42	4.1	89
5000	17	4.5	50

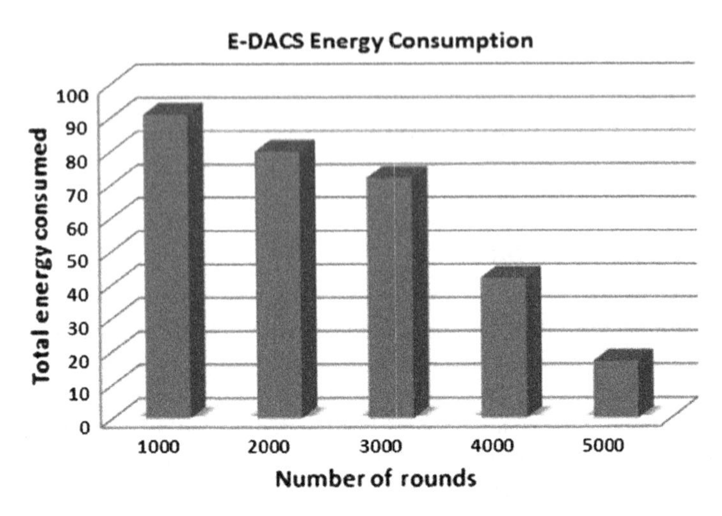

Fig. 6 Energy consumed versus number of rounds

5.4 *Network Efficiency*

Number of alive sensors during network lifetime and number of packets sent to base station are two important indicators that show network efficiency.

Figure 7 shows rapid increase in number of packets when the number of rounds increased. Through comparison, it is clear that a number of packets transmitted by 5000 rounds are 4 times greater than a network with 1000 rounds. Figure 8 presents the number of alive sensors during network lifetime. Generally, the network

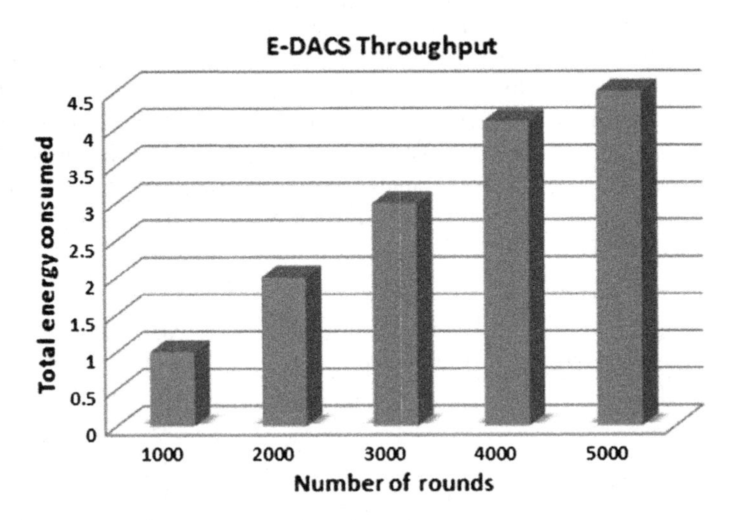

Fig. 7 Network throughput versus number of rounds

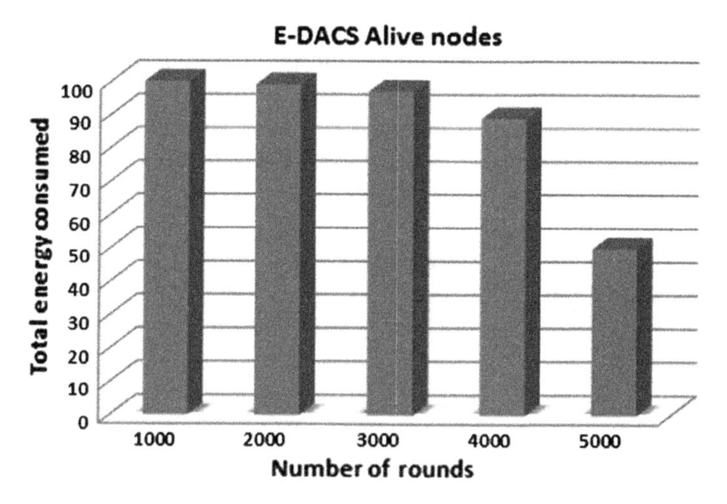

Fig. 8 Number of alive nodes versus number of rounds

remains active with a different number of rounds using E-DACS. It's clear that the number of remains alive sensor not less than 50 % of original number of sensors. Such networks can be used widely once their efficiency is high and can remain alive for more time.

5.5 *Comparison with Known Mechanisms*

To prove the efficiency of the proposed mechanism, it is compared with mechanisms described in [12]. The comparison factors are data packets transferred to the base station, energy consumed and the number of sensors that remains alive during network execution time.

First, Table 4 shows the results of the number of alive sensors for different mechanisms with different network rounds. The results represented in Fig. 10 show

Table 4 Comparison between EDEEC, EDDEEC, and proposed method (E-DACS) in terms of alive nodes during network lifetime

No. of rounds	E-DACS	EDDEEC	EDEEC
1000	100	99	96
2000	99	98	80
3000	97	96	75
4000	89	70	50
5000	50	30	20

E-DACS in different rounds have better results than other mechanisms, due to the optimal clustering and routing path; this means that a network can transmit a large amount of data before it dies compared to other methods (Fig. 9).

Energy consumption factor is the important factor that affects WSNs lifetime. Less energy consumption means the network will stay alive for a longer time.

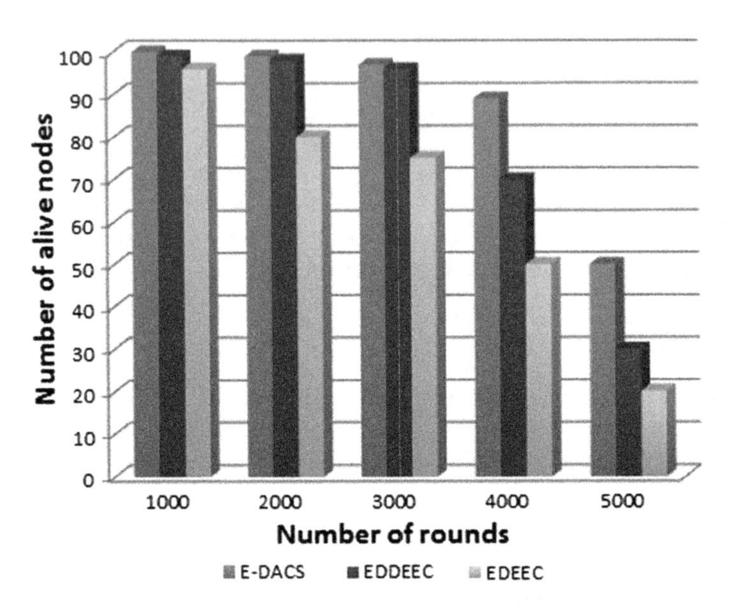

Fig. 9 Comparison between EDEEC, EDDEEC, and proposed method (E-DACS) in terms of alive nodes during network lifetime

Table 5 Comparison between EDEEC, EDDEEC, and Proposed Method (E-DACS) in Terms of energy consumed (J)

No. of rounds	E-DACS	EDDEEC	EDEEC
1000	91	98	92
2000	80	95	78
3000	72	90	70
4000	42	85	47
5000	17	50	18

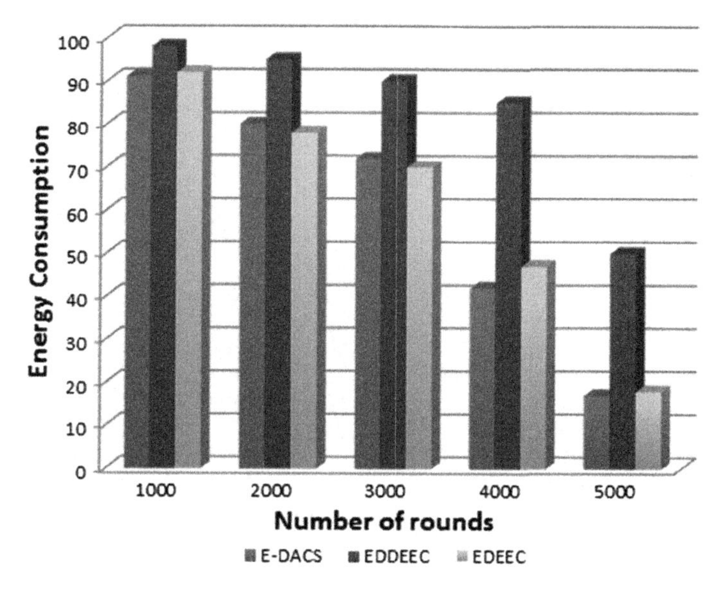

Fig. 10 Comparison between EDEEC, EDDEEC, and proposed method (E-DACS) in terms of energy consumption

Table 6 Comparison between EDEEC, EDDEEC, and proposed method (E-DACS) in terms of packets sent to base station (Bytes)

No. of Rounds	E-DACS	EDDEEC	EDEEC
1000	1	1	1
2000	2	2	2
3000	3	2.9	2.8
4000	4.1	3.9	3.7
5000	4.5	4	3.9

Results are shown in Table 5 and graphically represented in Fig. 10 prove that generally E-DACS consumes less energy than other methods for different rounds.

The last comparison factor is the number of packets transferred to the base station during network lifetime. As shown in Table 6 and Fig. 11, E-DACS has better performance than other mechanisms with a high number of rounds.

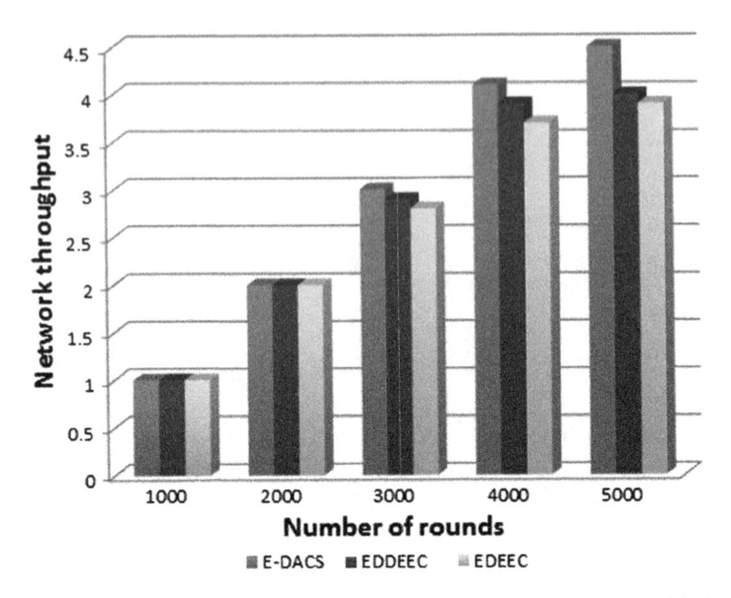

Fig. 11 Comparison between EDEEC, EDDEEC, and proposed method (E-DACS) in terms of packets sent to base station

6 Results Discussion

Simulation results demonstrate that E-DACS achieved significant energy savings and improved network lifetime compared to EDEEC and EDDEEC. We present that E-DACS obtains better network efficiency than other clustering methods. As a result, this mechanism can be used with different WSNs sizes with high performance. From the results of the simulation and comparison with known techniques, we can summarize the contribution of this paper in the following points:

- Simulation results have shown that total energy used to transmit data in the proposed model does not exceed 100 J for different network rounds.
- Comparing EDEEC, EDDEEC methods with E-DACS shows E-DACS has better results than other mechanisms, due to the optimal clustering and; this means that the network can transmit a large amount of data before it dies rather than other methods.
- In general, the energy consumption of the E-DACS is much less compared to other methods; this reduced energy consumption helps in maintaining the network for a longer time so that it can perform the task of transferring packets to the base station without affecting the computing power.
- E-DACS mechanism reduces the number of dead sensor nodes in a network so many sensor nodes will remain active till the network lifetime.

7 Conclusions

The energy constraints and limited computing resources of the sensor nodes present major challenges in transmitting and processing data within the network. In this paper, we focused on how to minimize the node energy consumption since it has been proven to be the important area in wireless sensor networks. By considering the limitation of the wireless sensor networks, E-DACS mechanism proposed to select network CHs in a way to prolonging the lifetime of the network. The main idea of the mechanism is its capability to be energy and distance aware during selecting CHs.

In E-DACS cluster head selected by considering residual energy, distances between elected node and other members in the cluster and distance between assigned cluster head and the base station. Simulation results prove the quality of mechanism and its efficiency rather than other methods in the same area. Energy consumption is minimum and a number of died sensors through network execution time is less compared to other mechanisms.

As future work, scheduling scheme enhancement can be applied by using mobile sensors to solve the problem of isolated clusters after some sensors have died which will increase network productivity.

References

1. Alkalbani, A.S., Tap, A.M., Mantoro, T.: Energy consumption evaluation in trust and reputation models for wireless sensor networks. In: Information and Communication Technology for the Muslim World (ICT4M), Morocco, pp. 1–6 (2013)
2. Shin, J., Suh, C.: CREEC: chain routing with even energy consumption. J. Commun. Netw. **13** (1), 17–25 (2011)
3. Chang, J.H., Tassiulas, L.: Maximum lifetime routing in wireless sensor networks. IEEE/ACM Trans. Netw. **12**(4), 1126–1137 (2004)
4. Zytoune, O., Fakhri, Y., Aboutajdine, D.: A fairly balanced clustering algorithm for routing in wireless sensor networks. Sensor Review. Emerald. **30**(3), 242–249 (2010)
5. Almomani, I., Almashakbeh, E.: A power-efficient secure routing protocol for wireless sensor networks. WSEAS Trans. Comput. J. **9**(9), 1042–1052 (2010)
6. Al-Karaki, J.N., Kamal, A.E.: Routing techniques in wireless sensor networks: a survey. In: IEEE Conference in Wireless Communications, vol. 11, no. 6, pp. 6–28 (2004)
7. Ramakant, K.K.S., Dalal, S.: Clustering in WSNs: A Review (2014)
8. Sharma, N., Nayyar, A.A: Comprehensive review of cluster based energy efficient routing protocols for wireless sensor networks. Int. J. Appl. Innov. Eng. Manage. **3**(1) (2014)
9. Daniel, R., Rao, K.N.: An optimal power conservation cluster based routing algorithm using Fuzzy verdict mechanism for wireless sensor networks. In: IEEE International Conference on Electrical, Electronics, Signals, Communication and Optimization (EESCO), pp. 1–9 (2015)
10. Kumar, D: Performance analysis of energy efficient clustering protocols for maximizing lifetime of wireless sensor networks. Wirel. Sensor Syst. IET. **4**(1), 1, 9–16 (2014)
11. Qing, L., Zhu, Q., Wang, M.: Design of a distributed energy-efficient clustering algorithm for heterogeneous wireless sensor networks. Comput. Commun. **29**(12), 2230–2237 (2006)

12. Javaid, N., Qureshi, T.N., Khan, A.H., Iqbal, A., Akhtar, E., Ishfaq, M.: EDDEEC: Enhanced developed distributed energy-efficient clustering for heterogeneous wireless sensor networks. Procedia Comput. Sci. **19**, 914–919 (2013)
13. Sha, K., Shi, W.: Modeling the lifetime of wireless sensor networks. Sensor Lett. **3**(2), 126–135 (2005)

Fast Tracking Algorithm for Designed Marker

Teera Siriteerakul and Rutchanee Gullayanon

Abstract Robust and real-time object tracking is a vital part of many application including the navigation of mobile robot and unmanned aerial vehicle. In such environment, the computational power and battery are limited. Hence, most of the state of the art algorithms will not be able to display their full potential. Thus, we propose an improved tracking algorithm, with a designed marker, to perform robustly in real-time in complex environments. Our algorithm based on a simple binarization and finding candidate connected components. Then, with physical and geometric property of the designed marker, we can filter out all the other components until we only have the marker. This scheme has been tested on Raspberry PI, a limited-power computing unit, and demonstrate sufficient robustness and speed.

Keywords Object tracking · Binarization · Connected component · Limited-power computer

1 Introduction

Object tracking is one of the fundamental task of many vision-based autonomous system. As stated in [1], object tracking is applicable to motion-based recognition, automated surveillance, video indexing, human-computer interaction, and many other systems. However, the tracking task can be difficult due to the loss of information from 3D world to 2D image, noise, object's shape/motion, partial and full occlusions, scenes illumination, and other factor including the real-time requirement from certain systems.

T. Siriteerakul (✉) · R. Gullayanon
Faculty of Engineering, King Mongkut's Institute of Technology Ladkrabang, Bangkok, Thailand
e-mail: teera@tboxteam.com

R. Gullayanon
e-mail: ler.gullayanon@gmail.com

P. Meesad et al. (eds.), *Recent Advances in Information and Communication Technology 2016*, Advances in Intelligent Systems and Computing 463,
DOI 10.1007/978-3-319-40415-8_28

There are many object tracking algorithms simply search the whole scene for an object and record object's location for tracking. However, both speed and robustness can be improved when we only search in a predicted area around the previously detect object area. For example, Kanade-Lucas-Tomasi (KLT) tracking [2, 3] is used in [4] for predict the search space of pedestrian in the next frame. This has been reported to be fast and stable tracking system for multiple objects. However, their approach utilize some advance features for detection which require a sufficient computing power.

Our target is to perform object tracking robustly in real-time in a limited computing power environment. Unfortunately, there are only limited attempts in this area. For example, [5] rely on chromatic information to perform fast tracking. However, the color information is not reliable when there are changes in illumination. On the other hand, [6–8] track object using MeanShift and the improved CamShift which would fail in clustered environments.

The tracking system in [9] followed example from Unmanned Aerial Vehicle (UAV) community and designed a tracking marker along with a tracking algorithm. In their approach, the marker is a white patch consists of four black circles with known geometric property (sizes and relative distances). Then, the tracking is done by searching potential marker's circles using relaxed Hough Circle Transform. Lastly, the actual marker's circles is identified by filtering out non-black circles and incorrect geometric properties. The approach is proven to be robust and run in sufficient time. However, the speed of this approach can be improved.

In their approach, the relaxed Hough Circle Transform is the major time consuming task. Thus, rather than performing Hough Circle Transform, we proposed that we can perform binarization based on Otsu's thresholding [10]. This can be done effectively since the marker is designed to be pure black circles on pure white area. Next, the potential circles can be found by searching for target size connected components of black areas which can be done in O(n) where n is the number of pixel.

The contribution of this paper lies on the improvement of potential circles detection part of [9]. By utilizing Otsu's thresholding and connected component search, we can improve their frame rate from approximately 10 frame per second to approximately 30 frame per second.

The rest of this paper is organized as follows. In the next section the proposed frame work of [9] is reviewed, along with our improve detection method. Then, in Sect. 3, the empirical results are described and discussed. Finally, the last section provides the final remarks and some suggestion for future improvement.

2 Tracking Framework

2.1 The Marker

We adopted the marker from [9] (see Fig. 1) where four black circles are placed in white area of 8×8 square-inches. The circles' radii are designed to be a ratio of 3:4:5:6 which translate to the size of 1.5, 2, 2.5, and 3 inches in diameter. The placement of the centers of the four circles form a square of 3×3 square-inches in the middle of the patch.

2.2 Tracking Algorithm

As stated in the introduction, our approach for improve tracking improvement is to substitute Hough Circle Transform with Otsu's thresholding and connection components detection. To make this work, we have to make some minor adjustments which to be described in this section.

Circle Detection The potential circles are detected in two steps. First, the image are transform into grayscale and then discretized into black and white using Otsu's thresholding. The fact that the marker is consist of pure black circles on pure white patch is helping with the thresholding operation.

To ensure that each circle separate nicely, we perform morphology erosion follows by dilation to the binalized image (aka. morphology opening). Then, we separate pixels in the image into 8-connected component which can be done in linear time.

Non-Circle Filtering This filter is in addition to [9]. Since what we are detecting are connected component, some of them may not even close to be a circle. In this filter, we filter out components where their bounding box are not close to squares (hence, not likely to enclosing a circle). Furthermore, the areas that are too large are also filtered out in this step as well.

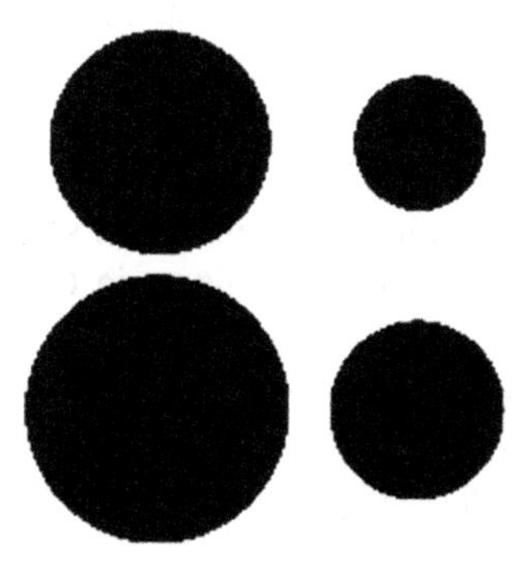

Fig. 1 The designed marker from [9]

Black Circle Filtering as mentioned in [9], only a few sub-sampling pixel are enough to ensure that we have black circle. However, in our case, since we are not actually detecting a circle, we need to have higher number of sub-sampling to ensure that every place in the connected component area is black.

The rest of filter which are Adjacency Filtering, Four Corners Filtering, Diameter Ratio Filtering, and The Third and The Forth Circle Verification can be done in the same way as [9]. Keep in mind that the last step where Black Circle Filtering is to perform again will need a higher number of sub-sampling as well.

3 Experiments

3.1 Experiment Setup

We implement the algorithm the same way as in [9] which consists of Raspberry PI 2 Model B with a 900 MHz quad-core ARM Cortex-A7 and 1 GB of Ram. The camera unit used is the PiCam NoIR. Dataset used is taken, with permission, from [9]. The dataset is 100 images sequence of size 640 $\times$ 480 taken from this setup hardware. An example of these images along with experimental results is illustrated in Fig. 2.

3.2 Experiment Result

Out of 100 images, we only missed 1 image where the marker is about to leave the scene which cause the Otsu's thresholding to perform poorly. This is a major improvement comparing to 12 misses by [9]. Apparently, our approach lessen the effect of motion blur by perform opening to make two circles separate. As shown in Fig. 2b, the motion blur cause Otsu's thresholding to join two large circles in the marker. Then, the morphology opening make them separate again (Fig. 2c). Figure 2d shows the connected component that has an almost square bounding box (potential circle). The circles which survive black circle filter is shown in Fig. 2e. Lastly, the final marker detection (Fig. 2f) is found after apply the rest of the filtering.

The average error of circle size detection is less than 1 pixel (comparing to 3 pixel in [9]). Although it seems to be an improvement, Fig. 3 shows that the circle size is not reliable as in the case of [9]. For position of the center of the largest detected circle, the graph in Figs. 4 and 5 demonstrate that the position detection is as robust as before (less than 1 pixel error in average). Lastly, the speed performance of our approach is more than 3 time faster than the Hough Circle Transform approach. The frame rate improves from approximately 10 to 30 frame per second.

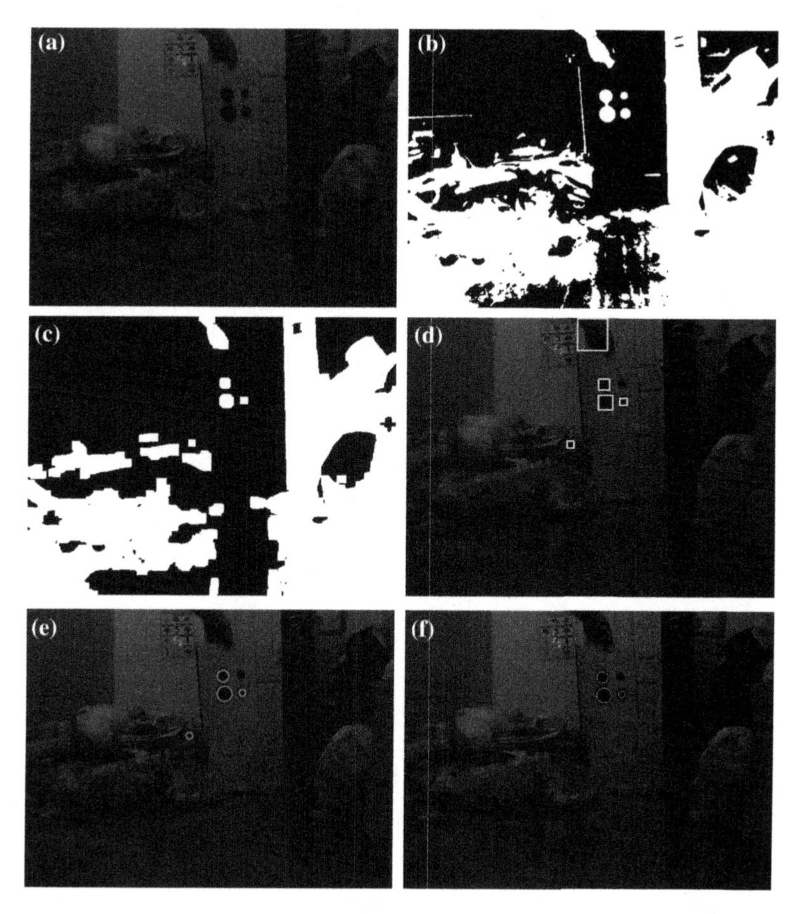

Fig. 2 **a** Complex scene with the marker. **b** Scene after Otsu's thresholding. **c** After morphology opening. **d** Detected connected components. **e** After *black circle* filtering. **f** Final marker detection

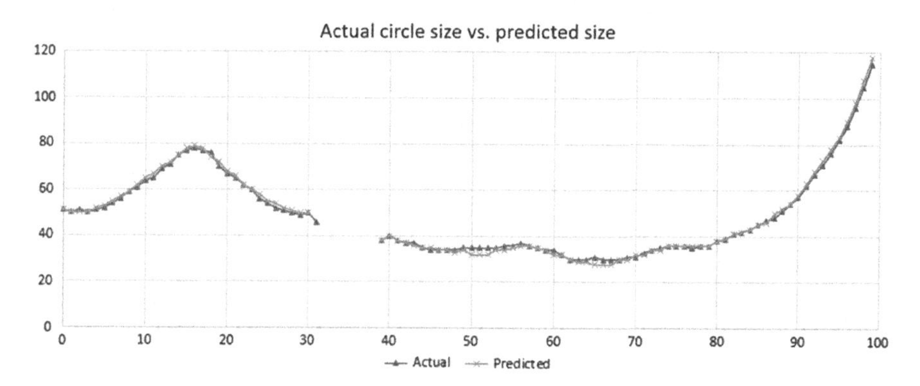

Fig. 3 Circle size detection

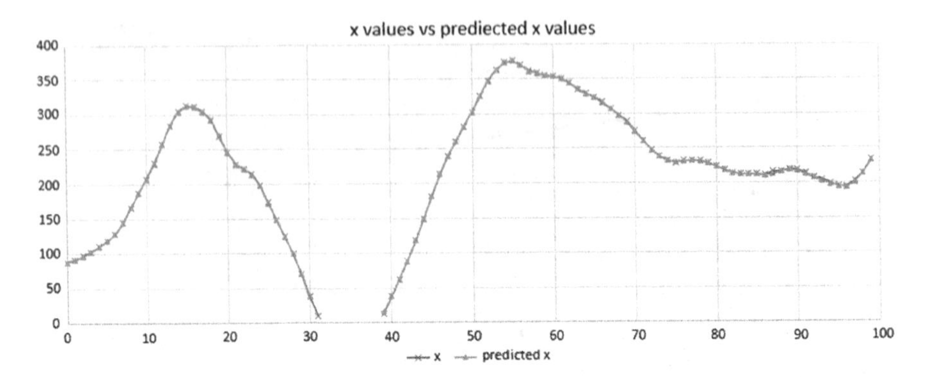

Fig. 4 x-value detection

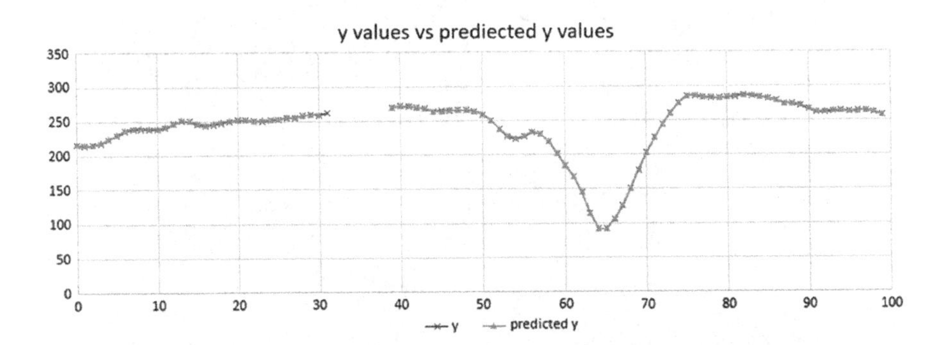

Fig. 5 y-value detection

4 Conclusion

This paper improves upon [9] and substitute Hough Circle Transform by a simple task of binarization and connected component detection. This substitution shown to increase the speed performance 3 times from approximately 10 to 30 frame per second. This would allow computation time for other tasks on the computing unit. Again, this paper is still not consider any tracking mechanism but still consider searching from the whole scene.

References

1. Yilmaz, A., Javed, O., Shah, M.: Object tracking: a survey. ACM Comput. Surv. (CSUR) **38** (4), 13 (2006)
2. Lucas, B.D., Kanade, T.: An iterative image registration technique with an application to stereo vision. In: IJCAI, pp. 674–679. William Kaufmann (1981)

3. Tomasi, C., Kanade, T.: Detection and tracking of point features. Technical Report CMU-CS-91-132, Carnegie Mellon University (1991)
4. Benfold, B., Reid, I.: Stable multi-target tracking in real-time surveillance video. In: Computer Vision and Pattern Recognition (CVPR), pp. 3457–3464. IEEE (2011)
5. Rasmussen, C., Toyama, K., Hager, G.D.: Tracking objects by color alone. DCS RR-1114, Yale University (1996)
6. Fukunaga, K., Hostetler, L.D.: The estimation of the gradient of a density function, with applications in pattern recognition. IEEE Trans. Inf. Theory **21**(1), 32–40 (1975)
7. Comaniciu, D., Meer, P.: Mean shift: A robust approach toward feature space analysis. IEEE Trans. Pattern Anal. Mach. Intell. **24**(5), 603–619 (2002)
8. Allen, J.G., Xu, R.Y., Jin, J.S.: Object tracking using CamShift algorithm and multiple quantized feature spaces. In: Proceedings of the Pan-Sydney Area Workshop on VISUAL INFORMATION PROCESSING, pp. 3–7 (2004)
9. Siriteerakul, T., Gullayanon, R.: Robust tracking algorithm with designed marker for limited-power computer. In: Proceedings of International Conference on Knowledge and Smart Technology (KST) (2016)
10. Otsu, N.: A threshold selection method from gray-level histograms. Automatica **11**(285-296), 23–27 (1975)

Real-Time Control of Mobile Communication Channels

Mario M. Kubek and Herwig Unger

Abstract Mobile devices such as mobile phones offer a wide range of communication channels that differ regarding offered speed, costs and availability. The transmission of sensitive data in the field of mobile control and monitoring technology requires an online control with respect to a high reliability and fault tolerance, goals that are reachable using an appropriate and combined selection of such channels. A system for this purpose along with its architecture and its modes of real-time operation is introduced that additionally to the required fault tolerance can increase the security of data transmitted.

Keywords ChannelSwitcher · Mobile application · Mobile ad hoc network · Hotspot · Communication channel · Fault tolerance

1 Introduction

Current solutions in the field of mobile control and monitoring technology frequently rely on wireless, radio-based communication channels that neither guarantee a sufficient transmission quality nor—in an extreme case—the needed degree of fault tolerance.

Mesh and ad hoc networks as described in [1, 2] have been a first step to manage geographically distributed functionalities in this regard. However, they usually offer only limited functionalities such as an adaptive routing [3] using just one communication channel. Therefore, an adaptive and context-specific management and usage of resources is hardly possible using them. To address these problems, the authors have developed a hardware platform—with a special focus on high availability—that primarily relies on the usage of mobile phones that typically offer a

M.M. Kubek (✉) · H. Unger
Chair of Communication Networks, FernUniversität in Hagen, Hagen, Germany
e-mail: mario.kubek@fernuni-hagen.de

H. Unger
e-mail: herwig.unger@fernuni-hagen.de

P. Meesad et al. (eds.), *Recent Advances in Information and Communication Technology 2016*, Advances in Intelligent Systems and Computing 463,
DOI 10.1007/978-3-319-40415-8_29

high connectivity due to many included radio interfaces such as NFC (Near Field Communication), Bluetooth and Wi-Fi and that are able to connect to 2G, 3G and 4G networks. However, their combined or alternative usage for the transmission of connected data streams is generally not provided or implemented.

One of the first security-related applications have been mobile surveillance cameras that, as autonomous devices, are distributed over a specific area and in contrast to off-the-shelf Wi-Fi network cameras such as the Plug & Play network camera C903IP.2 [4] can work independently and battery-powered over a longer period and send taken pictures or videos to a security server typically not located in the monitored area. Usually, the data to be transmitted is first sent wirelessly to a Wi-Fi router which forwards them through a wired connection such as DSL to the security server. Another application of this kind are autonomously working passenger information systems such as the dynamic passenger information display "DFI LCD" developed by EPSa [5] which is equipped with a GSM/GPRS-based radio interface.

However, deficiencies exist in terms of fault tolerant communication: An automatic switch between communication channels only occurs between Wi-Fi and GSM/3G protocol suites if the Wi-Fi connection is interrupted [6]. This mode of operation only considers the complete loss of connection to the local router, the quality of the Internet connection such as the guarantee of a minimum transmission rate is not checked or taken into account here. The consideration of server response times (see Fig. 1) clearly shows that using this transmission path, significant and time-dependent fluctuations of a factor of 5 to 10 in transmission quality occur that could possibly comprise a complete loss of the connection e.g. in case of an unreachable DNS server. In [7], the usage of alternative channels to speed up data transfers is proposed. However, due to the bottleneck in form of the Wi-Fi/DSL router, this can only work when the Wi-Fi connection available.

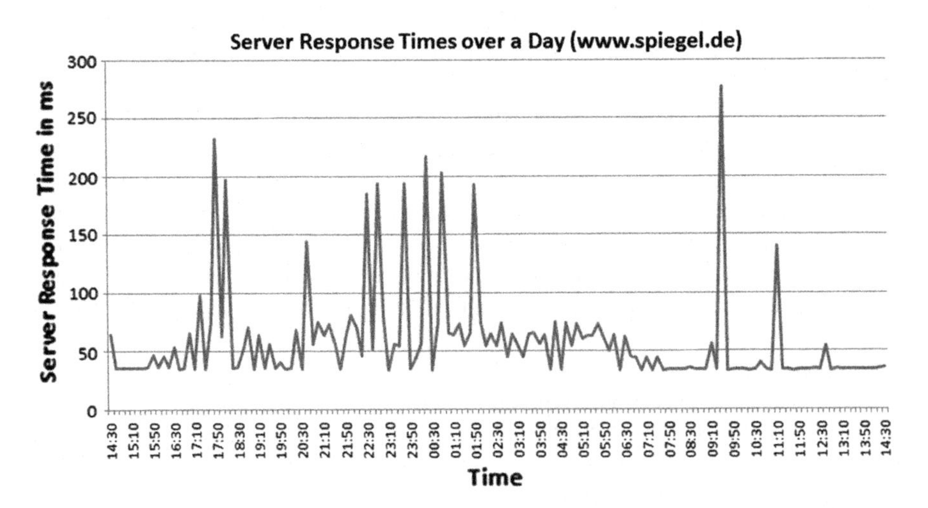

Fig. 1 Time-dependent fluctuations of the transmission rate

Newer mobile applications, that typically include a self-organising, context- and location-dependent service management, however, require increased attention to security aspects and to the management of user roles and rights. Here, the main problem is to uniquely identify users and devices that have never communicated before, to provide them a secure access to given communication channels and to offer the right information and possible means of action at the right time. A first system service for this purpose is the adaptation of access rights over a secure, alternative channel as well as the secure management of passwords, WPA-keys and access data. Here, in the easiest case, SIM cards can be a means of unique user authentication. The service Paybox [8] is an example for this approach that uses a secure second network for this purpose. The herein introduced mobile application (short: App) ChannelSwitcher [9] applies and extends this concept. Additionally, a communication concept for ChannelSwitcher will be described such that a stationary agent on each mobile device can monitor ongoing communications between connected devices and can, where appropriate, in cooperation with neighbouring devices manage these connections. For this purpose, this agent can monitor all available communication channels, heterogeneously organise local networks and switch communications appropriately using them.

2 The Mobile App ChannelSwitcher

The Android App ChannelSwitcher for mobile phones is introduced in this section.

2.1 Specification

ChannelSwitcher runs in the background of a mobile device as a service with a low power consumption and offers the following functionality:

1. Periodical analysis of the availability of communication channels and check for the parameters and costs of the current communication channel, in the simplest case by measuring the speed and stability of the current connection e.g. by determining the mean round-trip times of data packets (ping times).
2. Decision-making on when to switch between available communication channels, e.g. between Wi-Fi and GSM/3G in case the current connection is lost or its service quality (e.g. connection speed) is below a predefined threshold. Here, basically a multicriterial decision taking into account costs, speed and availability is made.
3. Organising of an adaptive and scalable communication infrastructure, e.g. by autonomously generating a mobile access point using the mobile phone's hotspot function for neighbouring phones without GSM/3G-connection including:

- access conflict resolution for neighbouring phones
- performance of Wi-Fi network access tests and tests of the functioning of the generated hotspot
- real-time negotiation on whether to add further network access points on neighbouring phones as well as to turn on additional communication channels such as Bluetooth on the local device

4. Activation and deactivation of communication interfaces of neighbouring phones on-demand (e.g. via Wake-on-BT, SMS)

2.2 Implementation

The first version of ChannelSwitcher only offers the most basic functionality to guarantee a high stability during autonomous operation. Besides its manual operation mode in which users can turn on and off Wi-Fi or the hotspot function by themselves, it offers in auto-mode an adaptive and automatic switching between Wi-Fi and GSM/3G based on the user-defined parameter for the Wi-Fi switching time. The core functionality of ChannelSwitcher is presented in the flowchart of Fig. 2.

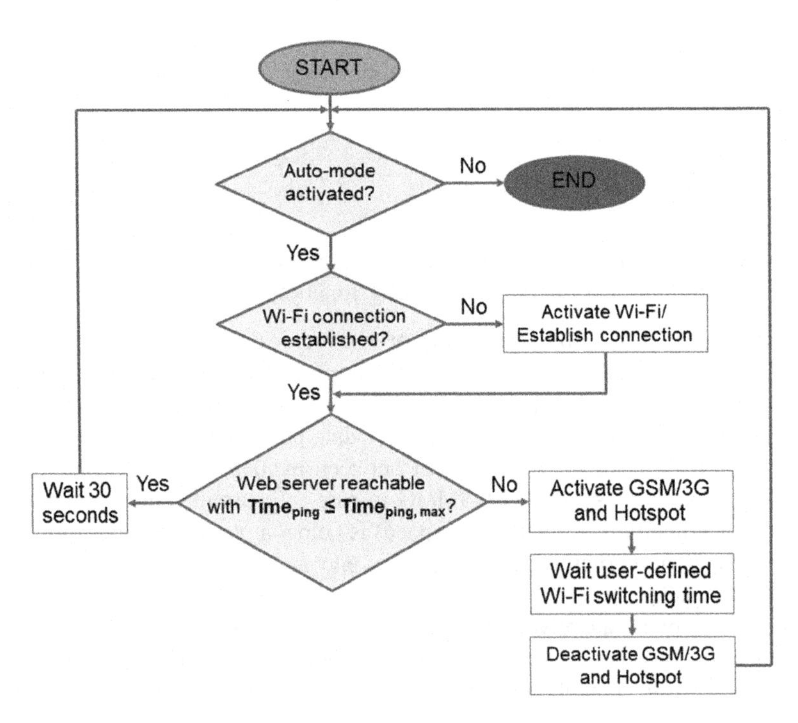

Fig. 2 Flowchart of "ChannelSwitcher"

Fig. 3 Screenshot of the App "ChannelSwitcher"

Figure 3 shows the graphical user interface (GUI) that is seen when the App is in the foreground. Normally, however, this GUI is hidden and the current state of the communication system is documented in the mobile phone's status bar using icons and notification messages. ChannelSwitcher has been developed for SAMSUNG mobile phones, yet it can run on all devices with Android version 2.3 or higher. It is also available in the Google Play Store [9].

2.3 *Extensions*

It is easy to be seen that an adaptive P2P-operation is basically realisable in order to construct a dynamic ad hoc network, too. Here, several mobile phones connect with each other, whereby they can be sender, receiver and also router of data. This way, also larger areas can be covered than by using just one router or hotspot. However, only a few manufacturers offer this P2P-mode of communication interfaces, e.g. using the standard Wi-Fi Direct [10]. That usually means that the normal Wi-Fi client mode and the hotspot function cannot be active at the same time as they are mutually exclusive. Future versions of ChannelSwitcher will largely extend the functionality of mobile phones in this regard and will make them useable as data relays while relying on these P2P-modes. At the same time, a mutual and periodical exchange of status information on the quality of the current connections as well as a

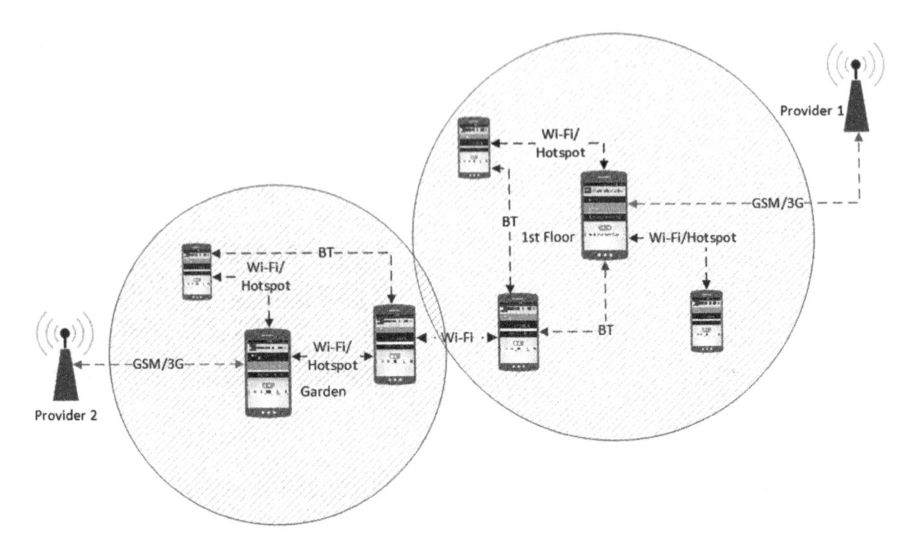

Fig. 4 Heterogeneous communication architecture

hierarchical routing of control signals between neighbouring mobile phones will be implemented. That means that first a short-range or mid-range connection will be established before a GSM/3G-connection in conjunction with the hotspot function is activated. In doing so, a heterogeneous network can be established. This means that devices in close proximity will be connected via Bluetooth and Wi-Fi, larger areas, however, will be covered using GSM/3G-based hotspots (see Fig. 4).

The advantages are that

- devices can be connected with a minimum amount of transmission power,
- a speedup of the communication by using multiple communication channels and an optimised routing can be achieved,
- a never-before-seen degree of fault tolerance can be guaranteed and
- the selection of communication channels which is non-transparent to an attacker can lead to an increased security.

3 Service Management

In a flexible, highly dynamic network, the real-time management of services and communication channels plays an important role. Here, an application such as ChannelSwitcher could be of central importance because it has knowledge of all running applications as well as all available hardware-related resources due to its closeness to the Android OS. This allows for the realisation of additional services such as:

1. Alternative (third) communication channels of various providers could be used to securely exchange keys. This way, a mechanism can be implemented that realises a function for key (e.g. WPA-keys) distribution using GSM/3G connections.
2. Permanent data streams (when detected) could be buffered to compensate short Internet connection failures or during the time of switching communication channels.
3. Data storages on the different connected mobile phones could be dynamically created when needed. As a side effect, due to the unknown location of data in advance, the probability of an unauthorised access to it is lowered, too.
4. Interactions with other applications to determine the right time for switching communication channels and to positively adapt their functioning (e.g. by reducing the amount of data transmitted using data compression techniques or by lowering the resolution of pictures or videos) could be further sensible use cases for ChannelSwitcher.

Further considerations involve the usage of a central server to control instances of ChannelSwitcher on the mobile phones remotely in order to dynamically modify access rights of users and devices that are then propagated across the generated network.

4 Conclusion

A simple, yet highly extensible controller for mobile phones has been introduced that can manage effectively and simultaneously the available communication channels on the devices. For this purpose, the most suitable way of transmitting data is selected before establishing or switching a connection. Numerous extensions allow for a construction of an adaptive and scalable communication infrastructure. However, future research is necessary regarding their applicability. A first real-life application for ChannelSwitcher is its usage for mobile video surveillance solutions such as "Mobile PowerCam" [11] that require a stable and reliable Internet connection at any time.

References

1. Akyildiz, I.F., Wang, X., Wang, W.: Wireless mesh networks: a survey. Comput. Netw. ISDN Syst. **47**(4), 445–487 (2005)
2. Basagni, S., Conti, M., Giordano, S., Stojmenovic, I.: Mobile Ad Hoc Networking: The Cutting Edge Directions. Wiley-IEEE Press (2013)
3. Lertsuwanakul, L.: Multiple criteria routing algorithms in mesh overlay networks. Ph.D thesis, FernUniversität in Hagen (2011)

4. Website of the ELRO Plug & Play network camera C803IP.2, http://www.elro.eu/en/products/cat/security/network-camera/fixedcamera/plug-play-network-camera3 (2014)
5. EPSa GmbH: Dynamic passenger information display "DFI LCD", www.epsa.de/de/produkte6.php?fid=e7f6bec1b8afe9dba533331035a217b2 (2014)
6. Google: Android 3.0 User's Guide, Android mobile technology platform 3.0, www.google.com/help/hc/pdfs/mobile/AndroidUsersGuide-30-100.pdf (2011)
7. Wulff, M., Unger, H.: Message chains and disjunct paths for increasing communication performance in large networks. In: Distributed Communities on the Web. Lecture Notes in Computer Science, vol. 1830, pp. 122–132. Springer Berlin-Heidelberg (2000)
8. Website of Paybox, https://www.paybox.at/ (2014)
9. App "ChannelSwitcher" in Google Play Store, https://play.google.com/store/apps/details?id=com.sws.channelswitcher (2014)
10. Wi-Fi Alliance: Wi-Fi Direct, http://www.wi-fi.org/discover-wi-fi/wi-fi-direct (2014)
11. App "Mobile PowerCam" in Google Play Store, https://play.google.com/store/apps/details?id=com.smallworldsec.mobilepowercam (2014)

E-Model Parameters Estimation for VoIP with Non-ITU Codec Speech Quality Prediction

Tuul Triyason and Prasert Kanthamanon

Abstract The aim of this research is to the improve performance of the E-model, which is one of the most successful non-intrusive speech quality prediction models for voice communication over a packet based network. However, the E-model still has limitations. The calculation method of the E-model is restricted to a set of voice codecs from ITU-T. This paper proposes a method to estimate two codec-related parameters that used to calculate the E-model, which are called equipment impairment factor I_e and packet loss robustness factor *Bpl* of the non ITU-T codec. The process to estimate both parameters uses a curve fitting method to calculate I_e values from PESQ results under various levels of network packet loss. The set of I_e and *Bpl* of eight narrowband codecs (G.711, G.729, GSM, AMR, iLBC, Speex, Silk, and Opus) are presented. Statistical analysis was also performed for model validation. The results show that the E-model with our I_e and *Bpl* parameters achieved a good accuracy and a good correspondence with PESQ MOS among the eight codecs.

Keywords VoIP · E-model · QoE · Codec

1 Introduction

The fundamental theorem of communication speech quality prediction has been researched and developed over many years to keep up with the rapid growth of telecommunication technology. Nowadays, packets switch networks, such as IP has become the main communication channel to carry both delay and non-delay sensitive data. Previous separate technologies such as voice, video and data commu-

T. Triyason (✉) · P. Kanthamanon
School of Information Technology, King Mongkut's University of Technology Thonburi, Pracha-utid Road, Bangmod, Toongkru, Bangkok, Thailand
e-mail: tuul.tri@sit.kmutt.ac.th

P. Kanthamanon
e-mail: prasert@sit.kmutt.ac.th

P. Meesad et al. (eds.), *Recent Advances in Information and Communication Technology 2016*, Advances in Intelligent Systems and Computing 463,
DOI 10.1007/978-3-319-40415-8_30

nication can now share resources over the same IP network. However, IP is a best-effort network that has not been designed to guarantee a quality of service for real time communication. The network characteristics such as packet loss, delay, and jitter play a crucial role for controlling a user's quality of experiences. In voice communication, speech quality is one of the key factors that map to a user's quality of experience. Therefore, an appropriate speech quality measurement method is a basis for maintaining a good quality of service. Communication speech quality measurement methods have been developed with either subjective or objective approaches. The subjective measurement is an assessment done by humans and the quality is rated by Mean Opinion Score (MOS) as defined in ITU-T Recommendation P.800 [1]. The overall procedures are to ask the testers to rate the quality of a set of speech sample in a controlled environment on a five-point scale (Bad, Poor, Fair, Good, and Excellent). The main drawbacks of the subjective method are that it is expensive, time consuming, lack of repeatability, and not useful for real-time speech quality monitoring [2]. These drawbacks have led to the development of objective methods.

The objective measurement method can be classified as intrusive and non-intrusive. Intrusive methods typically compare two input signals. These are a reference signal and a degraded signal which are recorded at the end of the communication chain. The ITU-T P.862 Perceptual Evaluation of Speech Quality (PESQ) is the most widely used intrusive method for narrowband VoIP applications [3]. Since PESQ is mainly designed for listening-only measurement, the objective MOS scores of PESQ are referred to as objective listening quality MOS (i.e., MOS-LQO) [4]. Non-intrusive methods do not need a reference signal. They typically predict voice quality based on the network (e.g., delay and packet loss) and communication relevant parameters (e.g., codec type and communication noise). The ITU-T G.107 E-model is the most successful non-intrusive speech quality prediction method for VoIP applications [5]. It can be used to predict MOS score directly from network relevant parameters. Non-intrusive methods are less accurate than intrusive methods, but are appropriate method for real-time speech quality monitoring because they do not need to record a speech signal from a network under tested.

Although the E-model is the most widely used method for VoIP speech quality prediction, it is only applicable to a very restrict set of codecs from ITU. The reason is that ITU only provides the model parameters that related to their set of codecs. This prevents the application of the E-model in new VoIP applications with non-ITU codecs. To address this problem, ITU provides two methodologies for the derivation of model parameters through subjective testing and the objective intrusive method [6, 7]. However, both methods require a lot of additivity checks by tandem operation with ITU codecs. This may not be appropriate for practical VoIP speech quality prediction. In this paper, the method to the estimate equipment impairment factor and packet loss robustness factor which are relevant parameters in the E-model calculation is proposed. The idea of the proposed method is to try to estimate both parameters from PESQ MOS by comparing with the E-model prediction MOS. The parameters obtained from this method are generic and can be used with the standard E-model seamlessly.

The remainder of this paper is structured as follows. In Sect. 2, a brief survey of related works is provided. In Sect. 3, the details of the E-model speech quality prediction methods are introduced. In Sect. 4, the method of how to derive the E-model relevant parameters for non-ITU codecs is presented. The accuracy of the E-model with new parameters is analyzed by comparing it with the PESQ method in eight codecs. The results are presented in Sect. 5. Section 6 concludes the paper.

2 Related Works

To the best of our knowledge, there are only a few work on this issue. This is due to the fact that most research in VoIP speech quality prediction field are related to the ITU codecs. The new framework for equipment impairment factor of the E-model is proposed by the inventor of the E-model for wideband (WB) and super-wideband (SWB) in [8, 9]. However, both bodies of research only present a method of how to obtain the equipment impairment factors of the ITU codec in WB and SWB codecs. Cole and Rosenbluth [10] and Sun and Ifeachor [11], have proposed a method to derive the equipment impairment factors of the AMR, G.729, G.723.1, and iLBC codecs from a curve fitting technique of PESQ MOS. The results show that the equipment impairment factor can be described in the form of a logarithmic function. However, the equipment impairment factor estimated by this method still do not conform to the factors defined by ITU. Raja et al. proposed a methodology to derive equipment impairment factor for a mix narrowband and wideband codec by using genetic programming evolve the impairment model from a modest set of initial parameters [12]. The impairment model of this method is also defined by a logarithmic function as presented in [10, 11]. In [13], the authors proposed a perceptual quality model for adaptive VoIP applications. They followed the method as presented in ITU-T recommendation P.833 to derive the equipment impairment factor of AMR codec but used PESQ instead of subjective listening test. Unfortunately, the result of derived impairment factor did not present in this paper. In [14], the authors presented wideband and narrowband SILK codec speech quality prediction model for VoIP application. They also presented a prediction model for the equipment impairment factor and packet loss robustness factor of the SILK codec in the form of an exponential decay function. Assem et al. proposed a genetic algorithm for mid-call audio codec switching [15]. They tested their algorithm on five codecs (G.711, SILK, iLBC, GSM, and Speex) and used E-model MOS as the indicator for codec switching. The paper also presented the method to estimate the equipment impairment factors of non-ITU codec in this research. In [16], the authors proposed a non-intrusive speech quality prediction model for Opus based codec VoIP. In this work, the model is not based on the E-model and the method to obtain MOS is by subjective tests. However, it could be the very first work that develops a non-intrusive model for a new Opus codec.

3 The E-Model

The E-model is a computation model that is used for network planning [5]. It was developed by the European Telecommunication Standards Institute (ETSI) work group. The E-model is a non-intrusive measurement technique that does not require a reference speech signal. Generally, the model calculates the basic quality value determined for a network, and later subtracts with the transmission impairment factors. The E-model combines the effects of each transmission impairment factor into one rating factor scale R, ranging from 0 (poor) to 100 (excellent). The R factor can be further mapped into MOS score by a transform function. The R factor can be calculated as Eq. (1).

$$R = R_0 - I_s - I_d - I_{e-eff} + A \tag{1}$$

where, R_0 represents the basic signal to noise ratio at 0 dBr (decibels relative to reference level). I_s is the sum of all impairments that may occur in a voice transmission. I_d is the impairment factor due to delay. I_{e-eff} is the effective equipment impairment factor caused by low bit rate coding and packet loss. A is the advantage factor or expectation factor (in conventional systems, A equals 0). Since,I_s is a function of communication over circuit switching and it depends on the method of VoIP access, we rely upon the default values of the ITU-T recommendation [17]. Assuming the system uses conventional communication; therefore the advantage factor A equals zero [10]. With the default values, the R factor can be calculated as Eq. (2).

$$R = 93.2 - I_d - I_{e-eff} \tag{2}$$

where, I_d is a function that is related to a one-way delay of the network. It can be calculated through a simplified delay model in [10] or a 6th order polynomial equation in [11]. However, for a network delay less than 50 ms, the value of I_d is insignificant (≤ 1) and can be dropped from calculation [10]. I_{e-eff} is the packet loss dependent equipment impairment factor. According to the recommendation G.107, I_{e-eff} can be expressed as follows:

$$I_{e-eff} = I_e + (95 - I_e)\frac{Ppl}{\frac{Ppl}{BurstR} + Bpl} \tag{3}$$

I_e is the equipment impairment factor which represents impairment factors caused by a low bitrate codec. *Ppl* represents a percentage of packet loss. *Bpl* is the packet loss robustness factor which is a codec-specific value. *BurstR* is the burst ratio. When packet loss is bursty, the value of *BurstR* is beyond 1. When packet loss is random, *BurstR* equals 1. Since I_e and *Bpl* are codec-specific values, ITU-T proposed the values of both parameters for ITU codecs in recommendation G.113 [17]. The values of I_e and *Bpl* of ITU codecs are summarized in Table 1.

Table 1 The values of I_e and *Bpl* of ITU codec in ITU-T G.113

Codec	I_e	*Bpl*
G.723.1 + VAD	15	16.1
G.729A + VAD	11	19.0
GSM-EFR	5	10.0
G.711	0	4.3
G.711 PLC	0	25.1

Once the *R* factor is calculated, the MOS score can be obtained from a transform function as follows:

$$\begin{aligned}&\text{For } R<0: && MOS=1\\&\text{For } 0<R<100: && MOS=1+0.035R+R(R-60)(100-R)\cdot(7\times10^{-6})\\&\text{For } R>100: && MOS=4.5\end{aligned} \quad (4)$$

4 Method for Deriving I_e and *Bpl* for Non-ITU Codecs

Since, ITU-T recommendation G.113 does not provide the I_e and *Bpl* values for non-ITU codecs such as iLBC, Speex, Silk, and Opus; in order to obtain these values, PESQ is used to estimate the MOS of non-ITU codecs by comparing reference and degraded signals of theses codecs. Then PESQ MOS can be transformed by using the 3rd order polynomial transform function from [11]. The transform function can be expressed as follows:

$$R=3.026MOS^3-25.314MOS^2+87.06MOS-57.336 \quad (5)$$

Since the PESQ MOS has listening-only quality, delay factor is not take into consideration. Therefore, Eq. (2) can be reduced into more simplified form and I_{e-eff} can be obtained from Eq. (6) as follows:

$$I_{e-eff}=93.2-R \quad (6)$$

In order to derive the I_e and *Bpl* values of a non-ITU codec, a simple experiment testbed is carried out as shown in Fig. 1.

Since the set of I_e and *Bpl* parameter of ITU was created from unknown set of speech signal in 2001 and that original set of speech signal is rarely available in present, therefore the reference signals in this experiment were taken from ITU-T recommendation P.564 [18]. The ITU-T recommendation P.564 speech corpus is a set of speech sample for narrowband VoIP transmission measurement purposes. It includes speech material as a set of four eight seconds speech files, which represent speech from four talkers, two males and two females in American English Spanish, French, and UK English languages. The VoIP softphone using in this testbed was developed from the PJSIP library [19]. In order to compare the new set of I_e and *Bpl*

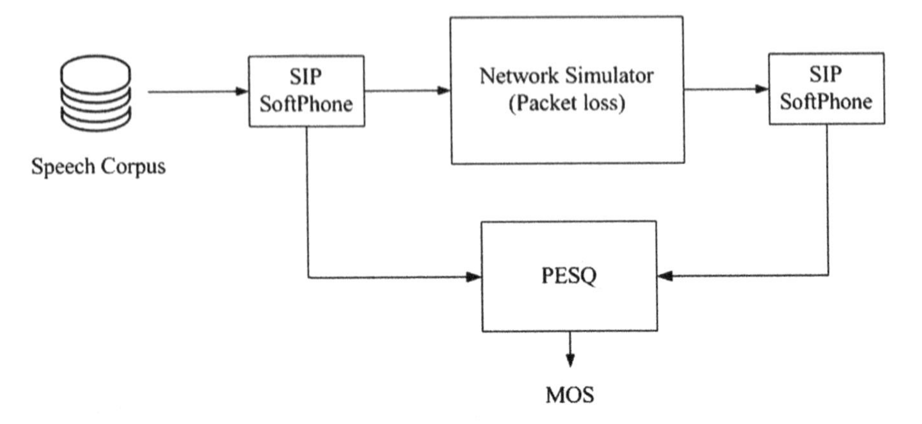

Fig. 1 Experimental testbed diagram

parameter with the ITU, eight VoIP codecs were chosen from both ITU codec (G.711, G.729, and GSM) and non-ITU codec (AMR, Speex, Silk, iLBC, and Opus). Since this research is limited to only narrowband codecs, all codecs were set to operate at 8000 Hz sampling rate only. 0–20 % packet loss levels were emulated by network simulator hardware [20]. PESQ algorithm was applied to each speech signal to measure the MOS score. In order to avoid packet loss location bias [11], the MOS score was averaged from 30 times measurements per packet loss conditions. The I_{e-eff} factor of eight codecs were calculated with Eqs. (5) and (6) from the PESQ MOS. The results are shown in Fig. 2 with packet loss on the x-axis and I_{e-eff} on the y-axis.

The MATLAB R2015b software was used to run the curve fitting method by randomize enter the candidate I_e and *Bpl* into Eq. (3). The curve was adjusted until the E-model $I_{e\text{-}eff}$ curve closely matched with PESQ curve in Fig. 2 (95 % confidence interval). The results of I_e and *Bpl* parameter that generated the E-model curve closely matched with the PESQ curve are listed in Table 2.

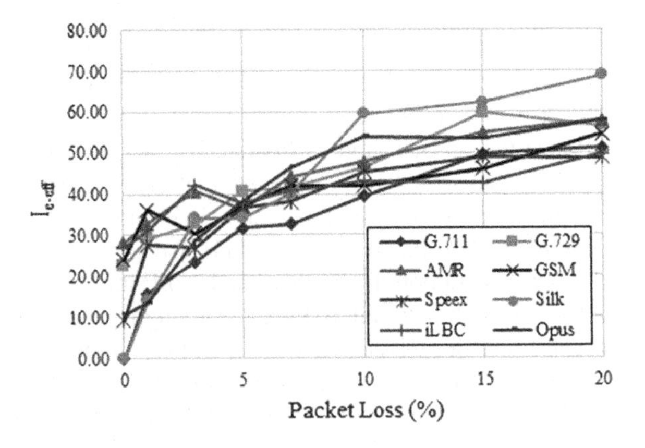

Fig. 2 The $I_{e\text{-eff}}$ of eight codecs

Table 2 Estimated values of I_e and *Bpl*

Codec	I_e	*Bpl*
G.711	7	16
G.729A	24	20
AMR (NB)	30	26
GSM	28	34
Speex (NB)	18	21
Silk (NB)	2	8
iLBC	32	50
Opus (NB)	11	12

5 Accuracy Analysis

In order to evaluate the accuracy of the estimated I_e and *Bpl* values in the E-model, eight codecs were selected to make a VoIP call in network under test. In this experiment, MicroSIP and Phonerlite software were used as a VoIP application for voice quality prediction [21, 22]. Both of them are Windows OS based software. The calls made by first seven codecs (G.711, G.729, GSM, AMR, iLBC, Speex, and Silk) were made by MicroSIP, whereas the Opus codec's calls were generated by Phonerlite software. The emulated WAN link with two hosts was constructed as shown in Fig. 1 [20]. Both host feature Intel based X64 architecture. Microsoft Windows 10 was installed to both hosts. The speech sample files were taken from ITU-T recommendation P.564 [18]. Since there is no appropriate solution to inject speech sample files to both VoIP softphones, the virtual audio cable software were used to connect audio inputs and outputs between a media player software and VoIP softphone [23]. Once the call session started, the speech files were streamed with eight codecs and packet loss level was emulated between 0 and 40 %. The voice packets were decoded at the endpoint machine and speech signals were recorded with media recording software [24]. The PESQ algorithm was applied to measure the MOS scores. Since a network delay in this testbed was very low (<1 ms), the speech quality of each codec can be predicted by the E-model with the new I_e and *Bpl* from Eq. (6). Figure 3 shows the MOS between the PESQ and the E-model in Opus codec between 0 and 40 % packet losses.

Taking the MOS values obtained by the PESQ as references; the accuracy evaluation can be performed with two criteria. The first one is how the MOS of the E-model with estimated I_e and *Bpl* correlated with the PESQ MOS. The second one is magnitude of prediction error occurred between the E-model compared to PESQ. In order to achieve this, the Pearson's correlation coefficient and the Mean Absolute Percent Error (MAPE) are chosen to satisfy both criteria. The result is presented in Table 3. The average correlation coefficient of E-model with estimated I_e and *Bpl* of eight codecs was 0.95. The E-model with a new set of parameter showed a good correspondence among predicted MOS and PESQ measured MOS. The average

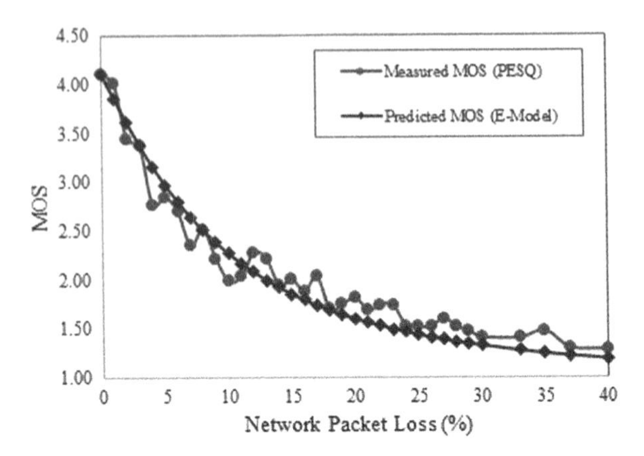

Fig. 3 The Opus MOS values of the PESQ and the E-model in packet loss 0–40 %

Table 3 Accuracy analysis of the E-model with estimated I_e and *Bpl* parameters

Codec	Pearson correlation coefficient	MAPE
G.711	0.98	3.98
G.729A	0.98	4.11
AMR (NB)	0.97	3.85
GSM	0.93	5.20
Speex (NB)	0.95	5.74
Silk (NB)	0.98	6.15
iLBC	0.86	5.33
Opus (NB)	0.98	7.60

MAPE error rate of eight codecs was 5.25 %. It can be seen that a high prediction accuracy of predicted MOS was achieved by using the new estimated I_e and *Bpl* parameters. This method is generic and can be used in the standard E-model equation of ITU-T for VoIP online-monitoring with non-ITU codec.

6 Conclusion

In this paper a method to estimate equipment impairment factor I_e and packet loss robustness factor *Bpl* of the E-model for non-ITU codecs is proposed. The method exploited an advantage of PESQ objective intrusive measurement method to obtain a MOS curve of non-ITU codecs. A curve fitting method was applied to estimate I_e and *Bpl* parameters from the E-model equation. The E-model with a new parameter set was validated by Pearson Correlation analysis and MAPE error analysis among eight codec. The E-model with estimated I_e and *Bpl* parameters achieved a high correlation with low error rate between the predicted and measured results. The I_e and *Bpl* parameters obtained by this method are generic and can be enhanced with the traditional version of the E-model equation to predict a speech quality of non-ITU codecs. The extracted parameters can be implemented in the E-model to monitor and tuning for optimal user quality of experience in real time. The future work will be extended for WB and SWB to have a complete parameter table for non-ITU codec.

References

1. ITU-T Recommendation P.800: Methods for subjective determination of transmission quality (1996)
2. Ebem, D.U., et al.: The impact of tone language and non-native language listening on measuring speech quality. J. Audio Eng. Soc. **59**(9), 647–655 (2011)
3. ITU-T Recommendation P.862: Perceptual evaluation of speech quality (PESQ), an ojective method for end-to-end speech quality assessment of narrowband telephone networks and speech codecs (2001)
4. ITU-T Recommendation P.800.1: Mean Opinion Score Terminology (2006)
5. ITU-T Recommendation G.107: The E-model: a computational model for use in transmission planning (2015)
6. ITU-T Recommendation P.833: Methodology for derivation of equipment impairment factors from subjective listening-only tests (2001)
7. ITU-T Recommendation P.834: Methodology for the derivation of equipment impairment factors from instrumental models (2015)
8. Möller, S., Raake, A., Kitawaki, N., Takahashi, A., Wältermann, M.: Impairment factor framework for wide-band speech codecs. IEEE Trans. Audio Speech Lang. Process. **14**, 1969–1976 (2006)
9. Waltermannn, M., et al.: Extension of the E-model towards super-wideband speech transmission. In: IEEE International Conference on Acoustics Speech and Signal Processing (ICASSP), pp. 4654–4657 (2010)
10. Cole, R.G., Rosenbluth, J.H.: Voice over IP performance monitoring, pp. 9–24. ACM SIGCOMM, Comput. Commun. Rev. (2001)
11. Sun, L., Emmanuel, I.C.: Voice quality prediction models and their application in VoIP networks. IEEE Trans. Multimedia **8**(4), 809–820 (2006)
12. Raja, A., et al.: A methodology for deriving VoIP equipment impairment factors for a mixed NB/WB context. IEEE Trans. Multimedia **10**, 1046–1058 (2008)
13. Hoene, C., Holger, K., Adam, W.: A perceptual quality model intended for adaptive VoIP applications. Int. J. Commun. Syst. **19**, 299–316 (2006)
14. Goudarzi, M., Lingfen, S., Emmanuel, I.C.: Modelling speech quality for NB and WB SILK codec for VoIP applications. In: 5th International Conference on Next Generation Mobile Applications, Services and Technologies (NGMAST). (2011)
15. Assem, H., Merabet, A., Brendan, J., David, M., Jonathan, D., Pat, O.: A generic algorithm for mid-call audio codec switching. In: IFIP/IEEE International Symposium on Integrated Network Management (IM 2013), pp. 1276–1281 (2013)
16. Orosz, P., Tamás, S., Zoltán, N., Tamás, L.: A no-reference voice quality estimation method for Opus-based VoIP services. Int. J. Adv. Telecommun. (2014)
17. ITU-T Recommendation G.113: Transmission Impairment due to Speech Processing (2007)
18. ITU-T Recommendation P.564: Conformance testing for voice over IP transmission quality assessment models (2007)
19. PJSIP–Open Source, SIP: Stack and Media Stack for Presence, Im/instant Messaging, and Multimedia Communication (2008)

20. IPNetSim-(IPNetwork/WANEmulator—100Mbps, 1Gbps, 4x1Gbps). http://www.gl.com/ipnetsim.html
21. MicroSIP lightweight VoIP SIP softphone for Windows. http://www.microsip.org
22. PhonerLite. http://phonerlite.de/index_en.htm
23. Virtual Audio Cable (VAC). http://www.fox-magic.com/vac.php
24. Audacity. http://www.audacityteam.org

ICT Implementation and Infrastructure Deployment Approach for Rural Nepal

Babu Ram Dawadi and Subarna Shakya

Abstract Information and Communication Technology (ICT) is the basic tools of our daily lives to optimize the resources, improve work performance and efficient service delivery. Due to diverse demographic and geographic situation of Nepal which has remote areas, government itself is not able to properly provide services to the citizens at Rural. Empowering rural people by means of ICT services in service delivery is becoming quite challenging. Nepal, after adopting the liberalization policy to encourage involvement of private sectors (Government of Nepal, http://nta.gov.np/en/legislation/policies, [1]), there have been optimum competitions in ICT sectors which help to reduce the digital divide and considerably increased the ICT penetration rate (Nepal Telecommunications Authority, http://nta.gov.np/en/mis-reports-en, [2]). But the services are more squeezed to city/urban areas. Varieties of challenges at rural Nepal raised the operability and sustainability issues of ICT centers. We have analyzed the survey data carried on the rural ICT centers to visualize its status and proposed the sustainable ICT implementation approach for rural Nepal.

Keywords Rural · ICT · Access centers · Policies · Sustainability · ADSL · Wi-MAX

1 Introduction

Within the last decade, we realized that there have been tremendous improvements in Information Technology (IT) and the telecommunications sector of Nepal. The global ICT business is increasing day by day with the development of smart applications, user friendly and rapid deployment of infrastructure. For the

B.R. Dawadi · S. Shakya (✉)
Institute of Engineering, Tribhuvan University, Pulchowk, Nepal
e-mail: drss@ioe.edu.np

B.R. Dawadi
e-mail: baburd@ioe.edu.np

P. Meesad et al. (eds.), *Recent Advances in Information and Communication Technology 2016*, Advances in Intelligent Systems and Computing 463,
DOI 10.1007/978-3-319-40415-8_31

developing countries like Nepal, Information and Communication Technology can play a remarkable role in human development through community development, economic growth, improvement in education and literacy, natural resource management, governance and health sector development. ICT accessibility areas like libraries, tele-centers, cyber-cafes, common service centers offer more access and use of ICT for development among underserved populations who are especially the rural peoples for our case in Nepal.

However, due to several other factors (such as demographic condition, terrain, transportation, electricity, literacy and health status etc.) make the maintainability, operability and sustainability of the ICT centers and services quite challenging. This paper analyzed the status of 58 ICT centers and ICT services provided by other agencies to identify the challenges and issues for better approach to sustainable ICT deployment.

2 Literature Review

2.1 ICT Services for Rural People

ICT service centers are the public places where people can access computers, the internet and other technologies that help people gather information and communicate with others, at the same time they develop digital skills to support community and social development—reducing isolation, bridging the digital divide, promoting health issues, creating economic opportunities and reaching out to youths [3].

"ICT Literacy" means it is using digital technology, communication tools and networks to access, manage, integrate, evaluate and create information in order to function in a knowledge base society [4]. According to the Swedish Program for ICT in Developing Regions [5], ICT Access Centers will help to enhance good governance, promote empowerment by facilitating tele-education and learning, ensure good health care and improve quality of life. ICT Access Centers may help to promote tele-medicine, tele-education, e-commerce, e-banking to improve the life of people in rural and remote areas, if appropriate model and intervention strategies are devised. The services of ICT-ACs will support to empower community with inclusion and reduce poverty. Information and Communication Technology (ICT) is gaining much research attention in present time especially due to its ability towards empowering people, promoting governance, alleviating poverty and contributing overall economic and social development of the country. Especially for the developing countries having difficult terrain and diverse population with low literacy rate like Nepal, ICT is an important tool that has considerable impacts on the lives of rural communities.

ICT can support democracy and human rights by expanding citizens' opportunities to participate in political decision-making, by providing citizens with access to information, and facilitating dissemination of information, as well as enabling social mobilization. ICT is of crucial importance in implementing information health systems at institutional levels. Establishing electronic patient records provides easy sharing of medical history among health personnel and patient-centered services. Organizing doctor-patient remote sessions or on-site remote training in the case of emergency, are all examples where ICT has a crucial role. ICT enables remote doctor-to-doctor consultation as well as real time instructions and treatment via telemedicine. Rural Nepal is still considered to have a backward social life in which government services are not evenly distributed.

2.2 *ICT Service Initiatives for the Rural Nepal*

The time period 2000–2010 (A.D.) become the ICT decade for Nepal. During when, government has promulgated several ICT related policies and also perform the research consultation including pilot projects to increase ICT accessibilities in Rural Nepal. The government 10th plan, 2002–2007 has focused the involvement of private sector for socio-economic development with promotion of telecommunications sector in Rural Nepal. The plan also focused on the establishment of community information centers to realize the importance of IT in social development [6].

IT Policy [7] sets the objective to make IT reachable to general public with the targets to provide internet facilities to all Village Development Committees (VDC) of Nepal. It was targeted to have knowledge based society and knowledge based industries by putting IT as a prioritized sector. Telecom Policy [1] encouraged the universal connectivity and liberalization of telecom services by provisioning different licensing regime like Rural Telecom Operator with subsidy in licensing fee only to Rs 100 for ISPs. Similarly policy provisioned the formation of Rural Telecommunications Disbursement Funds (RTDF) management committee to utilize the 4 % revenue collected by all service providers in their annual profit to enable ICT services to rural areas. More than 10 billion Nepalese rupees as revenue has been collected till date. Nepal Telecommunications Authority (NTA) has initiated for the consultancy of smart business concept towards utilization of RTDF funds [8]. The Broadband Policy, 2014 [9] has set some of the targets for better ICT services in rural Nepal. Expansion of fiber connection to every district headquarter, expansion of wired/wireless infrastructure to every Village Development Committee, guarantee of broadband access to every home, encouragement of e-service on every VDC and concept of e-VDC of Nepal is imagined by broadband policy, 2014. Local content development and management, proper utilization of RTDF for ICT and broadband service expansion are the major focus of broadband and ict policy 2015 [9, 10]. After the policy consultation to study the possibilities to increase ICT accessibilities in Rural Nepal in 2006, NTA in support of Asian Development Bank (ADB) come up with the establishment of 58 community centers at 26 rural and peri-urban districts of Nepal. These 58 ICT centers were taken as the primary source of data collection for status and impact analysis on this paper. E-Government Master Plan (e-GMP) of Nepal enforces the networked government with the establishment of telecenters at the rural places and access to broadband network services at densly populated areas [11]. The government enterprise architecture [12] proposed the technical and citizen service framework from central Government to the VDC level focusing on wireless infrastructure at the rural places.

Similarly, With the objective to reduce the digital divide and enhancing the rural people lives by introucing e-Health(tele-medicine), e-Learning, e-Education, e-Commerce, e-Agriculture and many more, Nepal wireless project, a non-commercial organization started the rural internet service by setting wireless connection form Pokhara city to Nangi village of magdi in 2001 [13].

Looking into the ICT service connectivity and coverage scenarios by major operators, Nepal Telecom is the largest incumbent operator provides basic telephone, mobile as well as internet (ADSL, Lease, Wi-MAX, and GPRS) services. As being a market leader in wireless internet, NTC Wi-Max service is covering almost the whole Nepal [14]. NCELL Pvt. Ltd. is the largest mobile operator of the country provides high speed data service to individual and the enterprise. Based on the media news and comments, NCELL mobile and data service quality is comparatively better [15]. There are other four telecom operators licensed to provide services at different parts of rural and peri-urban areas of Nepal [16] under limited mobility license.

3 Research Methodology

This reserach was conceptualized to identify the status and impact of those ICT access centers. The NTA funded 58 ICT centers of 26 districts were chosen as primary source of data collection. Questionnaire, opinion surveys/interviews techniques were taken in local language with at least four personnel at one center to gather qualitative as well as quantitative data. The study followed multiple methods of data gathering: semi-structured interviews were complemented with short time on-site observations and surveys with quantified responses. Qualitative methods of data collection helps us analyze the institutional, individualized and other factors responsible for the operations and management of the centers.

Figure 1 presents the interconnection relationship between ICT centers and the rural people, communities. ICT access centers provide different ICT services which are accessed by rural communities. The use of ICT services shall have impacts on communities like costumes, culture, lifestyle etc.

In total, 290 peoples including community members, operators and users (peoples of different profession, age group, gender and caste depicted in Table 1) of the ICT Access Centers were involved in the questionnaire and discussion session. Most of the research questions in the questionnaire session focused on collection of quantitative data. Discussion with the management committee and service operators

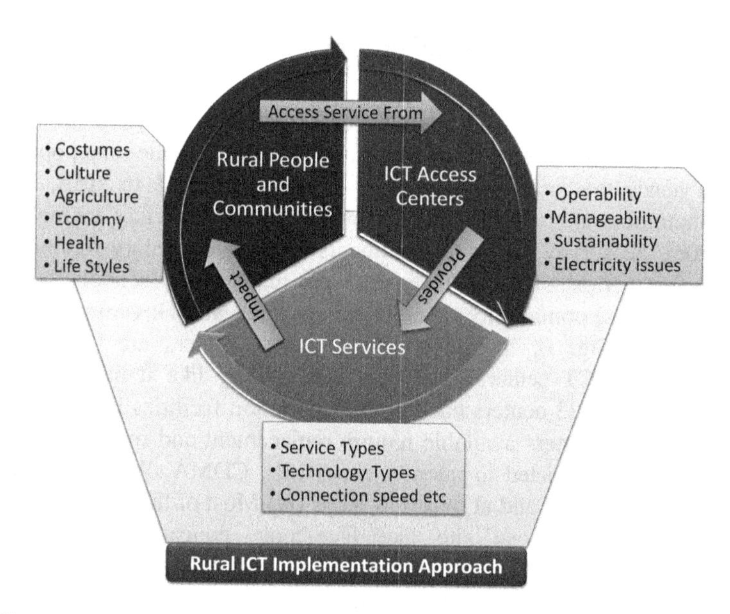

Fig. 1 Conceptual study

Table 1 Population samples for data collection by Gender, Caste, Age and Profession

Gender			Caste				
Male	Female	Total	Brahman	Chhetri	Dalit	Janajati	Total
191	99	290	68	61	41	120	290

Age (yrs)	Respondents
≤ 15	54
16 - 25	131
26 - 45	86
≥ 46	19
Total	290

Profession	Respondents
Business persons	31
Farmers	18
Housewives	3
Job Holders	26
Students	153
Teachers	59
Total	**290**

mostly focused on the challenges and issues with related positive and negative impacts in the society. Reports of other agencies regarding rural ICT activities were reviewed from secondary sources (previous reports of Ministry of Information and Communication, reports of NTA, ICT center establishment project reports carried out by partner organizations. Microsoft excel and SPSS packages were used to analyse the data collected from field survey. We mostly focused on the status of ICT centers towards its operability, resource manageability and power issues to predict its future sustainability.

4 Data Analysis and Visualizations

Figure 2 presents the voice and internet penetration rate of Nepal for last seven years [2]. The total active SIM cards distributed is more than the total population of Nepal and hence the total tele-density has reached 109.4 % in 2015, while the internet penetration rate is just crossing 46 %. The internet services are highly urban centric. From the survey data and documents study, the primary findings of this research were (i) Most of the ICT centers were managed by schools, community clubs, cooperatives, community member groups and Non-Governmental Organizations (NGOs) (Fig. 3), in which school access centers are operating more effectively (ii) 35 ICT centers have easy accessibility like it has transportation facility (road), while 23 centers have no transportation facilities like road (iii) The backup UPS and Inverters available require replacement and maintenance (iv) 43 ICT centers were connected to internet via ADSL, CDMA, Wireless Radio with internet speed of 64, 128 and at max 256 Kbps (iv) Most of the internet users have prioritized access to social site like face-book, education material, email, entertainment/games, health, job, agriculture and business information (v) School teachers are more benefited with the service and encouraged themselves to access latest information and knowledge in teaching/learning, that help the community students towards better education. Behind this, the service centers are facing different issues of operation and management leading to sustainability problems. The issues mostly faced are (i) Lack of equipments to operate the ICT services (ii) Unreliable internet connection (iii) Lack of skilled human resources to operate the services (iv) Unreliable energy supply, frequent load-shading and insufficient

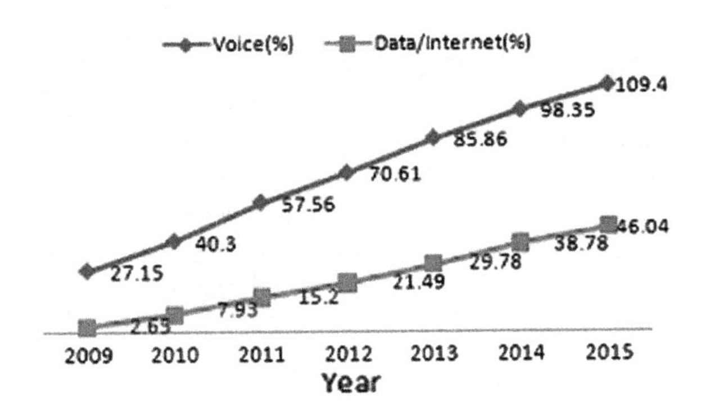

Fig. 2 Voice and data penetration of last seven years

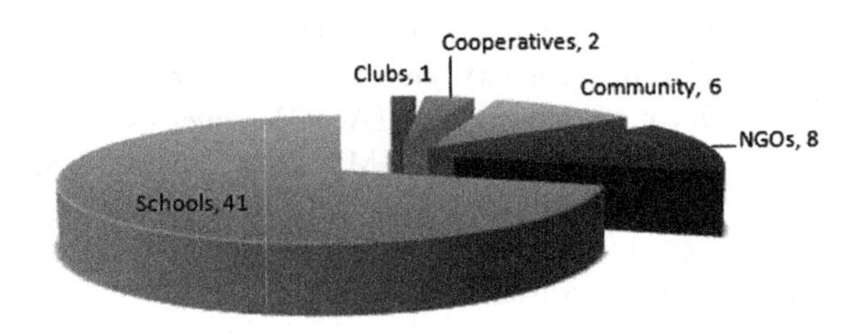

Fig. 3 Institutions for center management

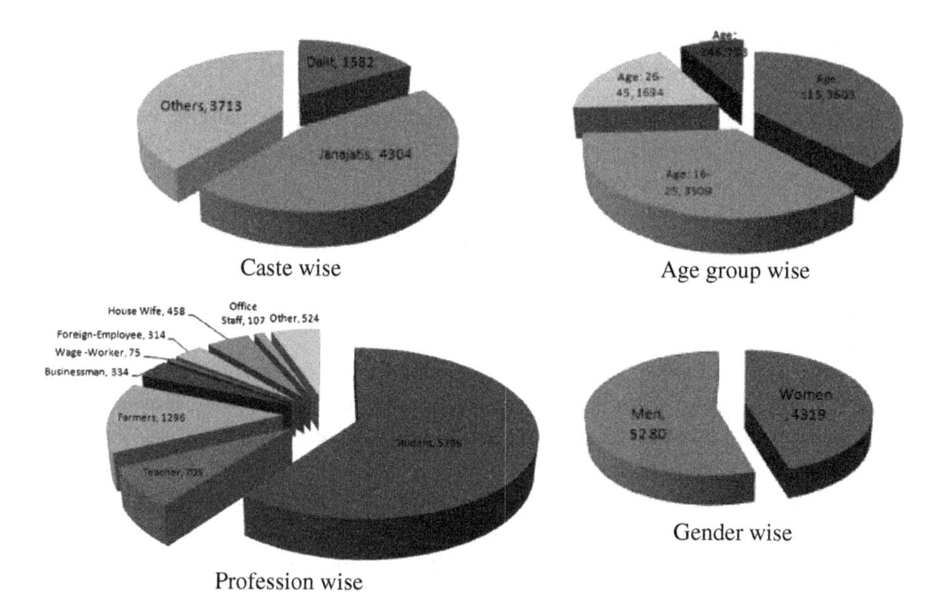

Fig. 4 Monthly status of ICT service users by different categories

backup making the service to be frequently interrupted (iv) Due to small market, income generation is difficult and only 32 centers are hardly in financial sustainability leading to service expansion problem.

The service centers over 26 districts cover in total 35,520 populations (population census-2010) as beneficiaries, in which the log records maintained by service centers show that in an average 9,599 people are served by the different ICT services per month. Figure 4 depicted the status of monthly ICT users by gender, age group, profession and caste. Dalits and Janajatis are remarked by new constitution of Nepal as backward societies. Their access to ICT services shall be considered as positive symbols for the rural development. Similarly women's participations help improve the gender imbalance and inclusive participation on every governmental and non-governmental sectors.

5 Policy Implication and ICT Implementation Challenges in Rural Nepal

Several attempts of government and non-government agencies with different policy provisions towards enabling ICT services in rural Nepal were reviewed in Sect. 2 of this paper. Government research documents and policies encouraged the provision of ICT services via tale-centers as well as the concept of rural ISPs and separate fund for rural ICT development has been provisioned. However, following were identified as challenges of ICT implementation in rural Nepal.

- Lack of feeling of ownership of service centers is a major challenge which may hinder smooth operation of services. This affects the quality of services and promotion of center activities.
- Due to the lack of various aspects such as (a) lack of trained operator (b) maintenance problem of equipments (c) insufficient backup power (d) low level of awareness in the community (e) unreliable internet connectivity (f) inactive management committee and operator, most of the centers are not operating in full fledge or closed down.
- Local wiring and Wi-Fi technologies are suitable for short range communication in the villages after from the service centers to distribute internet. However, the setup of backhaul network towards the different centers located at mountain and hilly zone is quite challenging.
- High cost of internet is another hindrance towards the service expansion in rural areas. Sparsely populated and low income rural people cannot afford the high cost of service which may require subsidy form the government.
- Financial sustainability is another major issue to properly operate the ICT centers in rural Nepal. The operation and maintenance cost is high due to lack of skilled human resources. In one hand the cost of internet is high and on the other hand rural people cannot afford high cost leading to the sustainability problems.
- However electricity on some of the centers are supplied by rural micro hydro, most of the centers have to rely on solar battery backup which may not be sufficient to run the equipments.
- Lack of awareness and access to local content could not attract the local users towards ICT services. Hence local content generation, access and connectivity are the major challenges of increasing ICT accessibilities in rural Nepal.
- Difficult and diverse terrain is another major concern of service providers hindering the reliable services. It leads to difficulty in network planning, optimization, maintenance, upgrade and network rollout.
- Security of ICT equipments and BTS tower at rural areas are also considered to be a challenge. The cause is the instability of countries political situation.

6 April, 2015 Earthquake Effects on Rural ICT Infrastructure and Services

The devastating April, 2015 earthquake having its epicenter at Barpak VDC of Gorkha district (the main author's birth place) and more than 400 bigger and smaller aftershocks [17] damaged the people's lives totally collapsing the homes including schools and ICT access centers in the affected zone. BTS, transmission towers, fiber backhaul, microwave links were damaged. NTA significantly run post-earthquake restoration activities to recover the damage and reduce impacts on ICT services throughout Nepal [18]. More than 8,500 peoples were died and more than 22,000 peoples were injured, similarly 500,000 houses were destroyed and 269,000 houses damaged by this earthquake in 14 highly affected districts of Nepal [19]. It hampers the solar backup system, telecom towers and computer centers of those districts. Peoples are suffering trouble living at their temporary plastic covered homes. The presence of ICT services helped early disaster identification to affected areas and facilitated for disaster post recovery in the emergency cases. This teaches us to require robust communication system in the rural areas for disaster preparedness and preparation of early rescue during emergency.

7 Sustainable Rural ICT Implementation Strategies and Approach

To fulfill the policy objectives, subsidized network expansion plan has to be promoted by the government. Optical fiber network up to federal and district head quarter is the mandatory. Wireless relay network (Wi-Fi) at district head quarter and city/urban areas using frequency of ISM band is feasible and these are the current solutions of ISPs providing the Wi-Fi service with affordable prices. For those people who can't handle and afford the smart device definitely be benefitted through services offered from ICT centers. The possible options for the rural Nepal network transport technologies are:

- Every VDC of Nepal is reachable by ADSL connection. Wi-Fi relay network shall be used locally in the villages beyond the ADSL connection. However reliability of the ADSL connection should be guaranteed.
- Wi-MAX service would be the best alternative approach in addition with ADSL service in the VDC, where the connection by ADSL shall be taken as backup connection. These services can easily be expanded in the home, school and community clubs in the villages.
- High range wireless communication with mountain top repeater shall be used like the concept implemented by Nepal Wireless [20]. But it could be better for pilot test only. For whole Nepal deployment, it is highly costly with respect to equipments, powers, security and maintenance as well.
- Government of Nepal shall utilize the satellite frequency like Ku and Ka band for high speed satellite broadband communication. To provide the high speed application services like VoIP phone calls, emergency communications, rural broadband Ka band spectrum has a potential to lunch those services [21]. Research on utilization of such satellite band with respect to operation and maintenance cost shall be required.

Following shall be the strategies to be followed for the successful and sustainable deployment of ICT service in the rural areas of Nepal.

- Establish ICT service centers at every VDC as a model centers and provide ADSL and Wi-MAX connections or other ISP broadband connection if any. The tentative total cost of equipments like ADSL and Wi-MAX receive routers, one 24-ports manageable switch, solar power backup, printer, one server and three client terminals shall be of USD 6,000. These minimal ICT equipments as seed shall be provided and established free of cost by the government.
- Every VDC centers should have at least one ICT volunteer whose task is ICT awareness and training, local content generation, management of ICT equipments and server which is connected in the WAN network so that it can act as a service gateway as a FOG server connected to government cloud.
- Encourage the private ISPs to deploy internet service. Provide subsidy by tax exemption on ICT equipments deployed in the rural areas and security guarantee of the ICT infrastructures. Also encourage the university researchers by providing grants under rural network deployment, FOG computing and Wireless Sensor Networks targeting to rural areas at different sectors.
- The RTDF is sufficient to facilitate village ICT centers to provide free internet service for two years. Once, people's awareness level is considered, targeted economic activities in the rural area shall be achieved, peoples will be habitual towards the ICT services, then thinking for business plan with income generation approaches would be suitable. If economic activity is increased, then source of income shall be identified for self sustainability of ICT centers.
- Identifying the far rural zones and provision of free SMS on telecom service with free access to government portals shall motivate the rural people towards the use of ICT services.
- A separate and dedicated rural ICT policy focusing on content generation and management, access technologies, services, infrastructure, research and development shall be fruitful for the case of Nepal to address the needs of rural people for their better life styles.
- The new constitution of Nepal proposed the federalism. Hence strategic review of the legislations like IT/ICT policies, e-Government Master Plan and Government Enterprise Architecture is mandatory to suit with the new geographic and government administrative structure.

Figure 5 outlines the network connection at VDC and distribution of service at the local level. The Wi-MAX/ADSL receive router connect the network at district or upper level, while the expansion from VDC is possible either through Wi-Fi or high range wireless relay network towards individual home, community centers or schools. Even wired connection shall be possible at local level based on the geographic distribution. VDC is the public access center where free internet service shall be provided with the access to local content and portal information of Nepal government. It shall acts as a FOG server where upper level administrative bodies can access local information and process for necessary action. Hence taking the reference of Government Enterprise Architecture [12] of Nepal, the network service deployment at the VDC level is proposed to be like depicted in Fig. 5. NTA has already planned and started documenting for the district optical fiber network. It is

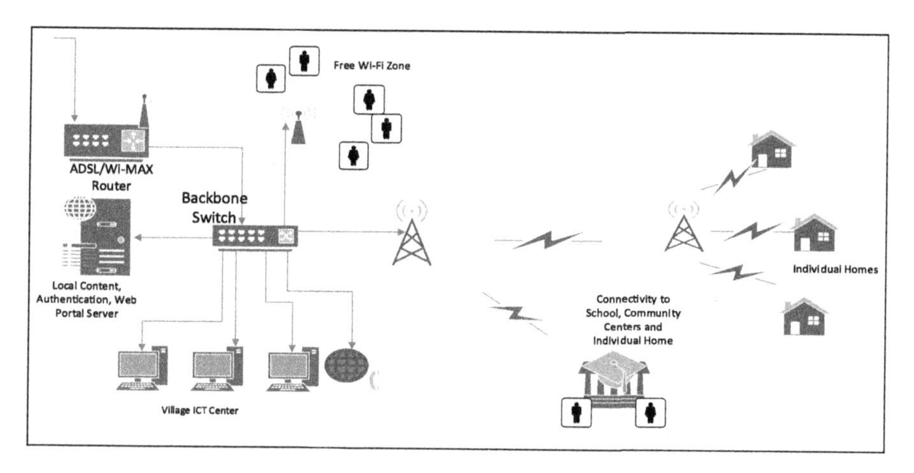

Fig. 5 ICT service expansion approach at rural VDCs of Nepal

the proposed project by Government of Nepal that connects all the 75 districts by high speed fiber backbone. The fiber network distributes form capital to district level and from district level, rural ISPs shall expand the wireless network towards far rural areas with ADSL and WiMAX service up to VDC level. Hence governmental and non-governmental interconnected network shall have presence at the VDC. This network deployment shall features the implementation of VillageCell [22], a localized voice connectivity by installing VoIP servers and enable for inter and intra-village communication. Al Hammond and John Paul [23] proposed the IP based satellite network for voice and data communication but the maintenance of this network shall be quite costly as compared with the existing ADSL/WiMAX service provided. Similarly research study in [24] proposed a Wi-Fi based Rural Extension (WiRE) architecture and analyzed the different issues for point-to-point and point-to-multipoint networks including Omni-directional antennas including the design of topology for wireless relay network for wider coverage in the rural areas of developing regions. For the situation of Nepal, we have to consider even low cost and reliable network where already available ADSL/WiMAX service shall be beneficial and point to multipoint approach at the village shall preferably be suitable.

8 Conclusion

The diverse geographic and demographic conditions of rural Nepal make the ICT accessibility service expansion quite a challenging task for all the ICT stakeholders. ICT centers or Community centers are one of the best options to increase ICT accessibilities in rural areas. However the factors like difficult and diverse terrain, low literacy rate, power issues, and lack of skilled resources raised the operability and maintainability issues of centers leading to sustainability problems. Increasing the economic activities creates business opportunities and makes the service sustainable and also attracts the private ICT service providers towards the remote. VDC office or community schools are proposed to be the best owners of the service centers. Also for the wider coverage, WiMAX service which is now easily provided by Nepal Telecom would be the best option. The WiMAX network shall be interconnected with backhaul district optical fiber network in which government shall provide subsidy to the service provider utilizing RTDF for free WiMAX connection to every ICT service centers at rural areas. We have identified the possible strategies with approach and necessary recommendations for suitable ICT service and infrastructure at rural Nepal with the requirement to change government legislation for federal state proposed by new constitution of Nepal.

Acknowledgement The pilot survey over 58 ICT service centers were conducted in support of Nepal Telecommunications Authority. We are thankful to Mrs. Roshani Ghimire, Lecturer of Purbanchal University with her active participation in collecting the survey data and relevant other documents for this research study.

References

1. Government of Nepal: Telecommunication Policy-2004. Policy document, http://nta.gov.np/en/legislation/policies
2. Nepal Telecommunications Authority: MIS Report-105, http://nta.gov.np/en/mis-reports-en
3. Definition of telecentre, http://en.wikipedia.org:/wiki/Telecentre
4. International ICT Literacy Panel: Digital transformation a framework for ICT literacy. ICT Report, ETS, USA (2007)
5. Grönlund, Å., Heacock, R., Sasaki, D., Hellström, J., Al-Saqa, W.: Increasing transparency & fighting corruption through ICT empowering people and communities. Universitetsservice US-AB, Stockholm (2010)
6. Chapagain, D.P.: A policy study on PPP Led ICT enabled services in rural Nepal, www.dineshchapagain.com.np
7. Government of Nepal: Information Technology Policy 2000, http://moste.gov.np/it_policy_2057_(2000_ad)
8. Chautari, M.: Stakeholders for universal connectivity in Nepal. Research Brief No. 15. Martin Chautari, Kathmandu (2015)
9. Government of Nepal: Broadband Policy-2014. Policy document, http://nta.gov.np/en/legislation/policies
10. Government of Nepal: Information and Communication Technology Policy-2015, Policy document, http://nta.gov.np/en/legislation/policies
11. Korean IT Industry Promotion Agency (KIPA): Nepal e-Government Master Plan Consulting Report 2006.8, http://nitc.gov.np
12. Government of Nepal: Government Enterprise Architecture, http://nitc.gov.np
13. Pun, M.: Bridging the digital divide: the case of Nepal wireless, www.unapcict.org/ecohub/bridging-the-digital-divide-the-case-of-nepal-wireless/at_download/attachment1
14. Nepal Telecom Website, https://www.ntc.net.np/WiMAX/ wimax.php
15. NCELL Website, www.ncell.com.np
16. STM Website, www.gramintel.com
17. Seismonepal website, http://seismonepal.gov.np/

18. Khanal, A.R.: Nepal's experience in responding to a disaster: a telecommunication/ICT Sector perspective. Technical report, www.itu.int/en/ITU-D/Regional-Presence/AsiaPacific/Documents/Events/2015/August-DF2015/Session-4/S4_Ananda_Raj_Khanal.pptx
19. United Nations Office for the Coordination of Humanitarian Affairs: Nepal Earthquake-2015 Humanitarian Response, www.humanitarianresponse.info
20. Nepal Wireless website, www.nepalwireless.net
21. Ka Band: the Future of Satellite Broadband, www.hughes.com/company/newsletters/spring-2011/ka-band-the-future-of-satellite-broadband
22. Anand, A., Pejovic, V., Johnson, D.L., Belding, E.M.: VillageCell: cost effective cellular connectivity in rural areas. In: ICTD2012, USA (2012)
23. Hammond, A., John Paul, J.: A new model for rural connectivity. In: World Resources Institute-Development through Enterprise (2006)
24. Subramanian, L.: A Low Cost Efficient Wireless Architecture for Rural Network Connectivity. New York University, USA (2014)

Author Index

P. Meesad et al. (eds.), *Recent Advances in Information and Communication Technology 2016*, Advances in Intelligent Systems and Computing 463,
DOI 10.1007/978-3-319-40415-8

Printed in the United States
By Bookmasters